KU-544-991

THE ECOLOGY BOOK

Withdrawn From Stock
Dublin Public Libraries

THE ECOLOGY BOOK

FOREWORD BY
TONY JUNIPER

Brainse Fhionnglaise
Finglas Library
Tel: (01) 834 4906

DK LONDON

SENIOR EDITORS
Helen Fewster, Camilla Hallinan

SENIOR ART EDITOR
Duncan Turner

ILLUSTRATIONS
James Graham

JACKET EDITOR
Emma Dawson

JACKET DESIGNER
Surabhi Wadhwa-Gandhi

JACKET DESIGN
DEVELOPMENT MANAGER
Sophia MTT

PRODUCER, PRE-PRODUCTION
Andy Hilliard

SENIOR PRODUCER
Meskerem Berhane

MANAGING EDITOR
Angeles Gavira Guerrero

MANAGING ART EDITOR
Michael Duffy

ASSOCIATE PUBLISHING DIRECTOR
Liz Wheeler

ART DIRECTOR
Karen Self

DESIGN DIRECTOR
Philip Ormerod

PUBLISHING DIRECTOR
Jonathan Metcalf

DK DELHI

SENIOR ART EDITOR
Ira Sharma

PROJECT ART EDITOR
Vikas Sachdeva

ART EDITORS
Shipra Jain, Sourabh Challariya,
Debjyoti Mukherjee

ASSISTANT ART EDITORS
Shreya Singal, Vidushi Gupta, Amrai Dua

SENIOR EDITOR
Janashree Singha

EDITOR
Aadithyan Mohan K.

ASSISTANT EDITORS
Rishi Bryan, Tanya Singhal, Nonita Saha

JACKET DESIGNER
Suhita Dharamjit

SENIOR DTP DESIGNERS
Harish Aggarwal, Jagtar Singh

DTP DESIGNERS
Mohammad Rizwan, Bimlesh Tiwary

PICTURE RESEARCHER
Vishal Ghavri

JACKETS EDITORIAL COORDINATOR
Priyanka Sharma

MANAGING JACKETS EDITOR
Saloni Singh

PICTURE RESEARCH MANAGER
Taiyaba Khatoon

PRE-PRODUCTION MANAGER
Balwant Singh

PRODUCTION MANAGER
Pankaj Sharma

MANAGING EDITOR
Soma B. Chowdhury

SENIOR MANAGING ART EDITOR
Arunesh Talapatra

TOUCAN BOOKS

EDITORIAL DIRECTOR
Ellen Dupont

SENIOR DESIGNER
Thomas Keenes

SENIOR EDITOR
Scarlett O'Hara

EDITORS
John Andrews, Alethea Doran, Sue George,
Guy Croton, Cathy Meeus, Abigail Mitchell,
Fiona Plowman, Dorothy Stannard,
Rachel Warren Chadd

ASSISTANT EDITOR
Isobel Rodel

INDEXER
Marie Lorimer

PROOFREADER
Richard Beatty

ADDITIONAL TEXT
Shannon Webber, Marcus Weeks

original styling by
STUDIO 8

First published in Great Britain in 2019
by Dorling Kindersley Limited,
80 Strand, London, WC2R 0RL

Copyright © 2019
Dorling Kindersley Limited
A Penguin Random House Company

Foreword © 2019 Tony Juniper

10 9 8 7 6 5 4 3 2 1
001–311040–Apr/2019

All rights reserved.
No part of this publication may be reproduced,
stored in or introduced into a retrieval system, or
transmitted, in any form, or by any means (electronic,
mechanical, photocopying, recording, or otherwise),
without the prior written permission of the
copyright owner.

A CIP catalogue record for this book
is available from the British Library.

ISBN: 978-0-2413-5038-6

Printed and bound in Malaysia

A WORLD OF IDEAS:
SEE ALL THERE IS TO KNOW

www.dk.com

CONTRIBUTORS

JULIA SCHROEDER, CONSULTANT

Julia Schroeder received her PhD in Animal Ecology from the University of Groningen in the Netherlands. From 2012 to 2017, she headed a research group at the Max Planck Institute for Ornithology in Germany, studying social behavioural ecology. Julia currently researches and teaches evolutionary biology at Imperial College London.

CELIA COYNE

Celia Coyne is a freelance writer and editor living in Christchurch, New Zealand. She is the author of *Earth's Riches* and *The Power of Plants* and writes and edits articles on science and natural history for magazines, newspapers, journals, websites, and books in the UK, Australia, and New Zealand. Her aim is to make scientific subjects accessible to lay readers.

JOHN FARNDON

The author of hundreds of books on science and nature for both children and adults, John Farndon studied geography at Cambridge University. He has written extensively on earth sciences and the environment, focusing in particular on conservation and ecology. His books include *The Oceans Atlas*, *The Wildlife Atlas*, *How the Earth Works*, and *The Practical Encyclopedia of Rocks and Minerals*.

TIM HARRIS

After studying Norwegian glaciers at university, Tim Harris travelled the world in search of unusual wildlife and extraordinary landscapes. He has explored the dunes of the Namib Desert, climbed Popocatépetl in central Mexico, camped in the Sumatran rainforest, and searched the frozen Sea of Okhotsk in Russia. He is a former Deputy Editor of *Birdwatch* magazine in the UK and has written books about nature for adults and children.

DEREK HARVEY

A naturalist and teacher with a particular interests in evolutionary biology, Derek Harvey graduated in Zoology from Liverpool University in the UK. He has taught a generation of biologists and led student expeditions to Costa Rica, Madagascar, and Australasia. Derek now concentrates on writing and consulting for science and natural history books.

TOM JACKSON

A writer for 25 years, Tom Jackson is the author of about 200 non-fiction books for adults and children and has contributed to many more. Tom studied zoology at Bristol University, UK, and worked in zoos and as a conservationist before turning to writing about natural history and all things scientific.

ALISON SINGER

Alison Singer is a PhD candidate in Community Sustainability at Michigan State University, US, where she studies storytelling and science communication. She has a broad educational background in writing, ecology, and the social sciences. Alison has worked as an educator for environmental charities, and for the US Environmental Protection Agency.

CONTENTS

THE VARIETY OF LIFE

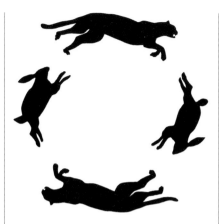

ECOSYSTEMS

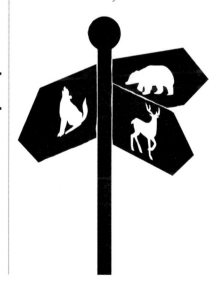

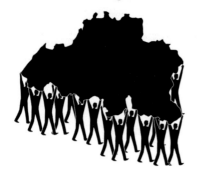

THE HUMAN FACTOR

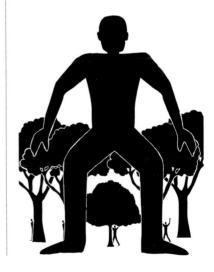

ENVIRONMENTALISM AND CONSERVATION

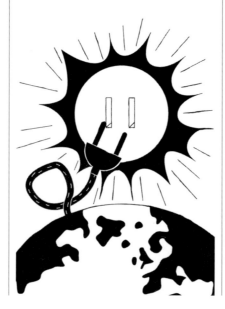

FOREWORD

As a small child, I was fascinated by nature – birds, butterflies, plants, reptiles, fossils, rivers, weather, and much else. My youthful passions set me on the path to being a life-long naturalist, and to working as an environmentalist, studying the natural world and promoting action for its conservation. I have worked as a field ornithologist, writer, campaigner, policy advocate and environmental advisor. All of these diverse interests and activities have, however, been linked by a single theme: ecology.

Ecology is a vast subject, embracing the many disciplines needed to understand the relationships that exist between different living things, and the physical worlds of air, water and rock within which they are embedded. From the study of soil microorganisms to the role of pollinators, and from research into the water cycle to investigating Earth's climate system, ecology involves many specialist areas. It also unites many strands of science, including zoology, botany, mathematics, chemistry, and physics, as well as some aspects of social science – especially economics – while at the same time raising profound philosophical and ethical questions.

Because of the fundamental ways in which the human world depends on healthy natural systems, some of the most important political issues of our age are ecological ones. They include climate change, the effects of ecosystem damage, the disappearance of wildlife, and the depletion of resources, including fish stocks, freshwater, and soils. All these ecological changes have implications for people and are increasingly pressing.

Considering the huge importance of ecology for our modern world, and the many threads of thought and ideas that must be woven to gain an understanding of the subject, I am delighted that Dorling Kindersley decided to produce *The Ecology Book,* setting out the key concepts that have helped shape our understanding of how Earth's incredible natural systems function. In the pages that follow readers will also discover something about the history of ecological concepts, the leading thinkers and the different perspectives from which they approached the questions they sought to answer.

One thing that sets this book apart is the manner in which the rich, memorable, and attractive content is presented. A huge body of information and insight is effectively conveyed by clear layout, graphics, illustrations, and quotes, enabling readers to quickly achieve an understanding of many important ecological ideas and the people behind them: James Lovelock's Gaia Hypothesis, Norman Myers's warnings about impending mass extinction,and Rachel Carson's work to expose the effects of toxic pesticides among them.

The diverse body of information found in the pages that follow could not be more important. For while the headlines and popular debate suggest it is politics, technology, and economics that are the vital forces shaping our common future, it is in the end ecology that is the most important context determining societies' prospects, and indeed the future of civilization itself.

I hope you find *The Ecology Book* to be an enlightening overview of what is not only the most important subject, but also the most interesting.

Tony Juniper CBE
Environmentalist

INTRODU

CTION

For the earliest humans, a rudimentary knowledge of ecology – how organisms relate to one another – was a matter of life and death. Without having a basic understanding of why animals grazed in a certain place and fruit grew in another, our ancestors would not have survived and evolved.

How living animals and plants interact with each other, and with the nonliving environment interested the ancient Greeks. In the 4th century BCE, Aristotle and his student Theophrastus developed theories of animal metabolism and heat regulation, dissected birds' eggs to discover how they grew, and described an 11-level "ladder of life", the first attempt at classifying organisms. Aristotle also explained how some animals consume others – the first description of a food chain.

In the Middle Ages (476–1500), the Catholic Church discouraged new scientific thought, and human understanding of ecology advanced very slowly. By the 16th century, however, maritime exploration, coupled with great technological advances, such as the invention of the microscope, led to the discovery of amazing new life forms and a thirst for knowledge about

them. Swedish botanist Carl Linnaeus developed a classification system *Systema Naturae*, the first scientific attempt to name species and group them according to relatedness. Throughout this time, essentialism – the idea that each species had unalterable characteristics – continued to dominate Western thought.

Great breakthroughs

Geological discoveries in the late 17th and early 18th centuries began to challenge the idea of essentialism. Geologists noted that some fossil species suddenly disappeared from the geological record, to be replaced by others, suggesting that

There are some 4 million different kinds of animals and plants in the world. Four million different solutions to the problems of staying alive.
David Attenborough

organisms change over time, and even become extinct. The Frenchman Jean-Baptiste Lamarck proposed the first cohesive theory of evolution – the transmutation of species by the inheritance of acquired characteristics – in 1809. However, some 50 years later it was Charles Darwin – influenced by his experiences on the epic expedition of HMS *Beagle* – and Alfred Russel Wallace, who developed the concept of evolution by means of natural selection, the theory that organisms evolve over the course of generations to adapt better to their environment. Darwin and Wallace did not understand the mechanism by which this happened, but Gregor Mendel's experiments on peas pointed at the role of hereditary factors later known as genes, representing another giant leap in evolutionary theory.

Making connections

The relationships between organisms and their environment, and between species, dominated ecological study in the early 20th century. The concepts of food chains and food webs (who eats what in a particular habitat) and ecological niches (the role an organism has in its environment) developed, and in 1935, Arthur

Tansley introduced the concept of the ecosystem – the interactive relationship between living organisms and the environment in which they live. Later ecologists developed mathematical models to forecast population dynamics within ecosystems. Evolutionary theories also advanced with the discovery of the structure of DNA, and the evolutionary "vehicle" provided by mutation as DNA is replicated.

New frontiers

Improved technology opened up new possibilities for ecology. An electron microscope can now make images to half the width of a hydrogen atom, and computer programs can analyse the sounds made by bats and whales, which are higher or lower than can be heard by the human ear. Camera traps and infra-red detectors photograph and film nocturnal creatures, and tiny satellite devices fitted to birds can track their movements.

In the laboratory, analysis of the DNA of faeces, fur, or feathers indicates which species an animal belongs to, and throws light on the relationship between different organisms. It is now easier than ever for ecologists to collect data, helped by a growing army of citizen scientists.

New concerns

Early ecology was driven by a desire for knowledge. Later, it was used to find better ways to exploit the natural world for human needs. As time went on, the consequences of this exploitation became increasingly evident. Deforestation was highlighted as a problem as early as the 18th century, and the problems of air and water pollution became obvious in industrialized nations in the 19th century. In 1962, Rachel Carson's book *Silent Spring* alerted the world to the dangers of pesticides, and six years later Gene Likens demonstrated the link between power station emissions, acid rain, and fish deaths.

In 1985, a team of Antarctic scientists discovered the dramatic depletion of atmospheric ozone over Antarctica. The link between greenhouse gases and a warming of Earth's lower atmosphere had been made as early as 1947 by G. Evelyn Hutchinson, but it was decades before there was a scientific consensus on the man-made causes of climate change.

The future

Modern ecology has come a long way since the science was first recognized. It now draws on many disciplines. In addition to zoology, botany, and their micro-disciplines, it relies on geology, geomorphology, climatology, chemistry, physics, genetics, sociology, and more. Ecology influences local and national government decisions about urbanization, transport, industry, and economic growth. The challenges posed by climate change, rising sea levels, habitat destruction, the extinction of species, plastic and other forms of pollution, and a looming water crisis pose serious threats to human civilization. They demand radical policy responses based on sound science. Ecology will provide the answers. It is up to governments to apply them. ∎

Even in the vast and mysterious reaches of the sea we are brought back to the fundamental truth that nothing lives to itself.
Rachel Carson

THE STO
OF EVOL

RY
UTION

James Hutton presents his theory that **Earth is much older** than was previously believed, and that Earth's crust is continuously changing.

1785

In his *Essay on the Theory of the Earth*, **Georges Cuvier** suggests that **fossils** are the remains of **extinct creatures** wiped out by periodic "catastrophic" events.

1813

HMS *Beagle* sets sail on a circumnavigation of the world, with **Charles Darwin** serving as the voyage's naturalist. The trip provides Darwin with the information that inspired his theory of **evolution by natural selection**.

1831

1809

Jean-Baptiste Lamarck publishes *Philosophie Zoologique*, where he argues that animals acquire characteristics as a consequence of use or non-use of different body parts, triggering **mutations** over generations.

1823

Amateur fossil hunter **Mary Anning** uncovers the first intact **plesiosaurus skeleton**.

Ancient myths, religions, and philosophy all reflect an enduring fascination with how the world began and man's place in the story of life on Earth. In the West, Christianity held that all animals and plants were the result of a perfect creation. On the chain or ladder of being, no species could ever move from one position to another. Species were immutable, an idea called essentialism.

The 18th-century Age of Enlightenment began to challenge orthodox Christian beliefs. French zoologist Jean-Baptiste Lamarck rejected the prevailing Bible-based notion of Earth being only a few thousand years old. He argued that organisms must have changed from simple life forms to more complex ones over millions of years, and that the "transmutation" of species was

the driving force behind this change. He speculated that characteristics acquired by animals during their lifetime were inherited by the next generation: giraffes, for example, became slightly longer-necked by stretching up to reach higher leaves, and passed this trait to their offspring; over many generations, giraffes grew longer and longer necks.

Fossil evidence of extinct life forms with features that resembled modern descendants, found by pioneering geologists such as Georges Cuvier, also suggested Earth had more ancient origins. Meanwhile James Hutton and Charles Lyell argued that geological features could be accounted for by the constant, ongoing processes of erosion, and deposition – a view called uniformitarianism. As these

processes take place slowly, Earth's history had to be much longer than was previously thought.

Natural selection

In 1858, Charles Darwin and Alfred Russel Wallace delivered a paper that would change biology forever. Darwin's observations on the epic voyage of the *Beagle* (1831–36), his correspondence with other naturalists, and the influence of Thomas Malthus's writings inspired Darwin's insight that evolution came about by what he called natural selection. He spent 20 years gathering supporting data, but when Wallace wrote to him with the same idea, Darwin realized it was time to go public. His subsequent book, *On the Origin of Species by Means of Natural Selection*, provoked outrage.

Gregor Mendel's paper "Experiments with Plant Hybrids" outlines findings from his pea plant experiments, laying the foundations for the field of **genetics**.

1866

The Selfish Gene by evolutionary biologist **Richard Dawkins** offers a **new perspective** on **evolution**, looking at the gene, as opposed to the species or group.

1976

1859

Darwin elaborates on his theories of evolution in *On the Origin of Species by Means of Natural Selection*, which is an instant sell out.

1953

In *The Eagle* pub in Cambridge, UK, **Crick and Watson** announce that they have discovered the **structure of DNA**.

2003

The **Human Genome Project** produces the first genetic blueprint of *Homo sapiens*.

Although the idea of evolution became widely accepted, the mechanism that made natural selection possible was not yet known. In 1866, an Austrian monk called Gregor Mendel made a huge contribution to genetics when he published his findings on heredity in pea plants. Mendel described how dominant and recessive traits pass from one generation to the next, by means of invisible "factors" that we now call genes.

The rediscovery of Mendel's work in 1900 initially sparked keen debate between his supporters and many Darwinians. At the time, evolution was believed to be based on the selection of small, blending variations, but Mendel's variations clearly did not blend. Three decades later, geneticist Ronald Fisher and others argued that the two schools of thought were complementary, rather than contradictory. In 1942, Julian Huxley articulated the synthesis between Mendel's genetics and Darwin's theory of natural selection in his book *Evolution: The Modern Synthesis*.

The double helix
Advances in technology such as X-ray crystallography led to more discoveries in the 1940s and '50s, and the foundation of the new discipline of molecular biology. In 1944, chemist Oswald Avery identified deoxyribonucleic acid (DNA) as the agent for heredity. Rosalind Franklin and Raymond Gosling photographed strands of the DNA molecule in 1952, then James Watson and Francis Crick confirmed its double helix structure the following year. Crick then showed

that genetic information is "written" on DNA molecules. The errors that occur when DNA copies itself create mutations – the raw materials for evolution. By the 1980s it was possible to map and manipulate the genes of individuals and species. In the 1990s, the mapping the human genome paved the way for medical research into gene therapy.

Ecologists also want to establish whether genes influence behaviour. Back in 1964, William D. Hamilton popularized the concept of genetic relatedness ("kin selection") to explain altruistic behaviour in animals. In *The Selfish Gene* (1976), Richard Dawkins further advanced the gene-centred approach. It is clear that aspects of evolutionary biology will still spark debate as long as ecologists continue to develop Darwin's theory. ∎

TIME IS INSIGNIFICANT, AND NEVER A DIFFICULTY FOR NATURE
EARLY THEORIES OF EVOLUTION

IN CONTEXT

KEY FIGURES
The Comte de Buffon
(1707–88), **Jean-Baptiste
Lamarck** (1744–1829)

BEFORE
1735 Swedish botanist Carl
Linnaeus publishes *Systema
Naturae*, a system of biological
classification that later helped
to determine species' ancestry.

1751 In "Système de la nature"
French philosopher Pierre
Louis Moreau de Maupertuis
introduces the idea that
features can be inherited.

AFTER
1831 Etienne Geoffroy Saint-
Hilaire writes that sudden
environmental change can
cause a new species to develop
from an existing organism.

1844 In *Vestiges of the Natural
History of Creation*, Scottish
geologist Robert Chambers
argues – anonymously – that
simple creatures have evolved
into more complex species.

Before the 18th century, most
people believed that plant
and animal species stayed
unchanged throughout time – a view
now known as essentialism. This
idea came under challenge as a
result of two developments: the
intellectual movement known as the
Enlightenment (*c.* 1715–1800), and
the Industrial Revolution (1760–1840).

The Enlightenment was marked
by scientific progress and increased
questioning of religious orthodoxy,
such as the claim that God created
Earth and all living things in seven
days. Then, as the Industrial
Revolution gathered pace, canals,
railways, mines, and quarries
cut through rock strata revealed
thousands of fossils, mostly of
animal and plant species that no
longer existed and had never been
seen before. These suggested that
life began long before the widely
accepted creation date of 4400 BCE,
deduced from biblical sources.

Animal adaptation
In the late 1700s, French scientist
Georges-Louis Leclerc, Comte de
Buffon, upset church authorities
by asserting that Earth was much
older than the Bible suggested. He
believed it was formed from molten
material, struck off the Sun by a
comet, that had taken 70,000 years
to cool (a huge underestimate, in
fact). As Earth cooled, species had
appeared, died off, and were finally
replaced by ancestors of those
known today. Noting similarities
among animals such as lions,
tigers, and cats, Buffon deduced
that 200 species of quadruped had
evolved from just 38 ancestors. He
also believed that changes in body
shape and size in related species
had occurred in response to living
in different environments.

In 1800, French naturalist Jean-
Baptiste Lamarck went further. In
a lecture at the Museum of Natural

*Nature is the system of laws
established by the Creator for
the existence of things and
for the succession of creatures.*
The Comte de Buffon

See also: Extinction and change 22 ▪ Uniformitarianism 23 ▪ Evolution by natural selection 24–31 ▪ The rules of heredity 32–33

History in Paris, he argued that traits acquired by a creature during its lifetime could be inherited by its offspring – and that a build-up of such changes over many generations could radically alter an animal's anatomy.

Lamarck wrote several books in which he developed this idea of transmutation. He argued, for instance, that the use or non-use of body parts eventually resulted in such features becoming stronger, weaker, bigger, or smaller in a species. For example, the ancestors of moles probably had good eyesight, but over generations this deteriorated because moles did not require vision as they burrowed underground. Similarly, giraffes gradually developed longer necks to enable them to reach leaves growing high up in trees.

Drivers of evolution

Larmarck's ideas about inherited acquired traits were part of a wider early theory of evolution. He also believed that the earliest, simplest forms of life had emerged directly from non-living matter. Lamarck identified two main "life forces" driving evolutionary change. One, he believed, made organisms

> … continuous use of any organ gradually strengthens, develops and enlarges that organ.
> **Jean-Baptiste Lamarck**

develop from simple to more complex forms in a "ladder" of progress. The other, via the inheritance of acquired traits, helped them to adapt better to their environment. When Charles Darwin developed his theory of evolution by means of natural selection, he would reject many of Lamarck's ideas, but both men shared the belief that complex life evolved over an immense period of time. ▪

Fossil finds changed ideas about how life began. The first example of an articulated plesiosaur – *Plesiosaurus dolichodeirus* – was discovered in 1823 by Mary Anning in Dorset, England.

Jean-Baptiste Lamarck

Born in 1744, Jean-Baptiste Lamarck attended a Jesuit college before joining the French army. Forced by an injury to resign, he studied medicine and then pursued his passion for plants, working at the Jardin du roi (Royal Garden) in Paris. Supported by the Comte de Buffon, Lamarck was elected to the Academy of Sciences in 1779. When the Jardin's main building became the new National Museum of Natural History during the French Revolution (1789–99), Lamarck was placed in charge of the study of insects, worms, and microscopic organisms. He coined the biological term "invertebrate" and often used the relatively simpler forms of such species to illustrate his "ladder" of evolutionary progress. However, Lamarck's work was controversial and he died in poverty in 1829.

Key works

1802 *Research on the Organization of Living Bodies*
1809 *Zoological Philosophy*
1815–22 *Natural History of Invertebrate Animals*

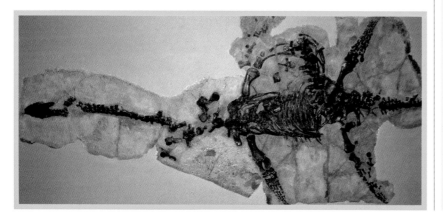

A WORLD PREVIOUS TO OURS, DESTROYED BY CATASTROPHE

EXTINCTION AND CHANGE

IN CONTEXT

KEY FIGURE
Georges Cuvier (1769–1832)

BEFORE
Late 1400s Leonardo da Vinci argues that fossils are the remains of living creatures, not just shapes spontaneously formed in the earth.

1660s English scientist Robert Hooke suggests that fossils are extinct creatures, since no similar forms can be found on Earth today.

AFTER
1841 English anatomist Richard Owen calls huge reptile fossils "dinosaurs".

1859 Charles Darwin's *On the Origin of Species* explains how evolution can occur through "natural selection".

1980 US scientists Luis and Walter Alvarez present evidence that an asteroid hit the earth at the time of the extinction of the dinosaurs.

In the early days of studying fossils, many people denied they could be extinct species. They failed to see why God would create and destroy creatures before humans ever appeared, arguing that unfamiliar fossil species might still be living somewhere on Earth. In the late 18th century, French zoologist Georges Cuvier looked into this by exploring the anatomy of living and fossil elephants. He proved that fossil forms such as mammoths and mastodons were anatomically distinct from living elephants, and so must represent extinct species. (It was highly unlikely that they still lived on Earth without being noticed.)

Cuvier believed that Earth had experienced a series of distinct ages, each of which ended with a "revolution" that destroyed existing flora and fauna. He did not, though,

believe that the evidence of fossil remains supported a theory of evolution. Nevertheless, Cuvier's central views have continued to win support, and modern evidence points to at least five catastrophic mass extinction events in Earth's past, including the one that wiped out the dinosaurs. Unlike Cuvier, however, today's scientists know that life is not recreated out of nothing after a catastrophe. Rather, when a mass extinction event kills off many species, those left will evolve and multiply – sometimes relatively quickly – to fill vacant ecological niches, as the mammals did after the age of the dinosaurs. ∎

Cuvier coined the name "mastodon" for its Greek meaning of "breast tooth", referring to the nipple-like patterns on the creature's teeth, which were unlike those of any living elephants.

See also: Evolution by natural selection 24–31 ▪ Ecological niches 50–51
▪ Ancient ice ages 198–199 ▪ Mass extinctions 218–223

NO VESTIGE OF A BEGINNING – NO PROSPECT OF AN END
UNIFORMITARIANISM

IN CONTEXT

KEY FIGURE
James Hutton (1726–97)

BEFORE
1778 The Comte de Buffon, a French naturalist, suggests that Earth is at least 75,000 years old – far older than most people believed at the time.

1787 German geologist Abraham Werner proposes that Earth's layers of rock formed from a great ocean that once covered the entire planet. His followers became known as Neptunists.

AFTER
1802 James Hutton's theory of uniformitarianism reaches a wider audience when Scottish geologist John Playfair publishes *Illustrations of the Huttonian Theory of the Earth*.

1830–33 *Principles of Geology*, by Scottish geologist Charles Lyell, supports and builds on the uniformitarian ideas of James Hutton.

… from what has actually been, we have data for concluding [what] is to happen thereafter.
James Hutton

Uniformitarianism is the theory that geological processes, such as the laying down of sediment, erosion, and volcanic activity, occur at the same rate now as they did in the past. The idea emerged in the late 18th century, as mining, quarrying, and increased travel brought ever more geological features to light, including unusual rock strata and previously unknown fossils, whose origins were then widely debated.

The generally accepted view that Earth was only a few thousand years old had been challenged by the Comte de Buffon, and in 1785 Scottish geologist James Hutton also argued for Earth's far greater antiquity. Hutton's ideas were formed during expeditions around Scotland to examine layers of rock. He believed that Earth's crust was constantly changing, albeit mostly slowly, and could see no reason to suggest that the complex geological actions of layering, erosion, and uplifting took place faster in the distant past than they did in the present. Hutton also understood that most geological processes happen so gradually that the features he was discovering must be astronomically old.

Uniformitarianism was not generally accepted at once, not least because it challenged a literal interpretation of the creation stories of the Old Testament. However, a new generation of geologists, such as John Playfair and Charles Lyell, threw their intellectual weight behind Hutton's ideas, which also inspired a young Charles Darwin. ■

See also: Evolution by natural selection 24–31 ▪ Early theories of evolution 20–21 ▪ Moving continents and evolution 212–213 ▪ Mass extinctions 218–223

THE STRUGGLE FOR
EXISTENCE

EVOLUTION BY NATURAL SELECTION

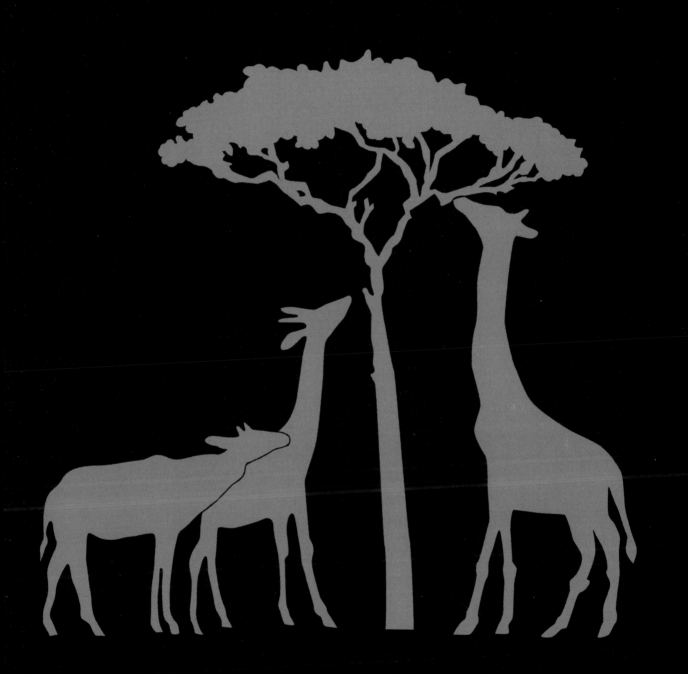

IN CONTEXT

KEY FIGURE
Charles Darwin (1809–82)

BEFORE
1788 In France, Georges-Louis Leclerc, Comte de Buffon, completes his 36-volume *Histoire Naturelle*, outlining early ideas about evolution.

1809 Jean-Baptiste Lamarck proposes that creatures evolve by inheriting acquired traits.

AFTER
1869 Friedrich Miescher, a Swiss doctor, discovers DNA, although its genetic role is not yet understood.

1900 The laws of inheritance based on the pea plant experiments of Austrian scientist Gregor Mendel in the mid-1800s are rediscovered.

1942 British biologist Julian Huxley coins the term "modern synthesis" for the mechanisms thought to produce evolution.

Natural selection, a concept developed by British naturalist Charles Darwin and set out in his book *On the Origin of Species by Means of Natural Selection* (1859), is the key mechanism of evolution in organisms, resulting in different survival rates and reproductive abilities. Those organisms that have higher breeding success pass on their genes to more of the next generation, so individuals with these characteristics become more common.

To the Galapagos

The young Charles Darwin first began to consider evolution during his pioneering scientific expedition around the world aboard HMS *Beagle* from 1831 to 1836. As a young man, Darwin accepted the orthodox interpretation of the Bible, that Earth was only a few thousand years old. However, while he was on board the *Beagle*, Darwin read Scottish geologist Charles Lyell's recently published *Principles of Geology*, in which Lyell demonstrated that rocks bore traces of tiny, gradual, and cumulative change over vast time periods – millions, rather than

> Natural selection is daily and hourly scrutinizing, throughout the world, the slightest variations.
> **Charles Darwin**

thousands of years. As Darwin looked at landscapes around the world that had been affected by processes of erosion, deposition, and volcanism, he began to speculate about animal species changing over very long time periods, and the reasons for such changes. By examining fossils and observing living animals, Darwin identified patterns; he noticed, for example, that extinct species had often been replaced by similar, but distinct, modern ones.

Darwin's field work on the islands of the Galapagos archipelago off South America in the autumn

---

Charles Darwin

Born in Shropshire, UK, in 1809, Darwin was fascinated with natural history from a young age. While at Cambridge University, he became friendly with several influential naturalists, including John Stevens Henslow. As a result, Darwin was invited to join the HMS *Beagle* expedition around the world. Henslow helped Darwin to catalogue and publicize his finds.

Darwin's research brought him fame and recognition – the Royal Society's Royal Medal in 1853, and fellowship of the Linnean Society in 1854. In 1859, his book *On the Origin of Species* sold out

instantly. Despite continuing ill-health, Darwin fathered 10 children and never stopped studying and developing new theories. He died in 1882.

Key works

1839 *Zoology of the Voyage of HMS* Beagle.
1859 *On the Origin of Species by Means of Natural Selection*
1868 *The Variation of Animals and Plants under Domestication*
1872 *The Expression of Emotions in Man and Animals*

See also: Early theories of evolution 20–21 ▪ The rules of heredity 32–33 ▪ The role of DNA 34–37 ▪ The selfish gene 38–39 ▪ The food chain 132–133 ▪ Mass extinctions 218–223 ▪ Population viability analysis 312–315

of 1835 provided especially strong evidence for his later theory of evolution by natural selection. Here, he observed that the shape of the carapaces (shells) of giant tortoises varied slightly from island to island. Darwin was also intrigued to find that there were four broadly similar, yet clearly distinct, varieties of mockingbirds, but that no single island had more than one species of the bird. He saw small birds, too, that looked alike but had a range of beak sizes and shapes. Darwin deduced that each group possessed a common ancestor but had developed diverse traits in different environments.

Darwin's conclusions

On Darwin's return to England, the differing beaks of the small birds he had found on the Galapagos, usually called "finches" although they are not in the true finch family, set him thinking. He knew that a bird's beak is its key tool for feeding, so its length and shape offer clues to its diet. Later research revealed that there are 14 different finch species on the Galapagos islands. The differences in their beaks are marked and significant. For example, cactus finches have long, pointed beaks that are ideal for picking seeds out of cactus fruits, while ground finches have shorter, stouter beaks that are better suited for eating large seeds on the ground. Warbler finches have slender, sharp beaks, which are ideal for catching flying insects.

Darwin speculated that the finches were descended from a common ancestral finch that had reached the archipelago from the mainland of South America. He concluded that a variety of finch

Comparison of Galapagos finch bill structure

Geospiza magnirostris
The short, sharp bill of the large ground finch, the biggest of Darwin's finches, enables it to crack nuts.

Geospiza fortis
The bill of the medium ground finch is variable, evolving rapidly to adapt to whatever size seeds are available.

Geospiza parvula
The stubby bill of the small tree finch, which forages in foliage, suits its diet of seeds, fruits, and insects.

Certhidea olivacea
The slender, probing bill of the green warbler-finch helps it to catch small insects and spiders.

populations had evolved in different Galapagos habitats, each group adapted for a more or less specialist diet by a process that he would later call "natural selection". Over time, the finch populations had become distinct species.

In the early 21st century, researchers at Harvard University uncovered new evidence of how this happens at a genetic level. Their findings, published in 2006, showed that a molecule called calmodulin regulates the genes involved in shaping birds' beaks, and is found at higher levels in longer-beaked cactus finches than in shorter-beaked ground finches.

Refining the theory

Darwin was influenced by Thomas Malthus's *An Essay on the Principle of Population* (1798), in which

Malthus predicted that population growth would eventually outstrip food production. This idea matched the evidence Darwin had observed of ongoing competition between individual animals and species for resources. This competitive aspect formed the backbone of Darwin's coalescing theory of evolution.

By 1839, Darwin had developed an idea of evolution by natural selection. He was, though, reluctant to publish because he understood that the theory would unleash a storm of controversy from those who would view it as an attack on religion and the Church. When, in 1857, he began receiving communications from fellow British naturalist Alfred Russel Wallace, who had independently arrived at very similar conclusions, Darwin realized he had to publish his » ideas. Papers by Darwin and Wallace were jointly presented at a meeting of the Linnean Society of London in July 1858, under the title "On the Tendency of Species to form Varieties; and on the Perpetuation of Varieties and Species by Natural Means of Selection".

The following year, Darwin published the theory in *On the Origin of Species by Means of Natural Selection*. It offended some scientists because it differed from Lamarck's ideas of transmutation, and also upset creationists who argued that it undermined a literal interpretation of the Bible. Others felt that the theory did not account for the huge range of characteristics in species and called it "unguided" and "non-progressive".

Darwin was confident. He knew that all individual organisms in a species show a degree of natural variation; some have longer whiskers, or shorter legs, or brighter colours, for instance. Because members of all species compete for limited resources, he deduced that those whose traits are best suited to their environment are more likely to survive and reproduce. He also argued that characteristics which

> I see no good reasons why the views given in this volume should shock the religious views of anyone.
> **Charles Darwin**

helped an individual organism to live longer and reproduce more successfully would be passed on to more offspring, while those that made the organism less successful would be lost. Darwin called this "natural selection" – a process that, over generations, enabled a population of any given species to adapt better and thrive in its chosen habitat.

Sexual selection

Darwin also developed a theory of sexual selection. First outlined in *On the Origin of Species*, this was developed further in *The Descent of Man, and Selection in Relation to Sex* (1871). This theory was distinct from natural selection, as Darwin recognized that animals select mates based on characteristics that do not simply favour survival. For example, when Darwin considered the spectacular but cumbersome tails of male peafowl (peacocks), he could not imagine the tail playing any role in helping the individual bird to survive. He concluded that they were designed to boost an individual's chance of reproductive success. Peahens choose males with the brightest tails, so the genetic material of these showy males is passed to the next generation. Bright tail feathers indicate that the bird is healthy, so choosing a mate with a bright tail is a good strategy for the peahen. However, Darwin's idea that females choose a mate came under fire; 19th-century society could accept that males competed to reproduce (intrasexual selection), but intersexual selection, where one sex (usually the female) makes the choice, was ridiculed.

Reproductive success is clearly essential for the future of a species. Natural selection is often described as "survival of the fittest", but longevity alone is not particularly

Natural selection

There is variation in traits.
For example, some beetles are pale and others dark.

There is differential reproduction.
No environment can support unlimited population growth, so some individuals lose out. Here, birds eat the pale beetles, so fewer of them reproduce.

There is heredity.
The dark beetles have more dark offspring because this trait has a genetic basis.

End result:
If darkness is the winning trait, producing more offspring, in time, all beetles will be dark.

The peacock with the most splendid tail will attract the most peahens. Its bright tail will be passed on to its male offspring, which will find it similarly easy to attract mates.

helpful. If individual A lives 10 times as long as individual B, but the latter produces twice as many offspring that then also breed, B will pass on more genes to the next generation than the longer-lived A.

Building on the theory

Many of Darwin's and Wallace's ideas have proved remarkably accurate, despite the fact that the workings of genetics were not understood at the time. Although Darwin himself had used the term "genetic" as an adjective to describe the as-yet-unknown mechanism of inheritance, it was British biologist William Bateson, in the early 20th century, who first used the term "genetics" in a

description of the scientific process. In 1930, British geneticist Ronald Fisher wrote *The Genetical Theory of Natural Selection*, which combined Darwin's theory of natural selection with the ideas of heredity that the 19th-century Austrian scientist Gregor Mendel had developed. In 1937, Ukrainian–American geneticist Theodosius »

Why do some die and some live?… the answer was clearly, that on the whole the best fitted live.
Alfred Russel Wallace

Kin selection

The term "kin selection" was first used by British biologist John Maynard Smith in 1964. It is the evolutionary strategy that favours the reproductive success of an organism's relatives, prioritising them above the individual's own survival and reproduction. It occurs when an organism engages in self-sacrificial behaviour that benefits its relatives. Charles Darwin was the first to discuss the concept when he wrote about the apparent paradox represented by altruistic non-breeding social insects, such as worker honeybees, which leave reproduction to their mothers. British evolutionary biologist William Donald Hamilton proposed that bees, for example, behave in an altruistic manner – assisting others in reproduction – when the genetic closeness of the two bees and the benefit to the recipient outweigh the cost of altruism to the giver. This is called Hamilton's Rule.

In honeybee colonies, female worker bees look after the queen bee. They build the honeycomb, gather nectar and pollen, and feed larvae, but they do not breed.

Albinism, as in this albino leopard gecko, is a mutation causing a lack of pigment. This mutation hinders the gecko's chances of survival, making it lighter coloured and sensitive to light.

Dobzhansky put forward the idea that regularly occurring genetic mutations are sufficient to provide the genetic diversity – and therefore different traits – that makes natural selection possible. He wrote that evolution was a change in the frequency of an "allele" in the gene pool, an allele being one of the alternative forms of a gene that arise by mutation.

A mutation is a permanent alteration in the sequence of deoxyribonucleic acid (DNA), the molecule that makes up a gene in one individual, resulting in a sequence that differs from that of other members of the species. Mutations may occur as the result of the miscopying of DNA during cell division, or they may be caused by environmental factors, such as damage resulting from the sun's ultraviolet radiation. One mutation might affect only the individual organism carrying it, whereas another might affect all its offspring and future generations.

Inherited mutations may or may not alter an individual's phenotype – its physical traits and behaviour. If mutations do affect the phenotype, they may be to its advantage or disadvantage, helping or hindering an organism's ability to survive and reproduce successfully. If they hinder, they are likely to disappear from the population; if they help an organism adapt better to its environment, they become more common over the course of generations. Over time, they may produce large enough divergences from the parent population for a new species to evolve – a process called speciation.

Mutation rates are usually very low, but the process is ever-present. The changes may be beneficial, neutral, or harmful. They do not occur in response to an organism's needs, and are, in that respect, random. However, some types of mutation occur more frequently than others. Scientists now know, for example, that evolution can take place very rapidly in bacteria because of their frequent mutations.

Different rates of evolution

The ancestors of all life on Earth were very simple organisms. Recent scientific research suggest that the earliest "biogenic" rocks – derived from early life forms – date back nearly four billion years. In that time, highly complex life forms have evolved, and later fossils of species that look more similar to those of today reveal what has

The vast majority of large mutations are deleterious; small mutations are both far more frequent and more likely to be useful.
Ronald Fisher

Seen in the light of evolution, biology is, perhaps, intellectually the most satisfying and inspiring science.
Theodosius Dobzhansky

occurred. For example, a fossil record stretches back 60 million years for ancestors of the horse. The earliest of these were dog-sized forest-dwelling animals with several toes on each foot. Evolution produced much larger horses with just a single hoof on each foot, adapted for life on open grasslands where they would often have had to outrun predators.

Peppered moths (*biston betularia*) reveal change over a shorter period. The moth is usually pale, providing camouflage against the bark of birch trees, but a mutation produces some black moths. Before the 19th century, most peppered moths were pale. During the Industrial Revolution (1760–1840), however, smoky air left deposits of soot on trees and buildings in British cities, and the black form became much commoner. By 1895, 95 per cent of peppered moths in Britain's cities were black, as paler moths were eaten by birds because their colouring provided no camouflage. This phenomenon continues to act as an example of Darwin's theory in action today, as the pale moth becomes common once more due to the declining soot concentrations in Britain's cities. ∎

Individuals within a species have a **variety of forms** of a **characteristic**.

↓

The **individuals** with the characteristic best **suited** to the **environment** are more likely to **survive** and **breed**.

↓

These characteristics are passed on to the next generation.

Two peppered moths exhibit evolution at work, the lower one an example of industrial melanism. The dark variety began to appear in British cities in the early 1800s.

Evolution in real time

Richard Lenski, a professor at Michigan State University, established the Long-term Experimental Evolution project in 1988. For more than 25 years, he studied 59,000 generations of the *E. coli* bacterium. During this time, he observed that the species used the glucose solution it lived in more efficiently, increasing in size but also growing faster. Also, a new species had evolved that was able to use a compound in the solution called citrate, which the parent bacterium could not. Evolving bacteria can pose a potential threat to humans. Increasing antibiotic use destroys many disease-causing bacteria, but not those with mutations that make them resistant to the drugs. As the non-resistant bacteria are killed off, the resistant strains become more dominant, multiplying and passing on their mutations to future generations. That is natural selection at work.

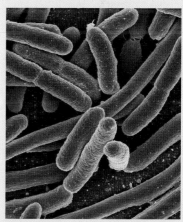

Escherichia (E.) coli bacteria can cause serious gut and other infections that will be increasingly difficult to treat as drug-resistant strains of *E. coli* multiply.

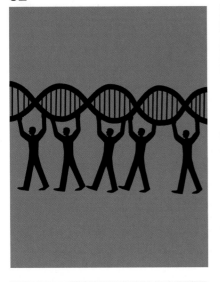

HUMAN BEINGS ARE ULTIMATELY NOTHING BUT CARRIERS FOR GENES
THE RULES OF HEREDITY

IN CONTEXT

KEY ECOLOGIST
Gregor Mendel (1822–84)

BEFORE
1802 French biologist Jean-Baptiste Lamarck suggests that traits acquired during the lifetime of an organism are transmitted to its offspring.

1859 Charles Darwin proposes his theory of evolution and natural selection in his book *On the Origin of Species by Means of Natural Selection*.

AFTER
1869 Swiss chemist Friedrich Miescher identifies DNA, which he terms "nuclein".

1953 Molecular biologists – including Briton Francis Crick and American James Watson – discover the structure of DNA.

2000s Researchers in the field of epigenetics describe inheritance by mechanisms other than through the DNA sequence of genes.

L‌ong before scientists cracked the genetic code, in 1866 an Austrian monk named Gregor Mendel was the first to show how traits are transferred through the generations. By means of much painstaking research, Mendel accurately predicted the basic laws of inheritance.

When Mendel began his experiments, scientists believed that the various traits seen in plants and animals were handed down through a "blending" process. However, Mendel noticed that this was not the case when he was working in his monastery garden. When he crossed a plant that always produced green peas with one that always produced yellow peas, the result was not yellowish-green peas – instead, all the peas were yellow.

Mendel's labours
During the course of his research (1856–63), Mendel grew nearly 30,000 pea plants over several generations and carefully recorded the results. He focused on traits

Mendel's pea experiment

Mendel's experiment with growing peas proved that the gene carrying the yellow colouration was dominant while the gene for green was recessive.

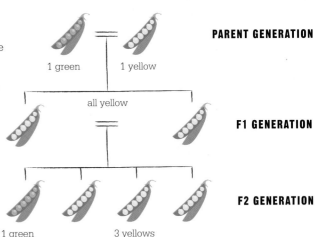

1 green   1 yellow   **PARENT GENERATION**

all yellow   **F1 GENERATION**

1 green   3 yellows   **F2 GENERATION**

See also: Early theories of evolution 20–21 ▪ Evolution by natural selection 24–31 ▪ The role of DNA 34–37 ▪ The selfish gene 38–39

> Heredity provides
> for the modification
> of its own machinery.
> **James Mark Baldwin**
> *American psychologist*

(phenotypes) that had only two distinct forms – for example, white or purple flowers. When examining the trait of yellow or green peas, Mendel took green pea plants and cross-pollinated them with yellow pea plants. The peas produced from this parent generation were all yellow and Mendel named them the F1 generation. He then cross-pollinated pea plants from the F1 generation with each other to produce the F2 generation. He found that some peas produced were yellow and some were green. The F1 generation showed only one trait (yellow), which Mendel called "dominant". However, in the F2 generation 75 per cent had the dominant yellow trait and 25 per cent displayed the non-dominant – or "recessive" – green trait.

Laws of inheritance

Mendel theorized that every pea plant has two factors controlling each trait. When plants are cross-pollinated, one factor is inherited from each plant. A factor can be

Pea plants provided the raw data that Mendel used to develop his theories explaining the transmission of traits from one generation to the next.

dominant or recessive. When both inherited factors are dominant, the resulting plant will show the dominant trait. With a pair of recessive factors, the plant will show the recessive trait. However, if one dominant and one recessive factor are present, the plant will show the dominant trait.

Pioneering geneticist

Mendel published his paper in 1866, but no one took much notice until 1900, when the botanists Hugo de Vries, Carl Erich Correns, and Erich Tschermak von Seysenegg rediscovered his work. Scientists then began proving Mendel's theories more widely.

Within just ten years, scientists named the pairs of factors "genes" and showed that they are linked on chromosomes. It is now known that inheritance is far more complex than Mendel recognized, but his meticulous research continues to form the basis for modern studies. ▪

Gregor Johann Mendel

Born Johann Mendel in 1822 on a farm in Silesia – then part of the Austrian Empire and now in the Czech Republic – Mendel studied philosophy and physics at the University of Olomouc (1840–43). At this time, he became interested in the work of Johann Karl Nestler, who was researching hereditary traits in plants and animals. In 1847 Mendel entered a monastery, where he was given the name Gregor. He then went on to study science further at Vienna University (1851–53).

When Mendel returned to his monastery in 1853, the abbot Cyril Napp gave him permission to use the gardens for his research into hybridization. Mendel himself became an abbot in 1868 and no longer had time for his experiments. Although he never received credit for his discoveries during his lifetime, he is widely regarded as the founder of modern genetics.

Key works

1866 "Experiments with Plant Hybrids", *Verhandlungen des naturforschenden Vereines in Brünn*

WE'VE DISCOVERED THE SECRET OF LIFE

THE ROLE OF DNA

IN CONTEXT

KEY FIGURES
Francis Crick (1916–2004),
Rosalind Franklin (1920–58), **James Watson** (1928–),
Maurice Wilkins (1916–2004)

BEFORE
1910–29 US biochemist
Phoebus Levene describes the
chemical components of DNA.

1944 US researchers Oswald
Avery, Colin Macleod, and
Maclyn McCarty show that
DNA determines inheritance.

AFTER
1990 British researchers,
led by embryologist Ian
Wilmut, successfully clone
an adult mammal – a sheep
named Dolly.

2003 Scientists complete
the mapping of the entire
human genome.

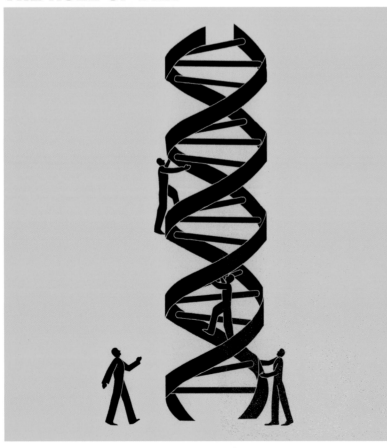

The discovery of the structure
of DNA (deoxyribonucleic
acid) in 1953 is one of
the most important scientific
breakthroughs to date. It offered
the key to understanding the very
building blocks of life and explained
how genetic information is stored
and transferred. Englishman Francis
Crick and American James Watson
famously celebrated their joint
discovery in a low-key fashion
at their local pub in Cambridge,
followed by a letter published in
the journal *Nature*. Their discovery
had enormous potential for scientific
advances and had an important
impact on many fields of research,
from medicine to forensic science,
taxonomy, and agriculture. The
ramifications of their work still
reverberate today, as methods of

See also: Early theories of evolution 20–21 ▪ Evolution by natural selection 24–31 ▪ The rules of heredity 32–33 ▪ The selfish gene 38–39 ▪ A system for identifying all nature's organisms 86–87 ▪ Biological species concept 88–89

Molecular biologists James Watson (left) and Francis Crick (right), pictured in 1953 with their double helix model of DNA. Watson called DNA "the most interesting molecule in all nature".

handling genetic material advance and we learn more about how individual genes operate.

Crick and Watson's breakthrough was the culmination of decades of research by numerous scientists, including Rosalind Franklin and Maurice Wilkins. While Crick and Watson worked with 3-D models to figure out how the components of DNA fitted together, at King's College, London, Franklin and Wilkins were developing methods of X-raying DNA to view its structure. Watson had seen examples of Franklin's work that hinted at DNA's helical shape shortly before he and Crick announced their breakthrough.

In 1962 Crick, Watson, and Wilkins were awarded the Nobel Prize for Physiology or Medicine. Franklin, who died in 1958, never received recognition for her part in the discovery during her lifetime, although Crick and Watson openly acknowledged that her work was essential to their success.

Double helix structure

DNA is a molecule featuring two long, thin strands that coil around each other to resemble a twisted ladder, in a shape known as a double helix. Using the ladder analogy, the sides of the ladder are made up of deoxyribose (a sugar) and phosphate, while the rungs of the ladder consist of paired nitrogenous bases, adenine (A), guanine (G), cytosine (C), and thymine (T). A always pairs up

DNA is like a computer program but far, far more advanced than any software ever created.
Bill Gates

with T to form base pair AT, and G always pairs with C to form base pair GC.

DNA is the blueprint for life. Sequences of bases along the DNA strand constitute the genes that provide the information that determines the complete form and physiology of an organism. A triplet of bases is known as a codon, and each codon specifies the production of one of 20 amino acids; the order in which the amino acids join together in a chain determines »

Genetic engineering

Understanding the structure of DNA has enabled scientists to change or "engineer" the genetic material in cells. It is possible to cut out a gene from one organism (the donor) and place it into the DNA of another organism. When this practice was first attempted in the 1970s it was both difficult and time-consuming, but technological advances – such as Clustered Regularly Interspaced Short Palindromic Repeats, or CRISPR, which has been particularly useful – have greatly simplified and accelerated the process. In theory, geneticists can now splice any gene with any other. They have attempted some intriguing combinations, such as the insertion of the gene for producing spider silk into goat DNA so that goats produce milk rich in proteins. Other substances that can be produced by modifying genes are hormones and vaccines.

In gene therapy, a genetically modified vector (often a virus) is used to carry a gene into the DNA of an organism to replace a faulty or unwanted gene.

A scientist analyses a sample of DNA. Genetic manipulation in medicine is standard practice and DNA profiling is a vital forensic tool.

Genetically modified food

In agriculture, crops may be engineered to enhance them in some way. A genetically altered crop is known as a genetically modified organism (GMO). Companies that operate in this sector may modify a plant's DNA so that it produces more of a certain nutrient or a toxin specific to a particular insect pest. The DNA of a plant may also be altered to become resistant to a particular herbicide, so that use of the chemical kills only the weeds and not the crop.

Some ecologists argue that there is a risk of genetically unmodified plants becoming contaminated by GMOs. They also point out that the long-term effects of eating such foods are as yet not properly understood. Another concern is that in the future large agrochemical companies could control the world's food supply by patenting the GMOs that they produce, to the detriment of poorer nations.

New kinds of rice are being developed through genetic modification. This may improve the nutritional value of the crop or its resistance to disease.

the type of protein they go on to make. For example, the combination GGA is the codon for glycine. Sixty-four possible triplets can be made from the four base pairs, and 61 of them code for a particular amino acid. The other three act as signals such as "start" and "stop", which govern how information is read by the cellular machinery. DNA is also organized into separate chromosomes, of which there are 23 pairs in the human cell.

Copying the code

When cells divide, DNA needs to be copied. This is achieved by the splitting of base pairs, which cuts

The structure of DNA

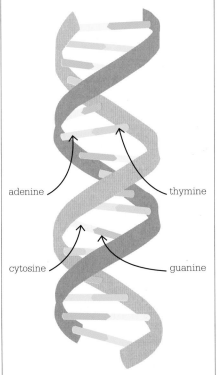

adenine

thymine

cytosine

guanine

A DNA molecule consists of a double helix formed by two strands, made up of sugars and phosphates, linked by paired base nucleotides: adenine and thymine or cytosine and guanine.

the ladder down the middle to produce two single strands. These act as templates for the production of a second complementary DNA strand on each of them by matching up the appropriate base pairs. The process results in two strands of whole DNA that are exactly the same as the original.

Since DNA remains in the nucleus of the cell, a related molecule called messenger ribonucleic acid (mRNA) copies segments of DNA coding sequence and carries the information to the regions of the cell where new proteins are made. RNA is chemically related to DNA, but the thymine base (T) is replaced by the base uracil (U) which is less stable but requires less energy to make. Stable living organisms benefit from having DNA genomes, but RNA makes up genomes of some viruses, where stability can be less advantageous.

DNA is found in all living things on Earth, from amoebae to insects, to trees, tigers, and humans. Of course, the sequence of base pairs varies, and this difference allows geneticists to trace relationships between different species.

Good and bad errors

DNA is a highly stable molecule, but sometimes mistakes, known as mutations, occur. These can be in the form of an error, duplication, or omission in the order of the nucleotides A, C, G, and T. Mutation can be spontaneous – the result of errors that occur when the DNA is copied – or may be induced by external influences such as exposure to radiation or cancer-causing chemicals. Some mutations have no effect, but others may change what the gene produces or inhibit the functioning of a gene. This can lead to problems in the organism as a whole.

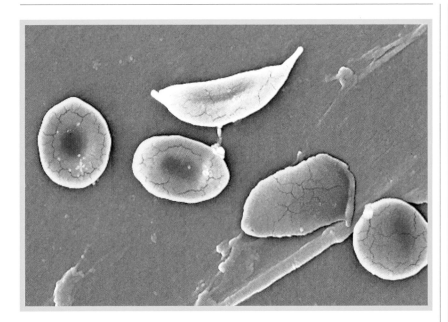

Examples of disorders caused by gene mutations include cystic fibrosis and sickle-cell disease.

Although many mutations are harmful, occasionally a mutation will confer an advantage on an individual, enabling it to survive in its environment better than others of the same species. This type of mutation may end up being passed on through the process of natural selection. Over many generations, mutation is a mechanism for diversification, survival of the fittest, and ultimately evolution.

The human genome
On 14 April, 2003, scientists completed the lengthy task of mapping (sequencing) the entire human genome. Geneticists worked out the precise position of all the base pairs in a chain of some three billion of the base nucleotides comprising an estimated 30,000 individual genes. This has allowed geneticists to identify new genes and the role they play in organisms.

Armed with this knowledge, an individual can find out if they have inherited a faulty gene from

Mutated blood cells occur in sickle-cell disease – a genetic disorder passed on when both parents carry the faulty gene. The condition can be painful and increases the risk of serious infections.

a parent. Additionally, with access to such data it is possible to screen embryos for known genetic disorders before implantation in the womb. By March 2018, the DNA of around 15,000 organisms had been sequenced. Such information can help to show how animals are related in the evolutionary line and how they have diversified.

While the discovery of the composition and structure of DNA has revolutionized the science of heredity, it is worth noting that the regions of DNA used for coding proteins account for just 2 per cent of the entire human genome. The nature of the other 98 per cent is not yet fully understood by geneticists, but it is believed that at least some of these regions involve the regulation of the way genes are expressed, or activated. It seems that many more discoveries await future geneticists. ∎

DNA barcoding
The idea of DNA barcoding was first raised in 2003 when a team at the University of Guelph, Canada, suggested that it would be possible to identify species by analysing a common section of their DNA. Led by Dr Paul Hebert, researchers chose a region in the gene known as cytochrome c oxidase 1 ("CO1"), made up of 648 base pairs. This region is quick to analyse, but the sequence is still long enough to differentiate between and within animal species. Different gene segments can be used for other forms of life.

The first part of the barcoding system involves cataloguing samples of known species. The DNA is extracted and organized into a sequence of base pairs, a process known as "sequencing". The sequence is then stored in a computer database, so that when a DNA sample from an unknown species is sequenced and entered into the database, the computer will match it with existing records. The barcoding technique has proved useful for taxonomy, helping to classify animals and plants.

With genetic engineering, we will be able ... to improve the human race.
Stephen Hawking

GENES ARE SELFISH MOLECULES

THE SELFISH GENE

IN CONTEXT

KEY FIGURE
Richard Dawkins (1941–)

BEFORE
1963 British biologist William Donald Hamilton writes about the "selfish interests" of the gene in *The Evolution of Altruistic Behaviour.*

1966 American biologist George C. Williams proposes in his book *Adaptation and Natural Selection* that altruism is a result of selection taking place at the level of the gene.

AFTER
1982 Richard Dawkins argues in *The Extended Phenotype* that the study of an organism should include analysis of how its genes affect the surrounding environment.

2002 Stephen Jay Gould critiques Dawkins' theory in *The Structure of Evolutionary Theory,* which revisits and refines the ideas of classical Darwinism.

The concept of the "selfish gene" was popularized by British evolutionary biologist Richard Dawkins in his 1976 book of that name. It states that evolution is fundamentally based upon the survival of different forms of a particular gene at the expense of others. The forms that survive are those that are responsible for the bodily types and behaviours (phenotypic traits) that successfully promote their own propagation. Supporters of the theory argue that because heritable information is passed through the generations by the genetic material of DNA, both natural selection and evolution are best considered from the perspective of genes.

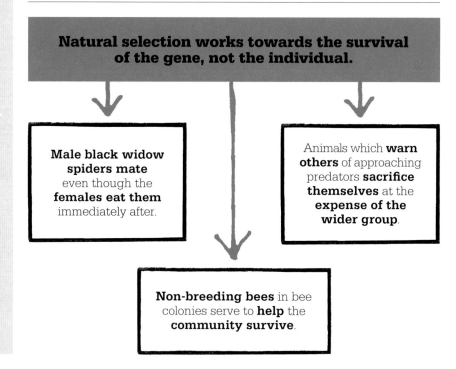

Natural selection works towards the survival of the gene, not the individual.

Male black widow spiders mate even though the **females eat them** immediately after.

Animals which **warn others** of approaching predators **sacrifice themselves** at the **expense of the wider group**.

Non-breeding bees in bee colonies serve to **help** the **community survive**.

See also: Evolution by natural selection 24–31 ▪ The rules of heredity 32–33 ▪ The role of DNA 34–37 ▪ Mutualisms 56–59

A male black widow spider gingerly approaches a huge female to mate. This genetically driven act will reproduce his genes but will lead to his death.

Dawkins was strongly influenced by the work of William Donald Hamilton on the nature of altruism and closely examined the biology of selfishness and altruism in *The Selfish Gene*. He argued that organisms were simply vehicles that supported their genes, or "replicators". Genes that help an organism to survive and reproduce tend also to improve those genes' own chances of being replicated.

Successful genes often provide a benefit to the host organism. For example, a gene that protects an animal or plant against disease thereby helps that particular gene to spread. However, the interests of the replicator and the vehicle may sometimes seem to be in conflict. Genes drive the male black widow spider to mate despite the risk of being eaten by her. However, the male's sacrifice nourishes the female and improves the prospect of his genes being passed on.

Selfishness and altruism

Gene selfishness usually gives rise to selfishness in the behaviour of an individual organism, but there are circumstances in which the gene can achieve its own selfish goals by fostering apparent altruism in the organism. One example is kin selection, the evolutionary strategy that favours the reproductive success of an individual organism's relatives, even at the cost of the individual's own reproduction or survival.

An extreme example of genetically based altruism is eusociality. Honey bees are a eusocial species. They live in colonies which include breeding and non-breeding individuals. By helping the colony survive, the many thousands of non-breeding worker bees ensure the reproduction of the genes they have in common with the sole breeding individual, the queen.

Critics of Dawkins' theory argue that since individual genes do not control behaviour, they cannot be said to be acting selfishly. Dawkins has maintained that he never meant to suggest that genes had their own conscious will. He later wrote that "the immortal gene" might have been a better title for both his concept and the book. ▪

The theory of evolution is about as much open to doubt as the theory that the Earth goes around the Sun.
Richard Dawkins

Richard Dawkins

Richard Dawkins was born in Kenya to British parents. After the family returned to the UK, he developed a strong interest in the natural world and studied zoology at Oxford University. While there, he was tutored by Nobel Prize-winner Niko Tinbergen, who was a pioneer of animal behaviour studies. After a brief spell at the University of California at Berkeley, Dawkins returned to Oxford to lecture in zoology.

Richard Dawkins is best known for his book *The Selfish Gene*, in which he argues that the gene is the principle unit of selection in evolution. His theory later triggered a series of fierce debates with Stephen Jay Gould and other evolutionary biologists. Dawkins is also known as a strong advocate of atheism and feminism.

Key works

1976 *The Selfish Gene*
1982 *The Extended Phenotype*
1986 *The Blind Watchmaker*
2006 *The God Delusion*
2009 *The Greatest show on Earth: The Evidence for Evolution*

ECOLOGI
PROCESS

CAL
E S

Joseph Grinnell publishes his research on the California thrasher bird, establishing the basis for the theory of **ecological niches**.

1917

Robert MacArthur's research on North American warblers shows how different species can **avoid directly competing** with each other in order to coexist.

1957

Dan Janzen observes the interdependence of acacia trees and the ants that reside on them, and concludes that the species evolved in a **mutualistic** manner.

1965

1925–26

The **Lotka-Volterra model** uses a mathematical equation to describe the interactions between **predator and prey**.

1961

Joseph Connell reveals that different types of barnacle **thrive in different tidal zones**, although they could, in theory, live in any of them.

1969

Robert Paine coins the term "**keystone species**" to describe species that play a crucial role in ecosystem functions.

In the 5th century BCE, the Greek historian Herodotus described watching crocodiles open their jaws for plovers to pick food from their teeth. He may have been the first to write about an ecological process – in this case a mutualistic relationship between reptiles and birds. Aristotle and Theophrastus observed many more interactions between animals and their environment in the 4th century BCE.

Over the next two millennia, countless other observations of the natural world were made, but a deep understanding or how organisms interacted with each other and the world around them was hampered by the inability to observe very small things, those that were active at night, or those living underwater. Additionally, few people with an interest in nature experienced much beyond their own local area. As technology improved and people began to travel the world, scientists such as Robert Hooke, Antonie van Leeuwenhoek, Carl Linnaeus, Alexander von Humboldt, Alfred Russel Wallace, Charles Darwin, and Johannes Warming became increasingly aware of ecological processes and laid the foundations of the science of ecology, even if they didn't use that word.

Mathematical models

It had long been understood that one of the most basic ecological processes is the struggle for survival: for herbivores to find food, predators to find prey, and prey to avoid being eaten. Predators do everything they can to hunt and eat prey, and the latter do all they can to avoid being eaten. In 1910, Alfred Lotka introduced one of the first mathematical models ever applied to ecology. Now known as the Lotka-Volterra model, its predator–prey equations help to predict the population fluctuations of these two groups.

In the early years of the 20th century, Joseph Grinnell conducted extensive research into animals' habitat needs in the western United States. He observed that species had different "niches" within a habitat – and that if two species have approximately the same food requirements, one will "crowd out" the other. Darwin had observed this on his travels aboard HMS *Beagle*, but Grinnell's axiom developed the idea further, as did subsequent research. In 1934, Georgy Gause demonstrated what he called the competitive exclusion principle in

Roy Anderson and **Robert May** demonstrate how **epidemics** affect animal population growth rates.

Research published by **Ronald Pulliam**, **Eric Charnov**, and **Graham Pyke** expands on the **optimal foraging theory** that animals try to gather resources while wasting as little energy as possible.

Robert Sterner and **James Elser** pioneer the study of **ecological stoichiometry** – how ratios of different chemicals within living organisms change with certain reactions.

1970s

1977

2002

1972

1991

Knut Schmidt-Nielsen publishes *How Animals Work*. The book hugely influences the field of ecophysiology.

Earl Werner publishes his findings about **non-consumptive effects** of predators on prey.

---

laboratory projects. As William E. Odum put it in 1959, "the ecological niche of an organism depends not only on where it lives, but also on what it does."

From field to lab
Laboratory experiments and field observations are the main methods of providing data for the study of ecological processes, but field experiments – in which a local environment is manipulated to test a hypothesis – were not conducted with scientific rigour until Joe Connell's work with barnacles in Scotland. His experiments – the results of which were published in 1961 – were meticulously planned and observed, and were repeatable.

Connell set the "gold standard" for fieldwork, but experiments in laboratories still have a vital role

to play, too – as Earl Werner demonstrated 30 years later. His work revealed the non-consumptive impact of predatory dragonfly larvae on the behaviour and physical development of their tadpole prey.

Since the mid-20th century, many new ideas on ecological processes have emerged. Work by Robert MacArthur and others on competition between species led to the development of optimal foraging theory, which seeks to explain why animals choose to eat some food items and not others. Mutualistic relationships became better understood through the research of biologists such as Daniel Janzen. Robert Paine's work with starfish and mussels also highlighted the concept of keystone species – those that have a disproportionate influence on their ecosystems.

New technology
Technological advances – including sophisticated chemical sampling techniques, satellites with remote sensing equipment, and computers capable of rapidly processing huge quantities of data – have opened up new areas of study.

Ecological stoichiometry, for example, studies the flow of energy and chemical elements throughout food webs and ecosystems, from the molecular level up. Like so many ideas in ecology, its origins can be traced back many years, but only took hold with Robert Sterner and James Elser's 2003 book *Ecological stoichiometry: The biology of elements from molecules to the biosphere*. New techniques such as this will undoubtedly continue to deepen our understanding of processes in ecology. ∎

LESSONS FROM MATHEMATICAL THEORY ON THE STRUGGLE FOR LIFE

PREDATOR–PREY EQUATIONS

IN CONTEXT

KEY FIGURES
Alfred J. Lotka (1880–1949),
Vito Volterra (1860–1940)

BEFORE
1798 British economist
Thomas Malthus shows that
the rate at which the population
changes increases as the size
of the population grows.

1871 In Lewis Carroll's novel
Through the Looking Glass,
the Red Queen tells Alice,
"you have to run just to stay
in the same place".

AFTER
1973 American biologist Leigh
Van Valen proposes the Red
Queen effect, which describes
the constant "arms race"
between predators and prey.

1989 The Arditi–Ginzburg
equations offer another model
of predator–prey dynamics
by including the impact of the
ratio between predator and prey.

Populations of **two species**,
one **predator**, the
other **prey**, interact.

The prey has **access to
food** and its **population
growth** is **exponential**.

When prey animals **meet
a predator**, they
are **eaten**.

Eating prey results
in **more predators**.

**More predators
results in less
prey, reducing the
number of predators.**

The predator–prey equations
are an early example of the
application of mathematics
to biology. Formulated in the 1920s
by American mathematician Alfred
J. Lotka and Italian mathematician
and physicist Vito Volterra, the
two equations – also known as
the Lotka–Volterra equations –
describe the way in which the
population of a predator species
and that of its prey fluctuate in
relation to each other.

Lotka proposed the equations
in 1910, as a way of understanding
the rates of autocatalytic chemical
reactions – chemical processes
that regulate themselves. In the
following decade, he applied
the equations to the population
dynamics of wild animals.

In 1926, Vito Volterra arrived
at the same conclusions. He had
become interested in the subject
after meeting Italian marine
biologist Umberto D'Ancona.
D'Ancona told Volterra how the
percentage of predatory fish
caught in nets in the Adriatic
Sea had greatly increased during
World War I. This change was
clearly linked to the drastic
reduction in fishing during the

Vito Volterra

Born in 1860 in Ancona, Italy, the
son of a Jewish cloth merchant,
Vito Volterra grew up in poverty.
Despite this, in 1883, aged just 23,
he secured a position as professor
of mechanics at the University
of Pisa and began a career as
a mathematician. Further
professorships at the universities
of Turin and Rome followed. In
1900, Volterra married, fathering
six children, although only four
survived to adulthood. He was
made a senator of the Kingdom
of Italy in 1905 and worked on the
development of military airships
during World War I. In 1931,

Volterra refused to swear loyalty
to Italy's fascist dictator Benito
Mussolini and was dismissed
from the University of Rome.
Forced to work abroad, he only
returned to Italy for a short time
before his death in 1940.

Key works

1926 "Fluctuations in the
Abundance of a Species
Considered Mathematically",
Nature
1935 *Les associations
biologiques au point de vue
mathématique*

A cheetah pursues a Thomson's gazelle. The predator–prey equations are able to model the way populations of both species will change in response to the activities of the other.

war years, but D'Ancona could not explain why less fishing did not produce more fish of all kinds in the nets. Using the same equations as Lotka, Volterra eventually explained the fluctuations in both the predator and the prey species.

Population principles
At the time Lotka and Volterra made their calculations, the science of population dynamics was still in its infancy, having barely moved on since the population studies of British economist Thomas Malthus in the late 18th century. According to Malthus's theory, a population grows or declines rapidly as long as the environmental factors for survival are constant, and the rate at which that population changes increases as the population grows. From this theory, Malthus predicted a catastrophic future for humanity. The number of humans was growing much more quickly than the amount of food that could be produced by the world's farmlands. Eventually, Malthus argued, a point would be reached when the human population would succumb to global famine and decline.

Malthus's bleak vision did not happen, thanks to technological advances in agriculture and the development of artificial fertilizers, but his population model became applicable to species populations within ecosystems. Every habitat, and the niche occupied by a species within its community of organisms, has a carrying capacity – the maximum population that can be supported by the resources available, such as water, space, food, and light. Any rise in population above this level is likely to be reduced by naturally occurring factors. As a result, wild populations should in theory be more or less static, fluctuating only around the carrying capacity, assuming the random impacts of catastrophic events are ignored.

However, this relative equilibrium did not always match up with observations – as in »

The food species cannot, therefore, be exterminated by the predatory species, under the conditions to which our equations refer.
Alfred J. Lotka

> Mathematics without
> natural history is sterile, but
> natural history without
> mathematics is muddled.
> **John Maynard Smith**
> *British mathematician
> and evolutionist*

D'Ancona's account of a sudden increase in the population of predatory sea fish. One theory to explain this discrepancy started from the premise that the population of predators is related to the size of the population of their food supply, such as prey species. The relationship suggests that when a lot of food is available, there will be a large predator population. The growing predator population should then begin to reduce the amount of prey, which will in turn lead to a drop in the number of predators. The size of both populations will rise and fall, but the ratio of predators to prey will remain stable.

Such a balanced theory was still at odds with species observations. Through mathematical modelling, Volterra was able to show that the average sizes of predator and prey populations do indeed oscillate but the rate at which each population is growing or declining is always changing and almost never matches the changes experienced by the other population. To eliminate variables, Volterra made a series of assumptions: first, that the prey and predator species have no reproduction limits and the rate of change in a population is proportional to its size; second, that the prey population – presumed to be a herbivore – is always able to find enough food to survive. Next, they assumed that the prey population is the predators' only source of nourishment, and that the predators never become full and never stop hunting. Finally, they assumed that environmental conditions, such as weather or natural disasters, had no impact on the process. The effect of the genetic diversity of the predators and prey animals on their ability to survive was not taken into account.

When plotted on a graph, the predator population trails the rise and fall of the prey population, and is still rising as the prey population starts to decline. This explained D'Ancona's observation of the larger proportion of predators after the prey population had been allowed to boom by a reduction in fishing.

The relative fluctuations of the populations depends on the relative reproductive rates of the two species and the predation rate. For example, oscillations in the size of an ant population and that of an anteater are barely noticeable because they reproduce at such different rates. The oscillations in the populations of species that breed at similar rates, such as the Iberian lynx and rabbit, are much more pronounced.

Nature's arms race

The predator–prey equations revealed that species are locked together in a never-ending struggle, swinging from near disaster and extinction to times of abundance and fertility. In this biological "arms race", the evolutionary pressure on the prey species is to escape predation and survive, so as to have more offspring. Meanwhile, the predator is under pressure to have a higher predation rate in order to provide food for more offspring. However, neither species is superior, responding instead to the adaptations of the other. The predator–prey relationship between even-toed hoofed mammals – such

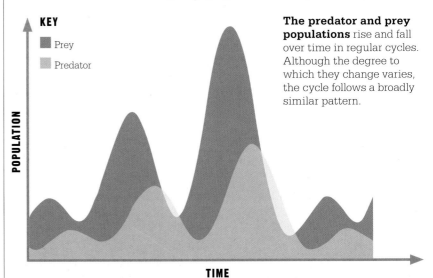

Predator–prey population cycles

KEY

Prey

Predator

POPULATION

TIME

The predator and prey populations rise and fall over time in regular cycles. Although the degree to which they change varies, the cycle follows a broadly similar pattern.

as antelopes and deer – and mammalian carnivores, like the big cats and wolves, is an example of this evolutionary arms race. The hoofed animals have long legs, extended by walking on the very tips of thickened and fused toe bones. This adaptation allows them to outrun and outjump their predators. In response, big cats – such as lions and tigers – have evolved speed and strength to bring down large, fleet-footed prey in surprise attacks. Wolves have evolved the stamina to run for long distances without stopping. This allows them to work as a team to chase down their prey and kill them when the exhausted prey collapse.

While the predator–prey equations offer an insight into the population dynamics of two species, the assumptions they rely on are rarely reflected in real life. Some predators do specialize in killing a single prey species, but other factors in the ecosystem also affect their populations.

Other applications

The Lotka–Volterra equations have been used to study the dynamics of food chains and food webs in which one species may be a predator of

Volterra was interested in a mathematical theory of 'the survival of the fittest'.
Alexander Weinstein
Russian mathematician

another species but also the prey species of a third. They have also been used to examine the relationship between host and parasite species, which bears some resemblance to that between prey and predator. Parasites often specialize in one host species – a relationship that should resemble the one described by the Lotka–Volterra equations. However, in practice the process of evolution is thought to interfere with this. A parasite does not usually kill its host (those that do are called parasitoids), but can reduce its fitness. The Red Queen evolutionary theory, proposed in the 1970s by Leigh Van Valen, describes how,

The parasitoid wasp lays its eggs in aphids (the smaller, yellow insects shown above). It is called a parasitoid because the wasp's larvae later eat the aphids as they grow.

thanks to beneficial genes, certain individuals in a host population are able to maintain their fitness despite the attacks from parasites. The parasites constantly evolve to exploit these seemingly immune individuals, and therefore the beneficial genes in the host population also change. In this way, evolution is happening all the time, as the parasite and host battle it out – although everything appears to stay the same. ■

EXISTENCE IS DETERMINED BY A SLENDER THREAD OF CIRCUMSTANCES

ECOLOGICAL NICHES

IN CONTEXT

KEY FIGURE
Joseph Grinnell (1877–1939)

BEFORE
1910 In a paper about beetles, Roswell Hill Johnson, a US biologist, is the first person to use the word "niche" in a biological context.

AFTER
1927 British ecologist Charles Elton stresses the importance of an organism's role as well as its "address" in his definition of an ecological niche in his book *Animal Ecology*.

1957 In an academic paper called "Concluding Remarks", British ecologist George Evelyn Hutchinson expands the theory of niches to embrace the whole of an organism's environment.

1968 A study by Australian D.R. Klein of the introduction, increase, and die-off of reindeer on St Matthew Island, Alaska, identifies the destructive niche.

An organism's niche is a combination of its place and its role in the environment. It encompasses how the organism meets its needs for food and shelter, as well as how it avoids predators, competes with other species, and reproduces. All its interactions with other organisms and the non-living environment are also part of what makes up its niche. A unique niche is an advantage for any animal or plant because this reduces competition with other species. For ecologists, a full knowledge of an organism's niche is vital to inform interventions to compensate for the environmental changes caused by habitat destruction and climate change.

The pioneer of the niche concept was Joseph Grinnell, a US biologist who studied a bird called the California thrasher. In 1917, he published his observations, which showed how the bird fed and bred in the underbrush of a scrubby

There is **constant competition** for **food and resources**; better adapted species outcompete those less suited to the environment.

→ **Reducing competition** increases the chances of survival.

↓

Finding a **unique niche** is the circumstance that **removes competition**.

← **Existence of each species is determined by a slender thread of circumstances.**

See also: Competitive exclusion principle 52–53 ▪ Field experiments 54–55 ▪ Optimal foraging theory 66–67 ▪ Animal ecology 106–113 ▪ Niche construction 188–189

An ultra-specialist

Giant pandas occupy a very specialized ecological niche, as their diet consists mainly of bamboo. Bamboo is a poor food source, low in protein and high in cellulose. Pandas can digest only a small proportion of what they eat, which means they have to eat a lot of bamboo – as much as 12.5 kg (28 lb) each day – and forage for up to 14 hours a day. It is unclear why pandas have become so dependent on bamboo, but some zoologists suggest it is because it is an abundant and reliable food source, and pandas are not skilled predators.

Pandas eat different parts of bamboo plants according to the seasons. In late spring, they prefer the first green shoots. They eat leaves at other times of the year, and stems in winter when little else is available. Pandas have evolved muscular jaws and a pseudothumb to manipulate bamboo stems. Their digestive tract is inefficient at processing large quantities of plant material because it remains similar to that of its carnivorous ancestors, although digestion is helped by the bacterial fauna in their gut.

habitat known as chaparral, and how it escaped predators by running through the underbrush. The thrasher's camouflage, short wings, and strong legs were perfectly adapted for life in this environment. Grinnell saw the chaparral habitat as the thrasher's "niche". His idea also allowed for "ecological equivalence" in plants and animals, whereby species distantly related and living far apart could show similar adaptations, such as feeding habits, in similar niches. In the Australian outback, for instance, babbler bird species forage in the scrubby vegetation in a similar way to the unrelated thrasher. Grinnell also identified "vacant" niches – habitats that a species could potentially occupy, but where it was not present.

Widening the niche

In the 1920s, ecologist Charles Elton looked beyond a simple habitat definition for "niche". For him, what an animal ate and what it was eaten by were the primary factors. Thirty years later, George Evelyn Hutchinson expanded the definition yet further. He argued that a niche should take into account all of an organism's interactions with other organisms and its non-living environment, including geology, acidity of soil or water, nutrient flows, and climate. Hutchinson's work encouraged others to explain the variety of resources used by a single organism (niche breadth), the ways in which competing species

[A niche] is a highly abstract multi-dimensional hyperspace.
George Evelyn Hutchinson

coexist (niche partitioning), and the overlap of resources by different animals and plants (niche overlap).

The importance of habitat

Ecological niches depend on the existence of a stable habitat; small changes can eradicate niches that organisms once filled. For example, dragonfly larvae only develop within a certain range of water acidity, chemical composition, temperature, and prey, and with a limited number of predators. The right vegetation is needed by adult females for egg-laying, and by larvae for metamorphosis. The dragonfly also impacts its environment: its eggs are food for amphibians; its larvae, which are both predators and prey, add nutrients to the water; and the adults prey on insects. These requirements and impacts define its ecological niche. Hutchinson argued that for a species to persist, conditions had to be within the required ranges. If conditions moved outside the niche requirements, a species could face extinction. ▪

COMPLETE COMPETITORS CANNOT COEXIST

COMPETITIVE EXCLUSION PRINCIPLE

IN CONTEXT

KEY FIGURE
Georgy Gause (1910–86)

BEFORE
1925 Alfred James Lotka first uses equations to analyse variations in predator–prey populations, as does mathematician Vito Volterra, independently, a year later.

1927 Volterra enlarges and updates his 1926 study to include various ecological interactions within communities.

AFTER
1959 G. Evelyn Hutchinson extends Gause's ideas and produces a ratio describing the limit of similarity between two competing species.

1967 Robert MacArthur and Richard Levins use probability theory and Lotka–Volterra equations to describe how coexisting species interact.

Competition is the driver of evolution; the need to be bigger, stronger, and better inevitably leads to adaptations that give a species an edge. When two species compete for identical resources, the one which has any advantage will outdo the other. As a result, the weaker of the two species either becomes extinct or adapts, so that it no longer competes. This proposition, known as the "competitive exclusion principle", was set out by Russian microbiologist Georgy Gause and is also known as Gause's Law.

Gause devised his principle from laboratory experiments, using cultures of microorganisms, rather than from observations in nature. In

How warblers coexist

Cape May
warbler

Blackburnian
warbler

Black-throated
green warbler

Bay-breasted
warbler

Yellow rumped
(Myrtle) warbler

Five species of warbler are able to share the same tree, because each inhabits its own "niche". Living in this way, without much overlap, the birds do not compete.

The red squirrel is smaller than the grey, and has a more restricted diet and habitat. Reds may also die from the squirrel parapoxvirus, which is carried by the greys but does not affect them.

nature, he proposed, there were too many variables to draw conclusions about how ecological mechanisms work. He argued that little progress had been made since Darwin's era in understanding how species compete for survival, whereas the experimental method had produced great advances in areas such as genetics. In fact, the competitive exclusion principle – although a useful theoretical model – is rarely seen in nature, simply because, in a bid to survive, a weaker competitor tends to quickly move on or adapt.

Avoiding competition

Most creatures can make the changes necessary for survival. A variety of birds can live in a garden during any one year because they all operate in different "niches". They have contrasting beak shapes and sizes that allow them to eat different types of food – the robin preferring insects, the finch eating seeds. Their choice of habitat and feeding times might also vary; this is known as resource partitioning.

In 1957, Robert MacArthur noted this phenomenon in North American warblers. The five species he observed, each with distinctive, colourful markings, flitted in and out of coniferous trees, feeding on bugs and other insects. They could coexist in one habitat because they did not try to feed in the same part of the tree but

Let us make for this purpose an artificial microcosm… let us fill a test-tube with nutritive medium and introduce to it several species of protozoa consuming the same food or devouring each other.
Georgy Gause

at different heights and depths of the foliage. In this way they avoid competing with each other.

An invasive competitor

Problems often arise if an exotic species is suddenly introduced to an ecosystem. Britain's red and grey squirrels provide a clear example. When the grey arrived from America in the 1870s, both squirrel species competed for the same food and habitat, which put the native red squirrel populations under pressure. The grey had the edge because it can adapt its diet; it is able, for instance, to eat green acorns, while the red can only digest mature acorns. Within the same area of forest, grey squirrels can decimate the food supply before red squirrels even have a nibble. Greys can also live more densely and in varied habitats, so have survived more easily when woodland has been destroyed. As a result, the red squirrel has come close to extinction in England. ▪

Types of competition

The Competitive Exclusion Principle covers two main types of competition. Intraspecific competition is between individuals of the same species and ensures the survival of the fittest, so that only the healthiest individuals – or those best adapted to a particular environment – will breed. The second type is interspecific: competition between two different species that rely on the same resources. The most important of these will be the "limiting resource", the one that both require in order to breed. Ecologists make a further two distinctions. Interference is when two organisms fight directly with each other over a limited resource, such as a mate or a preferred food. Exploitation is indirect competiton, such as stripping out a resource so there is none left for the competitor; this can be seen in plants, when a species' uptake of nutrients or water is more efficient than that of its neighbours.

POOR FIELD EXPERIMENTS CAN BE WORSE THAN USELESS

FIELD EXPERIMENTS

IN CONTEXT

KEY FIGURE
Joseph Connell (1923–)

BEFORE
1856 British scientists John Lawes and Joseph Gilbert start the Park Grass Experiment at Rothamsted, to test how different fertilizers affect the yield of hay meadows.

1938 Harry Hatton, a French ecologist, conducts one of the first marine ecology field experiments, on barnacles on the Brittany coast.

AFTER
1966 American ecologist Robert Paine removes *the starfish Pisaster ochraceus* from tide pools in a Pacific coast ecosystem, to test the effect of its absence on other species.

1968 The Experimental Lakes Area, comprising 58 freshwater lakes, is established in Ontario, Canada, to study the effects of nutrient enrichment (eutrophication).

Experimentation is crucial in ecology. Without it, our ideas about why organisms behave the way they do would be largely speculative. Rigorous observation is also essential, but, much of the time, experimentation is needed for a full understanding of those observations.

Three main types of ecological experiments are used to test theories: mathematical models, laboratory experiments, and field experiments. Each method has its merits, but it is only recently that the benefits of field experiments have been recognized. Before the 1960s, experiments outside of a laboratory were a rarity.

A laboratory, however, is an artificial environment, where organisms may not behave as they do in their natural habitat. For example, bats leaving a roost at dusk may follow different routes to their foraging areas in spring and late summer. The potential reasons for the switch – changes in prey distribution and predator threats; seasonal differences in tree cover; or human disturbance and light pollution – could not be established in a laboratory. Mathematical modelling might help to predict patterns, but would be less effective at identifying the causes of change. To understand the bats' behaviour, a study of their natural environment is crucial, and this is achieved only through field techniques.

Field experiments allow different factors to be manipulated to test their relevance. In the bat example,

Rainforest ecosystems are some of the most species-rich environments on Earth. This makes them especially valuable sites for ecologists to conduct experiments in the field.

See also: Ecological niches 50–51 ▪ A modern view of diversity 90–91 ▪ Animal behaviour 116–117 ▪ The ecosystem 134–137 ▪ Niche construction 188–189

street lights could be switched off to evaluate the impact of light pollution on their behaviour change.

Scottish barnacles

In 1961, American ecologist Joseph Connell published the results of his research on barnacles on the Scottish coast. Since free-swimming barnacle larvae can settle anywhere, Connell had tested why the lower part of the intertidal zone was colonized by *Balanus balanoides* barnacles and the upper part by *Chthamalus stellatus*. He wanted to know if this was due to competition, predation, or environmental factors.

Connell manipulated the local environment, and monitored it for over a year. In one area, he removed the *Chthamalus* barnacles. They were not replaced by *Balanus*, which suggested that *Balanus* could not tolerate the desiccation that occurred in the upper zone at low tide. Connell then removed the *Balanus* population from the lower zone, and found that *Chthamalus* barnacles did replace them. Both

[Connell's] studies ... have improved our understanding of the mechanisms that shape population and community dynamics, diversity, and demography.
Stephen Schroeter
Marine scientist

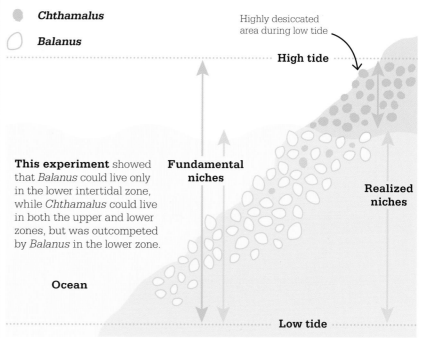

Joseph Connell's barnacle experiment

- *Chthamalus*
- *Balanus*

Highly desiccated area during low tide

High tide

Fundamental niches

Realized niches

This experiment showed that *Balanus* could live only in the lower intertidal zone, while *Chthamalus* could live in both the upper and lower zones, but was outcompeted by *Balanus* in the lower zone.

Ocean

Low tide

species could live in the lower zone, but only one could survive higher up. This suggested that *Chthamalus* was better able to deal with the harsh conditions of the upper zone, but was outcompeted by *Balanus* lower down. The "fundamental niche" of *Chthamalus* (where the species would normally be able to survive) encompassed both zones, but its "realized niche" (the actual area it inhabits) was more restricted.

Diversity experiments

In the early 1970s, Connell and American ecologist Daniel Janzen published an explanation of the degree of tree diversity in tropical forests: the Janzen–Connell hypothesis. Connell mapped trees in two rainforests in North Queensland, Australia, and found that seedlings tended to be less

successful when their nearest neighbour was of the same species. Each species is targeted by specific herbivores and pathogens, which will also eat or attack smaller, weaker individuals of the species nearby. This prevents "clumping" of one tree species.

In 1978, Connell proposed the intermediate disturbance hypothesis (IDH). This states that both high and low levels of disturbance reduce species diversity in an ecosystem, so the greatest range of species can be expected between those extremes. Several studies support IDH. One, carried out in waters off Western Australia, examined the effects of wave disturbance on diversity. Species diversity was found to be low both at exposed offshore sites and at sheltered sites. ▪

MORE NECTAR MEANS MORE ANTS AND MORE ANTS MEAN MORE NECTAR

MUTUALISMS

IN CONTEXT

KEY FIGURE
Dan Janzen (1939–)

BEFORE
1862 Charles Darwin proposes that an African orchid with a long nectar receptacle must be pollinated by a moth with an equally long proboscis.

1873 Belgian zoologist Pierre-Joseph van Beneden first uses the term "mutualism" in a biological context.

1964 The term "coevolution" is first used by American biologists Paul Ehrlich and Peter Raven to describe the mutualistic relations between butterflies and their food plants.

AFTER
2014 Researchers discover an unusual yet beneficial three-way mutualism involving sloths, algae, and moths.

I n biology, there are several kinds of interaction between organisms. One species in an ecosystem may lose out to another when competing for the same resources. A prey species may be eaten by a predator. There are also symbiotic relationships, in which one species benefits but not at the expense of the other, or where one organism does not benefit but still survives. In the relationship known as "mutualism", both organisms benefit from the relationship.

A tree and its ants
In the mid-1960s, Daniel Janzen, a young American ecologist, became fascinated by the amazing

See also: Evolution by natural selection 24–31 ▪ Ecological niches 50–51 ▪ Competitive exclusion principle 52–53 ▪ Animal ecology 106–113

Yuccas and their moths

In the hot, arid regions of the Americas, there is a remarkable mutualistic relationship between yucca shrubs and yucca moths. No other insects pollinate these plants, and no other plants host yucca moth caterpillars. A female yucca moth collects pollen from the flower of one yucca plant and deposits it in the flower of another yucca, fertilizing the plant as it does so. The moth then cuts a hole in the flower's ovary and lays an egg; she may lay several in the same flower. When the eggs hatch, the caterpillars feed on the seeds developing in the flower but do not eat them all, leaving enough for the plant to propagate. If too many eggs are laid in one flower, the plant sheds it before the caterpillars hatch – leaving those insects to starve. Without these moths, the yuccas would not pollinate and would soon die out. Without the yuccas, the moths would have nowhere to lay and nurture their eggs, and they too would not survive.

mutualistic relationship between acacia trees and ants in eastern Mexico. His research was one of the first in-depth studies of such an interaction. The two partners were the swollen-thorn acacia and the acacia ant, which lives in the bullhorn-shaped thorns of the tree. He found that queen ants sought out unoccupied shoots, cut a hole in one of the swollen thorns, and laid their eggs, sometimes leaving the thorn to forage on the tree's nectar. Larvae hatching from the eggs then fed on the acacia's leaf-tips, with their rich supplies of sugars and proteins. The larvae later metamorphosed into worker ants. In time, all the tree's thorns became occupied, with up to 30,000 ants living in a colony.

Janzen showed that, unless the acacia ants were present to defend it, the swollen-thorn acacia lost the ability to withstand damage caused by insects that ate its leaves, stems, flowers, and roots.

Ants and their larvae shelter inside the swollen thorn of an East African whistling thorn acacia tree. In return the ants swarm from their nests to protect the tree from herbivores.

Without the ants, a tree would be stripped of its leaves and die within six months or a year. Because it could not sustain growth, it was also likely to be shaded out by competing trees. Janzen clipped thorns and cut or burned shoots to remove ants from trees, and found that the ants moved back in when new thorns started to grow.

In return for food and shelter, the ants provided two services for the tree: they defended its foliage from leaf-eating insects and ate potentially competitive tree seedlings growing close by. Janzen described the acacias and their ants as "obligate mutualists", meaning that one species would die out without the other. If the ants were removed, the swollen-thorn »

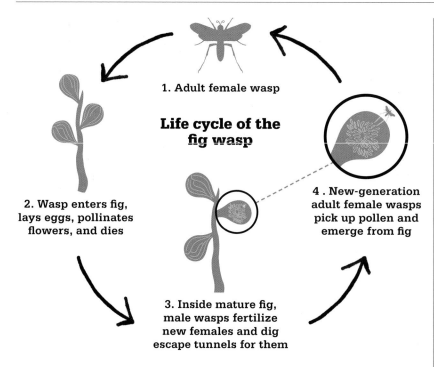

1. Adult female wasp

Life cycle of the fig wasp

2. Wasp enters fig, lays eggs, pollinates flowers, and dies

3. Inside mature fig, male wasps fertilize new females and dig escape tunnels for them

4 . New-generation adult female wasps pick up pollen and emerge from fig

The fig wasp and the fig share a complex service-resource mutualism, in which the wasp provides the service of pollination and the fig plants provide a secure environment for the wasp eggs to develop.

There is mutual aid in many species.
Pierre-Joseph van Beneden
Belgian zoologist

acacia would have no means of defending itself. And if the acacia trees were removed, the ants would have no home.

Benefits for all

There are two fundamental types of mutualism – service-resource and service-service relationships. They are defined by the nature of the relationship between the partner organisms, whether it is the provision of a service or the supply of a resource – both are usually key to survival. Service-resource relationships are common in nature, with the fertilization, or pollination, of flowers by butterflies, moths, bees, flies, wasps, beetles, bats, or birds the most widespread example. The resource (pollen) is provided by the flower, and the service (pollination) is provided by the animal. It is estimated that nearly three-quarters of flowering plants (some 170,000 species) are pollinated by 200,000 animal species. Typically, a pollinating insect is attracted to a flower by its colours or scent to drink nectar or collect pollen, and pollen attaches to part of the insect's body to be carried to the next flower, where it is deposited. The flower and its pollinator have evolved to make this mechanism work effectively.

Some plants have also evolved a service-resource relationship in which birds and mammals disperse their seeds, spores, or fruit. Seeds may become attached to the fur of a mammal browsing the plant's leaves; when the mammal wanders away, it disperses the seed. The vile odour of stinkhorn fungi attracts flies, which lick the fungi's slime and thence disperse their spores. When a bird swallows a fruit, it carries the seeds with it as it flies away; the indigestible seeds may be excreted in faeces far from where they were eaten. In all these situations, the plants provide a resource (food) and the mammals, flies, and birds provide a service (transport).

However, not all mutualistic relationships involve plants. In Africa, birds named oxpeckers and herbivorous mammals such as impalas and zebras practise another kind of service-resource mutualism. The oxpeckers pick ticks from the mammals' fur, removing irritation and a source of disease, while at the same time having a meal. Oxpeckers also make loud calls when they sense danger, alerting the mammal host as well as other oxpeckers.

In the insect world, some ants and aphids carry out a different form of service-resource mutualism. While the aphids feed on plants, the ants protect the aphids. Subsequently, the ants consume the honeydew that the aphids release, using a "milking" process on their smaller partners, by stroking them with their antennae.

Service-service mutualisms, in which both organisms offer each other protection, are far less common. One unusual relationship takes place in the western Pacific Ocean, between around 30 species of clownfish and 10 species of venomous sea anemones. The sea anemones' stinging, toxin-filled nematocysts, or capsules, on their tentacles kill most small fish that come close, but not the clownfish. Its thick layer of protective mucus provides immunity against the anemone's sting, allowing the fish to live within the tentacles. In

The clownfish and sea anemone
could both survive without the other's protection, but their coevolved mutual relationship gives them a much higher chance of survival.

return for the protection offered by the sea anemones' venomous tentacles, the clownfish deters predatory butterfly fish, removes parasites from its host, and also provides nutrients from its faeces.

Cooperative evolution

Relationships between service and resource providers have developed over millions of years in a process called "coevolution" – the evolution of two or more species that affect each other reciprocally.

The term coevolution was coined by American biologists Paul Ehrlich and Peter Raven in 1964, but a century before the word existed, the naturalists Charles Darwin and Alfred Russel Wallace were already aware of the concept, not least through their observation

of orchids. Like many other flowering plants, orchids rely on insects to pollinate them. Some have extraordinary structures in which to hold nectar and pollen. To lure the insect pollinators, the plants offer them a drink of energy-giving nectar. This fascinated Darwin, who was given a specimen of the Madagascar orchid in 1862. The flower stores its nectar in a hollow spur nearly 30 cm (12 in) long. Darwin and Wallace speculated that only a large moth could have a proboscis long enough to reach the nectar – a theory eventually proven in 1997. If the orchid's spur were shorter, a moth could drink without picking up pollen and so would not pollinate the flower. If the spur were longer, a moth would not visit. ■

WHELKS ARE LIKE LITTLE WOLVES IN SLOW MOTION

KEYSTONE SPECIES

IN CONTEXT

KEY FIGURE
Robert Paine (1933–2016)

BEFORE
1950s In Kenya, farmer and conservationist David Sheldrick introduces elephants to Tsavo East National Park, and discovers this results in a major increase in biodiversity.

1961 Fieldwork by American ecologist Joseph Connell on Scotland's rocky shores shows that removing predatory whelks alters the distribution of their barnacle prey.

AFTER
1994 In the US, a group of ecologists led by Brian Miller publishes a paper explaining the valuable role prairie dogs play as a keystone species.

2016 Fieldwork leads marine ecologist Sarah Gravem to conclude that organisms can be keystone species in some places but not in others.

A keystone species plays a crucial role in the way an ecosystem functions, even though it is often a small part of the overall biomass of the ecosystem. Because it exerts a disproportionately large effect on the environment relative to its biomass, if a keystone species disappears from an ecosystem, that ecosystem will change dramatically. The importance of keystone species was brought to light by the American biologist Robert Paine – who derived the term from the central "keystone" at the top of an arch that stops it from collapsing – in his 1969 article "A Note on Trophic Complexity and Community Stability".

The keystone concept

In the 1960s, Paine spent several years studying the animals of the intertidal zone of Tatoosh Island on the Pacific coast of Washington State. He removed the ochre starfish and watched its key prey, a mussel whose numbers had been kept in check by the starfish, dominate the zone, replacing other subordinate species. The removal of a single, keystone species had a

> Do you want an auto mechanic who… can name, list, and count all of the parts of your engine, or one who really understands how each part interacts with the others to make a working engine?
> **Robert Paine**

clear impact on many others. Paine developed the idea to include the concept of "trophic cascades" – the strong, top-down effects that ripple through an ecosystem and its organisms. Since Paine's work with starfish, several studies have demonstrated that there are

Black-tailed prairie dogs look out from their burrow in a field in Wyoming, USA. Study of this species has revealed its key role in fostering diversity in its native habitat.

See also: Predator–prey equations 44–49 ▪ Mutualisms 56–59 ▪ Animal ecology 106–113 ▪ Trophic cascades 140–143 ▪ Evolutionary stable state 154–155

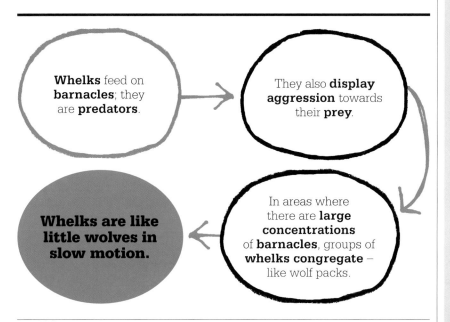

Whelks feed on **barnacles**; they are **predators**.

They also **display aggression** towards their **prey**.

In areas where there are **large concentrations** of **barnacles**, groups of **whelks congregate** – like wolf packs.

Whelks are like little wolves in slow motion.

Robert Paine

Born in 1933, in Cambridge, Massachusetts, Robert Paine studied at Harvard. After a stint in the US Army, where he was the battalion gardener, Paine focused his research on marine invertebrates. His study of the relationship between starfish and mussels on the Paciic coast led him to propose the concept of keystone species – the disproportionate impact that a single species can have on its ecosystem.

Paine worked for most of his career at the University of Washington, where he popularized field manipulation experiments, or "kick-it-and-see" ecology. He was awarded the International Cosmos Award by the National Academy of Sciences in 2013, and died in 2016.

Key works

1966 "Food Web Complexity and Species Diversity", *American Naturalist*
1969 "A Note on Trophic Complexity and Community Stability", *American Naturalist*
1994 *Marine Rocky Shores and Community Ecology: An Experimentalist's Perspective*

---

many other keystone organisms, and they each fulfil their role in different ways.

Ecological engineers

Prairie dogs in the American Midwest are a good example of a keystone species whose impact is the result of their "engineering" activities. Huge colonies of these small mammals dig networks of tunnels beneath the prairie grasslands. They sleep and raise their young in these extensive burrows, converting the grassland into a suitable habitat.

The prairie dogs' constant digging dramatically increases soil turnover and allows nutrients and water from rain and snow to penetrate deeper than would otherwise be the case. The damp, nutrient-rich soil encourages a diversity of plants, and birds such as mountain plovers feed and nest in the short grass. Predators like ferruginous hawks and black-footed ferrets are attracted to the area

to hunt for prey, and the ferrets and tiger salamanders use the burrows for shelter. Almost 150 species of plant and animal are known to benefit from prairie dog colonies. Although there are "losers" – notably vertebrates that favour tall vegetation – the prairie dogs' presence increases overall biodiversity. When colonies die out, scrubby patches of mesquite vegetation replace short grasses, plovers abandon the area, and predator numbers decline.

Coral cleaners

The princess parrotfish in the Caribbean is another keystone species, this time because of the consequences of its feeding. The fish lives around coral reefs, where corals fight each other for light, nutrients, and space. The parrotfish scrapes the surfaces of the corals to remove layers of algal seaweed to eat. If the parrotfish did not do this, clumps of seaweed would grow on the corals, smothering as well as » chemically damaging the reef. If the parrotfish was overfished or died out from disease, the health of the reefs would rapidly deteriorate.

Landscape managers

On African grasslands, elephants smash down small and medium-sized trees for food, helping to maintain savanna as grassland and opening up new areas that were formerly woodland. This destructive behaviour helps to maintain the feeding habitat for grazing animals such as zebras, antelope, and wildebeest. It also indirectly helps the predators that hunt the grazers – including lions, cheetahs, and hyenas – and the smaller mammals that burrow in grassland soils. Without the elephants, these animals would soon disappear. Elephants are also very important seed dispersers; undigested seeds pass through their gut, are then defecated, and later germinate. Up to one-third of all West African tree species depend on elephants for their seed dispersal. Elephants also dig and maintain waterholes, which benefit many other species.

Forest-dwelling Asian elephants have a similar role. In southeast Asia, they smash through gaps and clearings in woodland, opening up holes in the canopy. The new plants that grow in these unshaded areas add to the forest's plant and animal diversity and also help a broader range of animals to thrive there.

Keystone predators

The sea otter is a marine mammal that lives in the Pacific coastal waters of North America. In the 18th and 19th centuries, they were hunted extensively for their fur. By the early 20th century, they had been wiped out in many areas, and their total population was thought to be fewer than 2,000 individuals. Since 1911, legal protection has led to a slow increase in numbers.

Sea otters are important because they eat large numbers of sea urchins. These seafloor-dwelling invertebrates graze on the lower stems of kelp that grow up

Yellowstone wolf pack territories

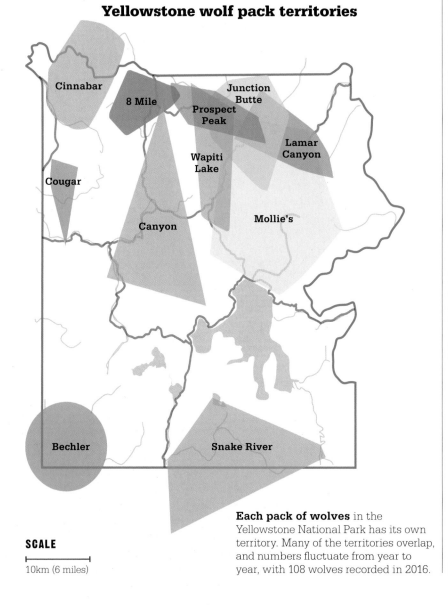

Each pack of wolves in the Yellowstone National Park has its own territory. Many of the territories overlap, and numbers fluctuate from year to year, with 108 wolves recorded in 2016.

SCALE

10km (6 miles)

Every species in the coastal zone is influenced in one way or another by the ecological effects of sea otters.
James Estes
American marine biologist

Reintroducing beavers to the UK

Beavers were wiped out in the UK 400 years ago, but the beneficial role of this keystone mammal is now better understood. Beavers are ecological engineers, building dams and canals, and their presence increases biodiversity.

In 2009, 11 beavers were reintroduced to Knapdale Forest, Scotland, and in 2011, the Devon Wildlife Trust introduced a pair to a fenced enclosure. Both schemes have been monitored to test their impact on the environment. In Knapdale Forest, the beavers' dams changed the water level of a loch, and Devon's beavers built several dams on the headwaters of the River Tamar, creating 13 new freshwater pools and making surrounding areas wetter.

In Devon, the damp areas created by beavers led to an increase in the number of bryophyte species (mosses and liverworts), and the range of aquatic invertebrates has risen from 14 to 41 species. Increased numbers of flying insects have also improved bat diversity, with two nationally rare bat species drawn into the area. More beaver reintroduction schemes are now planned in the UK.

---

from the seabed, causing it to drift away and die. If the kelp disappears, however, so do the many other marine invertebrates that graze on it. "Forests" of kelp also absorb large amounts of atmospheric carbon dioxide and, by slowing water currents, help to protect coastlines from storm surges. The protection that sea otters offer kelp along stretches of open coast is therefore particularly significant.

Unlike the sea otter, some keystone species are also "apex" predators at the top of the food chain, such as the grey wolf. Before 1995, there had been no grey wolves in Yellowstone National Park for at least 70 years. American elk were common in the park, but there was just a single colony of beavers. That year, 31 wolves were introduced to the park and by 2001 their numbers had increased to more than 100, largely due to the abundance of elk for food.

The presence of wolves in the park forced the elk to become more mobile. Rather than over-grazing willow, aspen, and cottonwood trees in favoured locations, the elk moved on, allowing plants to regenerate and provide a food source for other herbivores, such as beavers. Within 10 years, the number of beaver colonies had increased from one to nine. Beaver dams helped to revive wetlands, and wetland wildlife flourished. The increase in elk carcasses also benefited carrion-eaters – especially coyotes, red foxes, grizzly bears, golden eagles, ravens, and magpies – as well as several smaller scavengers.

Jaguars are apex predators in South and Central American forests, preying on more than 85 species. Although there are very few jaguars in any given area, their impact on the numbers of other predators – such as caimans, snakes, large fish, and large birds – as well as herbivores, such as capybaras and deer, has a significant ripple-down effect on their ecosystem. Left unchecked, the herbivores could devour most of the plants and destroy the habitat on which so many other species depend.

Keystone plants

Not all keystone species are animals. One example is the fig tree, of which there are about 750 species, mostly found in tropical forests. In this habitat, most fleshy-fruited plant species share one or two peaks of ripening each year. Fig trees bear fruit throughout the year, supporting many animals when other trees are fruitless. More than 10 per cent of the world's bird species and 6 per cent of mammals (a total of 1,274 species) are known to eat figs, as do a small number of reptiles and even fish. Fig trees therefore provide a vital support mechanism for fruit-eating species. Without them, fruit bats, birds, and other creatures would decline or disappear. ■

By protecting a keystone species such as the prairie dog, the public could be educated about the value of ecosystem conservation.
Brian Miller
American ecologist

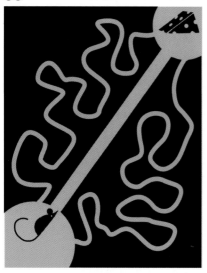

THE FITNESS OF A FORAGING ANIMAL DEPENDS ON ITS EFFICIENCY
OPTIMAL FORAGING THEORY

IN CONTEXT

KEY FIGURES
Ronald Pulliam (1945–),
Graham Pyke (1948–),
Eric Charnov (1947–)

BEFORE
1966 John Merritt Emlen,
Robert MacArthur, and Eric
Pianka outline the concept of
optimal foraging in two
articles published in the
American Naturalist magazine.

AFTER
1984 Argentinian–British
zoologist Alejandro Kacelnik
researches the foraging
behaviour of starlings to
illustrate the marginal
value theorem (MVT).

1986 Belgian ecologist Patrick
Meire investigates prey
selection by oystercatchers.

1989 Swiss environmental
scientists T. J. Wolfe and Paul
Schmid-Hempel examine how
the weight of nectar carried
by bees has an effect on the
bees' foraging behaviour.

Every plant and animal
on Earth needs resources
to survive. Plants obtain
their nutrients and water from soil,
and sunlight provides the energy
for photosynthesis. Animals
generally have to work harder to
find their food – they have to move,
and this uses extra resources.
Optimal foraging theory (OFT)
proposes that animals try to gather
resources in the most efficient way
to avoid using additional energy.
Searching for and capturing food
takes energy and time. The animal
needs to gain maximum benefit
for minimal effort in order to
achieve optimal fitness. OFT

Diets should be broad
when prey are scarce,
but narrow if food
is abundant.
Eric Pianka

helps to predict the best strategy
that an animal can use to achieve
this goal.

Foraging theories
The first theory of foraging by
wild animals did not emerge until
the mid-1960s, when Americans
Robert MacArthur and Eric Pianka
examined the question of why,
when a range of food was available
to them, animals often restricted
themselves to a few preferred types
of prey. They argued that natural
selection favoured animals whose
behaviour maximized their net
energy intake per unit of time spent
foraging. An animal's foraging time
includes searching for prey and the
killing and eating of the food
(handling time).
    These ideas were developed
by US ecologists Ronald Pulliam
and Eric Charnov and Australian
ecologist Graham Pyke. It seems
that OFT works best for mobile
foragers seeking immobile prey,
and some researchers believe it is
less relevant when prey are mobile.

Key choices
Animals must choose which types
of food to eat, which is rarely
straightforward. For example,

See also: Evolution by natural selection 24–31 ▪ Predator–prey equations 44–49 ▪ Competitive exclusion principle 52–53 ▪ Mutualisms 56–59

> The expected behaviour of animals with respect to available resources can be used to predict ... the biotic structure ... of communities.
> **Ronald Pulliam**

American ecologists Howard Richardson and Nicolaas Verbeek studied northwestern crows feeding on clams in the intertidal zone of British Columbia. The crows put lots of effort into digging clams out of the mud, opening the shells, and feeding on the animal inside. The ecologists noticed that smaller clams went unopened and concluded that the crows had to make an energy trade-off between handling time and edible food. The time and energy needed to open up small clams was better spent digging for another, larger clam. A similar study with oystercatchers and mussels found that the largest mussels were left – they had thicker, barnacle-clad shells, so opening them was more difficult. The oystercatchers benefited more by looking for thin-shelled mussels, despite their smaller size.

Animals also have to make choices about where and when to feed. The longer a starling spends in one patch of suitable grassland, for example, the harder it will become to find prey, so they have to decide when to abandon that patch and move to another – an example of what is known as the "marginal value theorem". Foraging animals also need to consider a range of other factors such as the presence of predators, the number of animals competing for the same food, and the impact of human activity. ▪

Oystercatchers, despite their name, are reliant on cockles and mussels as their primary food source. Without these shellfish, they are forced to forage further inland.

Echolocating bats

Technological advances have greatly helped research into the hunting strategies of animals. Insectivorous bats (also known as microbats) use echolocation in the dark to locate and pursue flying insect prey, such as moths and midges. A team of Japanese scientists set out to study the bats' feeding behaviour using microphone array measurements and mathematical modelling analysis. The researchers recorded the echolocation calls and flight paths of the bats and discovered that they often directed their sonar not just at their immediate prey but at the next target they were lining up as well.

The team also found evidence that the bats chose flight paths that would allow them to plan two steps ahead, rather like skilled chess players. Not only were the animals maximizing their energy input by targeting multiple prey items, but they were also minimizing their energy output by reducing the distance they flew in pursuit of insects. This behaviour fits in well with optimal foraging theory.

PARASITES AND PATHOGENS CONTROL POPULATIONS LIKE PREDATORS
ECOLOGICAL EPIDEMIOLOGY

IN CONTEXT

KEY FIGURES
Roy Anderson (1947–),
Robert May (1936–)

BEFORE
1662 English statistician John Graunt seeks to classify causes of death in London in *Natural and Political Observations made upon the Bills of Mortality.*

1927 Scottish scientists Anderson Gray McKendrick and William Ogilvy Kermack develop an epidemic model for infected, uninfected, and immune individuals.

AFTER
1996 American epidemiologist James S. Koopman calls for greater use of computational technologies to simulate disease generation and spread.

2018 A global team tracks the origins and spread of a fungus devastating frog populations.

Epidemiology is the study of how disease spreads through a population. Its initial application was to human diseases, but its methods have been recognized as an effective way of modelling populations of other organisms, too.

Ecologists have long known that the size of an animal or plant population and its growth rate depend on the availability of food, living space, and levels of predation. In the 1970s, British epidemiologist Roy Anderson and Australian scientist Robert May showed how parasites and infections from pathogens such as bacteria and viruses limited the

See also: The microbiological environment 84–85 ▪ Microbiology 102–103 ▪ The ubiquity of mycorrhizae 104–105 ▪ Biodiversity and ecosystem function 156–157

Map of deaths from cholera in London in 1854

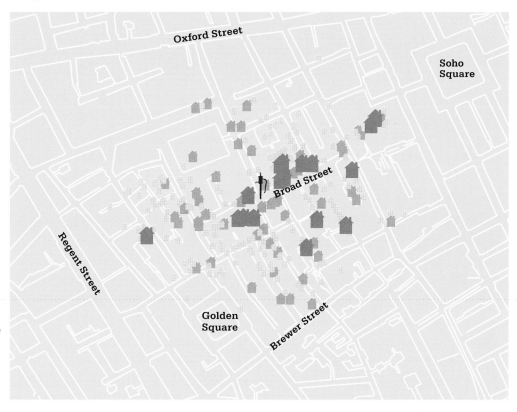

KEY

🏠 1–4 deaths

🏠 5–9 deaths

🏠 10–15 deaths

⚲ Broad Street pump

Fatalities in London's cholera outbreak of 1854 were linked to the central pump; its water was found to have been contaminated with infected sewage from a stricken family.

size of a population. In wild sheep, for instance, the chief cause of death is lungworms, while most wild birds die from viral infections.

In ecology, the effects of disease have wider implications. Up to 40 per cent of ocean bacteria are killed each day by viruses. This causes a "viral shunt", as nutrients that would otherwise flow up the food chain to consumers revert back to the bottom of the chain again.

Human beginnings

Epidemiology has its beginnings in the work of physician John Snow, who witnessed a cholera epidemic in the Soho district of London in 1854. At the time, disease was thought to be caused by miasma – a sort of poisonous vapour in the air – that spread from the bodies of the dead and dying. Snow was not the first to question this theory, but he was especially suspicious of it in the case of cholera.

In 1854, Snow plotted every case of cholera on a map of Soho, and found that afflicted households collected their water from a pump on Broad Street (later renamed Broadwick). He shut down the »

British doctor John Snow fought the establishment to gain acceptance for his belief that cholera was waterborne. The medical journal *The Lancet* finally conceded that he was right in 1866.

The role of drought in plant diseases

Like other disease-causing agents, a plant pathogen (disease-causing agent) needs a supply of susceptible individuals to infect. Periods of drought slow the rate of plant reproduction and growth, thereby reducing the prevalence of disease.

Aridity, however, also weakens plants and makes them susceptible to pathogens that thrive in dry conditions. These include various forms of fungi that attack the leaves of cereal crops, legumes, and fruits. These fungi are adapted to survive in a dormant state as hardened microscopic bodies in soil. They can exist for many years in dry soil, but when the soil becomes wet, the fungi must find a host within a few weeks or die. They do not necessarily kill their host. Recent research into chickpeas suggests that although infections from such fungi do rise during a dry spell, the mortality rate of the affected plants goes down during a drought.

A summer drought produces only sparse growth of young barley plants. Lack of moisture and too much heat reduce their resistance to fungi that attack their roots.

pump, and the epidemic soon ended. This showed that cholera was a waterborne disease that humans contracted through contaminated food and drink. A decade later, Louis Pasteur's "germ theory" proposed that diseases, as well as general rotting and decay, were the work of microorganisms.

Disease model

In their 1970s studies, Anderson and May focused first on building a mathematical model to show how a microorganism could affect a population. This led to a set of equations that they hoped would help to explain the real-life impact of different kinds of pathogens, from bacteria and viruses to parasitic worms and insect larvae.

In their model, a number of mice were divided into three groups: susceptible (uninfected) mice, infected mice, and mice that had survived infection and were now immune. Unlike many earlier epidemiological models, the total population was not a fixed number; mice could be added either by reproduction or by additions from other populations. Mice also died from natural causes. In the absence

> Sensibly used, mathematical models are no more and no less than tools for thinking about things in a precise way.
> **Roy Anderson and Robert May**

A ravaged tree in North Yorkshire, UK, shows the effects of Dutch elm disease, a fungus spread by elm bark beetles accidentally introduced to Europe and America from Asia.

of disease, the total would remain more or less the same, with the rate of added mice balancing that at which other mice died.

For simplicity, the model assumed that the diseases were transmitted by contact between infected and uninfected mice. Not all infected mice would die, so the model also included a recovery rate. Mice that recovered would be immune, at least initially. Immunity to viruses is more or less lifelong, but it is possible to become susceptible again to the same bacterial infection as time passes. Therefore, the calculations also included a rate of loss of immunity.

Putting all this together, Anderson and May produced a set of equations to predict the rate of population change in the three initial groups of uninfected but susceptible mice, infected mice, and the immune survivors. These equations could be added together to give the rate of change for the total mouse population.

> Diseases such as measles and rubella, with short infections and lasting immunity, will tend to exhibit epidemic patterns.
> **Roy Anderson**

From their calculations, they deduced that a disease will persist in a population whose equilibrium point (the rate of new additions, balanced by the natural death rate) is greater than the combined effects of natural mortality, disease deaths, recovery, and transmission rate. While the disease is present, that equilibrium point will be lower than if the population were disease free. If, however, the equilibrium point of a population affected by disease is lower than the combined effects of deaths, recoveries, and rate of transmission, the disease will die out. Once a population is disease free, its equilibrium point will return to its former level.

Matching the real world

Anderson and May needed to show that their model was an accurate predictor of a real-life population. They did so by using data from a study of laboratory mice infected with the bacterial disease pasteurellosis; the data included the impact on the population of adding individuals at different rates. The observed data confirmed their predictions, so the two scientists were able to consider the effects of hypothetical values. They found, for instance, that when the rate of added mice was highest, the disease had the greatest impact on population numbers. This suggests that species with high reproductive rates (introducing large numbers of uninfected offspring) are most likely to have endemic diseases within the population, and show depressed numbers compared with species that breed more slowly. They also explored the differing effects on populations of diseases of different intensities.

Unlike endemic diseases, in which the population's level of infection remains consistent, epidemics appear in populations when the growth rate of all infected and uninfected members is low compared to the death rate caused by the disease. Infection numbers rise sharply to a maximum, then drop away. Epidemics also occur when a disease is not particularly deadly but slows the population growth rate; this has occurred with human diseases such as measles and chickenpox.

Applying the theory

The characteristics of disease and its effects on animal and plant populations are of increasing ecological importance. Food producers, for example, benefit from studies into the nature of parasites and the dynamics of diseases that can affect crops and livestock. Conservationists also employ epidemiology to predict how exotic diseases and invasive parasites might affect fragile ecosystems. ∎

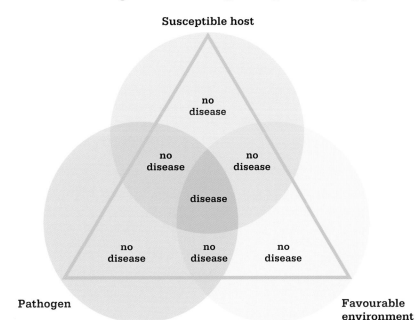

Venn diagram of ecological epidemiology

A pathogen strikes when it finds a suitable host in an environment that favours infection, as shown where the circles intersect. For instance, diarrhoeal diseases spread quickly among sick people in unsanitary conditions.

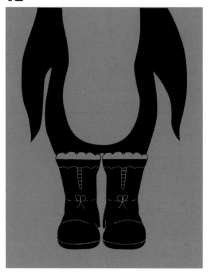

WHY DON'T PENGUINS' FEET FREEZE?

ECOPHYSIOLOGY

IN CONTEXT

KEY FIGURE
Knut Schmidt-Nielsen
(1915–2007)

BEFORE
1845 The explorer Alexander von Humboldt reveals that plants facing similar ecological factors also have many analogous features.

1859 Charles Darwin argues that organisms evolve because they are adapting to changed ecological conditions.

AFTER
1966 Australian biochemists Marshall Hatch and Charles Slack explain that the most widespread plants are the ones that photosynthesize most efficiently.

1984 Peter Wheeler, a British scientist, suggests that human bipedalism – the ability to walk on two legs – evolved as a thermoregulatory adaptation that reduces the body's exposure to direct sunlight.

The central principle of Darwinian evolution is that all organisms, from simple bacteria to complex mammals, are adapted by natural selection to survive in a particular niche and habitat. Ecophysiology – for which Knut Schmidt-Nielsen's book *Animal Physiology* (1960) was a vital inspiration – is the study of an organism's anatomy and the way it functions (its physiology), as well as how these characteristics relate to the challenges posed by its environment. It shows how an animal or plant's anatomy is linked to its ability to survive, as well as

From a physiological viewpoint, fresh water is no more freely available in the sea than in the desert.
Knut Schmidt-Nielsen

to its distribution, abundance, and fertility. Ecophysiology now plays an important role in helping scientists to understand how the stresses created by climate change impact on both wild ecosystems and cultivated environments.

Managing temperature
Ecophysiology has revealed a number of specific adaptations for different environments. For example, animals that live in colder regions tend to have larger bodies and smaller legs, ears, and tails than related species living in warmer climes. A larger body has a smaller surface-area-to-mass ratio, and therefore loses heat more slowly, while smaller appendages reduce exposure to frostbite.

In the most extreme cold, the feet of a warm-blooded animal are at risk of becoming frozen to the ground. Mammals in Arctic regions such as musk oxen and polar bears are adapted for life in these conditions by having thick hairs to insulate their feet.

In the Antarctic, the undersides of penguins' feet are insulated by a thick layer of fat. Penguins also have a heat-exchange (or counter-current) mechanism in

See also: Evolution by natural selection 24–31 ▪ Ecological niches 50–51
▪ Competitive exclusion principle 52–53 ▪ Ecological stoichiometry 74–75

their legs. The warm blood arriving from the body is cooled to near 0°C (32°F) by the chilled blood arriving from the feet, which warms to body temperature in the process.

Gazelles in Africa use a similar counter-current system to cool their body temperature. They are able to chill the blood entering their head, giving them an advantage over their predators, who often overheat. Camels have a heat-exchange system in their nasal cavity, which reduces the amount of water lost in their breath. Hot, dry air is inhaled and mixes with moisture inside the nose before travelling to the lungs. The exhaled air is much cooler than the air outside, so the moisture it carries condenses in the nose. This creates the cool, damp conditions needed to chill the next in-breath.

Future challenges
Today ecophysiology is becoming increasingly focused on plants, fungi, and microbes. Like animals, they have to adapt to survive – and studying them offers the possibility of vital discoveries for commercial and conservation purposes. ▪

Emperor penguins survive freezing Antarctic temperatures thanks in part to the way their bodies have evolved to adapt to the harsh environment.

Knut Schmidt-Nielsen

Knut Schmidt-Nielsen grew up in the Norwegian town of Trondheim. His interest in the way animal physiology related to habitat was inherited from his grandfather who, years before Knut's birth, had released thousands of flounder (marine fish) hatchlings into a freshwater lake. Although the fish thrived, they were unable to breed because their reproductive physiology was adapted for life in salt water.

Schmidt-Nielsen joined Duke University, North Carolina, in 1954. He built a climate-controlled space for keeping desert animals, where he considered the anatomy of camels, gerbils, and other species able to live for long periods without water. He also investigated the respiratory systems of birds and the buoyancy of fish. His 1960 textbook *Animal Physiology* is still a classic work.

Key works

1960 *Animal Physiology*
1964 *Desert Animals*
1972 *How Animals Work*
1984 *Scaling*
1998 *The Camel's Nose: Memoirs of a Curious Scientist*

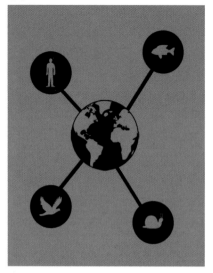

ALL LIFE IS CHEMICAL
ECOLOGICAL STOICHIOMETRY

IN CONTEXT

KEY FIGURES
Robert Sterner (1958–),
James Elser (1959–)

BEFORE
1840 German biologist and chemist Justus von Liebig asserts that the limitations on agriculture productivity are primarily chemical.

1934 US oceanographer Alfred Redfield measures the atomic ratio of carbon, nitrogen, and phosphorus (C:N:P) in plankton and seawater, and finds it to be relatively consistent in all oceans. The Redfield Ratio soon becomes a benchmark for such research in all habitats.

AFTER
2015 In "Ocean stoichiometry, global carbon, and climate", Robert Sterner highlights inconsistencies in C:N:P ratios in phytoplankton, which absorb more atmospheric carbon in low-nutrient, low-latitude ocean surface waters and adjust their ratios accordingly.

Every living organism – from tiny ocean algae to a mighty redwood – is made up of chemical elements in varying ratios. Ecological stoichiometry considers the balance of these elements, and how the ratios change during chemical reactions. Studying such ratios throws light on the way the living world operates, revealing how organisms obtain the nutrients and other chemicals they require for life from the resources in their environment.

Individual organisms also show differences in stoichiometry during their life cycles. Young organisms may have different compositions from older ones …
Robert Sterner and James J. Elser

The field of ecological stoichiometry was comprehensively described for the first time by American biologists Robert Sterner and James Elser; in *Ecological Stoichiometry* (2002), they used mathematical models to demonstrate the application at every level, from molecules and cells to individual plants and animals, populations, communities, and ecosystems.

Key chemicals
In ecological research, the three main elements examined are carbon (C), nitrogen (N) and phosphorus (P), because each plays a vital role. Carbon is a basic building block of all life and an important part of many chemical processes. Nitrogen is a major constituent of all proteins, while phosphorus is crucial for cell development and storing energy.

An organism's C:N:P ratio is not necessarily consistent. Plants have a variable ratio: they can adjust the balance of their elements according to their environment. For instance, the proportion of carbon in their chemical makeup can rise on a particularly sunny day because more photosynthesis occurs – the process by which they take carbon

See also: Ecophysiology 72–73 ▪ The food chain 132–133 ▪ Energy flow through ecosystems 138–139 ▪ The foundations of plant ecology 167

Controlling ecological stoichiometry ratios

A locust eats grass that may contain six times as much carbon as it needs. To get the right balance, it excretes carbon or breathes it out as CO_2. Locusts are widely used in research because they are easy to breed.

KEY

▪ Carbon ▪ Nitrogen

LOCUSTS
5:1

GRASS
33:1

dioxide from the air and use the sun's energy to convert it into the nutrients they require.

Higher up the food chain, animals have largely fixed C:N:P ratios, so they must deploy various mechanisms to deal with any imbalances of chemicals entering the body. If an insect or animal herbivore is getting too much carbon from its plant diet, for instance, it may adjust its digestive enzymes and excrete it, store it as fats, or raise its metabolic rate to burn it off, breathing out the excess carbon as CO_2. Overuse of such mechanisms to redress a high imbalance can, however, affect fitness, growth, and reproduction. An animal that eats other animals has less work to do, as its prey's C:N:P ratio already closely matches its own. However, the size of its prey population is still determined by the plants in its environment, as plant foods with a high carbon ratio can only support a small food chain of consumers.

Understanding our world

Food chains are one area of study; ecological stoichiometry covers just about everything and all the links in between. By discovering how the chemical content of organisms shapes their ecology, scientists are also learning how environments can be better managed. Their findings may significantly influence the future of life on Earth. ▪

The desert locust (*Schistocerca gregaria*) has to eat vast quantities of carbon-rich plants in order to get enough nitrogen and phosphorus to maintain its C:N ratio.

The Growth Rate Hypothesis

Cancer research is one area where stoichiometry is now being employed. Evidence is growing for a theory called the Growth Rate Hypothesis (GRH), which may help to explain why some cancerous tumours grow at faster rates than the rest of the body.

The hypothesis states that organisms with high C:P (carbon:phosphorus) ratios, such as fruit flies, have more ribosomes in their cells, which enables them to grow and reproduce more rapidly. Around half of all phosphorus in an organism is in the form of ribosomal RNA (rRNA); it is present in every cell, creating proteins to build new cells and grow the body. Applying biological stoichiometry, James Elser and his team have shown that fast-growing tumours have a much higher phosphorus content than normal body tissue. Such research may help scientists understand how tumour growth could be controlled.

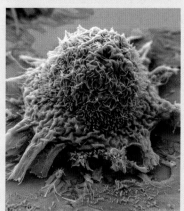

Malignant lung tissue (seen here) and cancerous colon tissue both had the highest phosphorus content in research exploring the rapid growth rates of tumours.

FEAR ITSELF IS POWERFUL

NON-CONSUMPTIVE EFFECTS OF PREDATORS ON THEIR PREY

IN CONTEXT

KEY FIGURE
Earl Werner (1944–)

BEFORE
1966 American ecologist Robert Paine conducts a series of groundbreaking field experiments to highlight the crucial effects of a predator on the community in which it lives.

1990 Canadian biologists Steven Lima and Lawrence Dill analysed the decision-making of organisms that are at the greatest risk of being preyed on by other creatures.

AFTER
2008 American behavioural biologist and ecologist John Orrock teams up with Earl Werner and others to produce mathematical models to explain the non-consumptive effects of predatory animals.

Many descriptions of ecosystems focus on predator–prey interactions in which predators kill and prey are eaten. However, American ecologist Earl Werner and others have shown that the mere presence of a predator affects the behaviour of prey.

Apart from apex predators, all animals must balance the need to sleep, reproduce, and feed with the risks of being eaten. The lethal role of predators is obvious, but their non-lethal (non-consumptive) role can have an even bigger impact on an ecosystem. Potential prey are forced to change their way of life in order to avoid being killed.

In 1990, Werner studied the effects of green darner dragonfly larvae on toad tadpoles. He noticed that when the predatory larvae

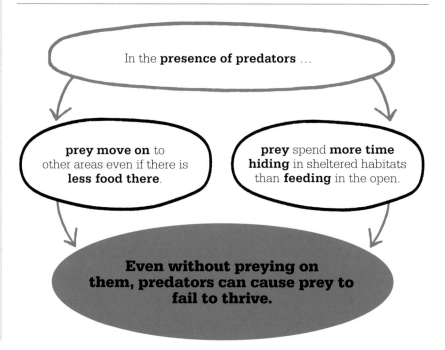

In the **presence of predators** …

prey move on to other areas even if there is **less food there**.

prey spend **more time hiding** in sheltered habitats than **feeding** in the open.

Even without preying on them, predators can cause prey to fail to thrive.

See also: Evolution by natural selection 24–31 ▪ Predator–prey equations 44–49 ▪ Ecological niches 50–51
▪ Competitive exclusion principle 52–53 ▪ Mutualisms 56–59 ▪ Optimal foraging theory 66–67

A green darner dragonfly laying its eggs in a pond. The larvae that hatch out are predators and have been shown to influence the behaviour of their tadpole prey.

were in the tank, the tadpoles were less active, swam to other parts of the tank, and metamorphosed into adults when they were smaller. The predator had changed the toads' morphology and their behaviour, just by being there.

In 1991, Werner investigated what happened when more than one prey species was involved. In the absence of a predator, bullfrog and green frog tadpoles grew at virtually identical rates. However,

when predatory dragonfly larvae were introduced to the tank, both prey species became less active and chose different places in which to swim. The bullfrog tadpoles grew more quickly than they had in a predator-free tank, but the green frog tadpoles decreased their feeding activity and grew more slowly. Werner concluded that for prey species there was a trade-off between the need to grow as fast as possible and the risk of predation. Growing more quickly requires more feeding activity, and this in turn increases the chances of being eaten by a predator. As the larvae's presence altered the behaviour of the prey species differently, the

bullfrog's new behaviour gave it a competitive advantage over the green frog by making it bigger.

Terrestrial animals

Early studies of non-consumptive effects (NCEs) were concerned with aquatic organisms under laboratory conditions, but more work has now been done in the wild with land-dwelling animals. German field research published in 2018 focused on lynx and their roe deer prey. When lynx were present, researchers found that the roe deer avoided areas they knew to be high-risk, both during the day and on summer nights when nocturnal predation is more common. The deer treated some grazing areas as out of bounds, presumably due to fear of being attacked by lynx.

Wherever there are predators, they exert NCEs. They also affect some sessile (non-moving) species, as well as mobile prey. This can happen when certain dominant competitors are displaced by predators and, in their new habitats, out-compete sessile animals for food. Small fish that are displaced, for example, could out-compete sponges for food. ▪

... species react [to predators] by reducing activity and altering space use.
Earl Werner

ORDERIN
NATURAL

G THE
WORLD

Aristotle's *History of Animals* groups living things based on their **species**, in a *scala naturae* that places organisms into 11 grades.

c.350 BCE

A private collection of natural history curiosities is displayed at Oxford University's Ashmolean Museum, the world's **first public museum**.

1683

The **Natural History Museum in London** opens its doors to the public, free of charge. It now houses 80 million specimens.

1881

1665 CE

Micrographia, the richly illustrated book by **Robert Hooke**, reveals **microscopic structures** to a wider audience.

1758

The 10th edition of *Systema Naturae* by **Carl Linnaeus** classifies a range of plant and animal species using his **binomial system**.

P eople have long marvelled at the variety of life, celebrating nature's gifts in prehistoric cave art that dates back 30,000 years or more. In Ancient Greece in the 4th century BCE, Aristotle made an early attempt to classify living organisms; his 11-grade *scala naturae* ("ladder of life") placed humans and mammals at the top, and descended through other, more "primitive" animals to plants and then minerals. A thousand years on, the medieval world still considered variations on Aristotle's system to be valid. There were several reasons for this. Without microscopes, nothing was known of cells and microorganisms. Without the means to explore underwater, science's knowledge of aquatic creatures was limited, and many parts of the world were still unknown to Western scientists.

In keeping with the prevailing ideas of the Catholic Church, the natural world was seen as static and unchanging.

An age of discovery

The age of great expeditions of discovery revealed previously uncharted regions and their animals and plants. In his *History of the Animals* (1551–58), Swiss physician and naturalist Conrad Gesner included some of the recent finds from the New World and the Far East, as well as relying on classical literature. The five-volume work reflected his division of animals into mammals; reptiles and amphibians; birds; fish and aquatic animals; and snakes and scorpions.

The invention of the microscope also had a major impact. English scholar Robert Hooke was quick

to adopt this new technology: his book *Micrographia* (1665) inspired others to do likewise. Able to view specimens magnified to 50 times their actual size, he made meticulous drawings of microsopic life, and also coined the term "cell" after examining plant fibres. Hooke also suggested a living origin for fossil fragments found in rocks.

Classifying variety

English vicar John Ray's *History of Plants* (1686–1704) was the botanical equivalent of Gesner's earlier work, listing some 18,000 species in three huge volumes. Ray also produced a biological definition of a species, remarking that "one species never springs from the seed of another". Swedish botanist Carl Linnaeus, the "father of taxonomy", first published *Systema Naturae* in 1735,

Carl Woese
establishes a new, third
category of organisms –
the **prokaryotes**.

**Norman Myers's
"biodiversity hotspots"**
concept identifies ten hotspots
where conservation efforts
should preserve rare species.

1977

1988

1942

1988

2018

Ernst Mayr develops the
biological species concept,
which categorizes species
based on their ability to
breed with each other.

Edward O. Wilson coins
the term **biodiversity**
and later identifies the
key **human threats**
to biodiversity.

The **IUCN Red List** shows that
more than **26,000 species** –
more than 27 per cent of all
those assessed – are at
risk of **extinction**.

---

but it is the 10th edition from 1758 that founded the modern zoological naming system. Two volumes of Linnaeus' work are devoted to plants and animals, which he divided into classes, orders, genera, and species. The binomial system, in which every species is given generic name followed by a specific name, is still in use today. Linnaeus also wrote a third volume on rocks, minerals, and fossils.

Species concepts

Building on Darwin's theory of evolution by means of natural selection, German-American evolutionary biologist Ernst Mayr cemented the biological concept of species in his *Systematics and the Origin of Species* (1942). He argued that a species is not just a group of morphologically similar individuals,

but a population that can breed only among themselves. Mayr went on to explain how if groups within a species become isolated from the rest of the population, they may start to differ from the rest, and over time, through genetic drift and natural selection, may even evolve into new species.

Modern technological advances, including electron microscopy and mitochondrial DNA analysis, have revealed much information – some of it surprising – about the number of species and the relationships between them. In 1966, striving to reflect the intricacies of evolution, German entomologist Willi Hennig proposed a new taxonomic system of clades – groups of organisms based on a common ancestor. In the 1970s, American biologist Carl Woese classified all life into three

new domains. As of 2018, about 1.74 million extant plant and animal species have been described, but estimates of the total number range from 2 million to 1 trillion.

The threat to diversity

By the late 20th century, however, alongside a growing knowledge of the scale and critical role of biodiversity – and of how evolution can destroy species as well as create them – American ecologist Edward Wilson and others made the world aware that human activity was responsible for causing a rapid acceleration in the extinction rate. Some have even warned that Earth could be on the verge of a sixth mass extinction. Many policies are now being proposed to counter this, including the protection of biodiversity hotspots. ■

IN ALL THINGS OF NATURE THERE IS SOMETHING OF THE MARVELLOUS

CLASSIFICATION OF LIVING THINGS

IN CONTEXT

KEY FIGURE
Aristotle (*c.* 384–322 BCE)

BEFORE
***c.* 1500 BCE** Different properties of plants are recognized by ancient Egyptians.

AFTER
8th–9th centuries CE Islamic scholars of the Umayyad and Abbasid dynasties translate many of Aristotle's works into Arabic.

1551–58 Conrad Gessner's *History of Animals* classifies the animals of the world into five basic groups.

1682 John Ray publishes his *History of Plants*, which lists more than 18,000 species.

1735 Carl Linnaeus devises a system of binomial names, the first consistent classification of organisms, according to which he names every species listed in his *Systema Naturae*.

From the beginning of recorded history, people have attempted to identify organisms according to their uses. Egyptian wall paintings from *c.* 1500 BCE show, for example, that people understood the medicinal properties of many plants. In the text *History of Animals*, written in the 4th century BCE, the Greek philosopher and scholar Aristotle made the first serious attempt to classify organisms, studying their anatomy, life cycles, and behaviour.

Features of classification

Aristotle divided living things into plants and animals. He further grouped about 500 species of animal according to obvious anatomical features, such as whether they had blood, were "warm-blooded" or "cold-blooded", whether they had four legs or more, and whether they gave birth to live offspring or laid eggs. He also noted whether animals lived in the sea, on land, or flew in the air. Most significantly, Aristotle used names for his groupings that were later translated into the Latin words "genus" and "species" – terms that are still used by modern taxonomists to this day.

Aristotle placed animals in a *scala naturae* (ladder of nature), with 11 grades distinguished by their mode of birth. Those in the top grades gave birth to live, hot, wet offspring; those in the lower grades to cold, dry eggs. Humans were at the very top of the scale, with live-bearing tetrapods (four-legged creatures), cetaceans, birds, and egg-laying tetrapods lower down. Aristotle placed minerals on the bottom grade of his scale, with plants, worms, sponges, larva-bearing insects, and hard-shelled animals on the levels above.

If any person thinks the examination of the rest of the animal kingdom an unworthy task, he must hold in like disesteem the study of man.
Aristotle

See also: The microbiological environment 84–85 ▪ A system for identifying all nature's organisms 86–87 ▪ Biological species concept 88–89 ▪ Microbiology 102–103 ▪ Animal behaviour 116–117 ▪ Island biogeography 144–149

An octopus blends in with its surroundings. The ability of these creatures to change colour was one of Aristotle's many accurate observations.

While Aristotle's system of classification was rudimentary, it was based largely on first-hand observations, many of which were made on the island of Lesvos. He recorded things that noone else had described, including that young dogfish grew inside their mothers' bodies, male river catfish guard eggs, and octopuses can change colour. Most of his observations were good – and some were confirmed only centuries later.

The great chain of being

Despite its limitations, Aristotle's method of classification heavily influenced every later attempt at grouping animals and plants until the 18th century. Medieval Christianity developed his *scala naturae* as a "great chain of being", with God at the top of a strict hierarchy, humans and animals beneath, and plants at the bottom.

The Swiss doctor Conrad Gessner wrote the first modern register of animals – also called *History of Animals* – in the mid-16th century. This monumental five-volume work was based on classical sources but included newly discovered species from East Asia. It covered the main animal groups as Gessner saw them: live-bearing quadrupeds (mammals); egg-laying quadrupeds (reptiles and amphibians); birds; fish and aquatic animals; and snakes and scorpions. In 1682, the English naturalist John Ray produced the equivalent register for botany with his *History of Plants*. Within little more than 50 years, the classification of living things would be completely transformed by Carl Linnaeus's *Systema Naturae*. ▪

Aristotle

Aristotle was born in Macedonia, ancient Greece. Both his parents died when he was young, and he was raised by a guardian. Aged 17 or 18, Aristotle joined Plato's Academy in Athens, where he studied for 20 years, writing on physics, biology, zoology, politics, economics, government, poetry, and music. Later, he travelled to the island of Lesvos with a student named Theophrastus to study the island's botany and zoology. Much of his *History of Animals* was based on observations he made there. Aristotle taught both the future scholar Ptolemy and King Alexander the Great. In 335 BCE, he established his own school at the Lyceum in Athens. After Alexander's death in 322 BCE, Aristotle fled the city, and died on the island of Euboea in the same year.

Key works

4th century BCE
History of Animals
On the Parts of Animals
On the Generation of Animals
On the Movement of Animals
On the Progression of Animals

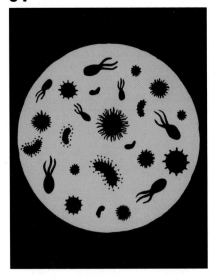

BY THE HELP OF MICROSCOPES NOTHING ESCAPES OUR INQUIRY
THE MICROBIOLOGICAL ENVIRONMENT

IN CONTEXT

KEY FIGURE
Robert Hooke (1635–1703)

BEFORE
1267 English philosopher Roger Bacon discusses the use of optics for looking at "the smallest particles of dust" in his *Opus Majus* Volume V.

1661 Microscopic drawings by English architect Christopher Wren impress Charles II, who commissions more drawings from Robert Hooke.

AFTER
1683 Dutch amateur scientist Antonie van Leeuwenhoek uses a microscope to observe bacteria and protozoa, and publishes his findings with the Royal Society of London.

1798 Edward Jenner, an English physician and scientist, develops the world's first vaccine – for smallpox – and publishes *An Inquiry into the Causes and Effects of the Variolae Vaccinae.*

Leafing through the pages of *Micrographia*, a 17th-century reader would have been astonished. Here, in English scientist Robert Hooke's seminal 1665 book, were many detailed illustrations of structures previously hidden from the human eye due to their minuscule size. Hooke's microscope magnified things by a factor of fifty, but the accuracy of his drawings also owes much to his painstaking approach. Hooke would make numerous sketches from many different angles before combining them into a single image.

… in every little particle… we now behold almost as great a variety of Creatures, as we were able before to reckon up in the whole Universe itself.
Robert Hooke

Although it is not known for certain who developed the first microscopes, they were certainly in use by the 1660s. The early instruments were unreliable – due to the difficulty of making the lenses – and scientists had to be inventive and work around the problem. At first, Hooke had difficulty seeing his specimens clearly, so he invented an improved light source, named a "scotoscope".

Hooke's book is more than just an accurate representation of what he saw through the lens; it also theorizes on what the images reveal about the workings of the organisms he studied. For example, when looking at a wafer-thin specimen of cork, Hooke saw a honeycomb-like pattern, the elements of which he described as "cells" – a term that is still used today.

Microscopic marvels
Micrographia inspired many other scientists to investigate the microscopic world. Following notes and diagrams from Hooke's book, Dutch scientist Antonie van Leeuwenhoek was able to construct his own microscopes. He achieved magnifications of more than 200 times actual size.

See also: Classification of living things 82–83 ▪ A system for identifying all nature's organisms 86–87 ▪ Microbiology 102–103 ▪ Thermoregulation in insects 126–127

Van Leeuwenhoek examined samples of rainwater and stagnant pondwater and marvelled at the multitude of life he saw there. He identified single-celled protozoa, naming them "animalcules", and went on to discover bacteria. He also made many observations of human and animal anatomy, including blood cells and sperm.

While van Leeuwenhoek examined water samples, fellow Dutchman Jan Swammerdam was placing insects under his own microscope. He published records of all manner of insects depicted in the finest detail and uncovered much about their anatomy. Swammerdam's most influential work was *Life of the Ephemera* (1675), which recorded in great detail the life cycle of the mayfly.

In England, Nehemiah Grew used microscopy to examine a wide range of plants. He was the first to identify flowers as being the sexual organs of plants. In *The Anatomy of Plants* (1682), Grew named the stamen as the male organ and the pistil as the female

[*Micrographia* is] … the most ingenious book that I ever read in my life.
Samuel Pepys
English diarist

organ. Grew also spotted pollen grains and noted that they were transported by bees.

Since the early days of microscopy, devices have grown in sophistication. The electron microscope, first used in 1931, uses beams of electrons – rather than light – to reveal objects, allowing scientists an even closer look. Electron microscopes provide views of up to one million times actual size – 600 times greater than most modern light microscopes. ▪

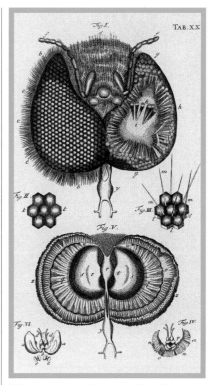

The compound eye and brain of a bee, drawn by Jan Swammerdam and published in *A Treatise on the History of Bees*, shows the eye exterior (left) and the eye dissected (right), with the brain cross-sectioned below.

Robert Hooke

Born on the Isle of Wight, England, Hooke showed an early interest in science. A small inheritance allowed him to attend the prestigious Westminster School, where he excelled, earning a place at Oxford University. There he assisted the natural philosophers John Wilkins and Robert Boyle. In 1662 Hooke became the first curator of experiments for the Royal Society of London. In 1665 he became Professor of Physics at Gresham College.

Like many scientists of his day, Hooke had a broad range of interests. His achievements include some early insights into the wave theory of light; the construction of some of the earliest telescopes; and the formulation of Hooke's Law. Hooke was also a respected architect, an activity that made him a wealthy man.

Key works

1665 *Micrographia*
1674 *An Attempt to Prove the Motion of the Earth*
1676 *A Description of Helioscopes and Some Other Instruments*

IF YOU DO NOT KNOW THE NAMES OF THINGS, THE KNOWLEDGE OF THEM IS LOST

A SYSTEM FOR IDENTIFYING ALL NATURE'S ORGANISMS

IN CONTEXT

KEY FIGURE
Carl Linnaeus (1707–78)

BEFORE
1682 John Ray, an English botanist, proposes that the plant kingdom be divided into trees and two families of herbaceous plants.

1694 French botanist Joseph Pitton de Tournefort publishes *Eléments de Botanique*. This beautifully illustrated book becomes the botanical classification benchmark for half a century.

AFTER
1957 Sir Julian Huxley is the first to use the term "clade" to describe a common ancestor and all of its descendants.

1969 Robert Whittaker, an American ecologist, argues for a five-kingdom categorization of life: Monera, Protista, Fungi, Plantae, and Animalia.

Before the 18th century, there was no consistent naming system for animals and plants. Botanists and zoologists often did not know if they were discussing the same organism. To overcome the problem, Swedish botanist Carl Linnaeus invented a revolutionary system, which is still in use today. He is known as the "father of taxonomy" – the science of naming and classifying organisms.

Linnaeus divided both the plant and animal kingdoms into classes, orders, genera, and species. Organisms were placed in these levels on the basis of shared characteristics, such as similarity of body parts, size, shape, and methods of getting food. Linnaeus also adopted a precise two-word (binomial) name for each species.

Early insights

By 1730, while still a student, Linnaeus began to have issues with the system for classifying plants developed by Joseph Pitton de Tournefort more than 30 years earlier. For Linnaeus, the characteristics of individual species needed to be analysed more closely in order to produce a more thorough taxonomic system.

Collaborative work is crucial for the advancement of **scientific knowledge**.

To **work together** over long distances, scientists need things to be **named** with **accuracy**.

Misunderstandings cause **discrepancies** in scientific knowledge.

If you do not know the names of things, the knowledge of them is lost.

See also: Classification of living things 82–83 ▪ Biological species concept 88–89 ▪ A modern view of diversity 90–91

> In natural science,
> the principles of truth
> ought to be confirmed
> by observation.
> **Carl Linnaeus**

In 1732, Linnaeus joined an expedition to Lapland, where he collected about 100 unidentified species. These formed the basis of his book *Flora Lapponica*, in which he aired his ideas about plant classifications for the first time.

Three years later, Linnaeus wrote about his idea for a new hierarchical classification of plants in a further book, *Systema Naturae*, and thereafter in arguably his greatest work, *Species Plantarum*, published in 1753, which covered 7,300 species. Previously, plants had been known by long impractical names – for example, *Plantago foliis ovato-lanceolatis pubescentibus, spica cylindrica, scapo tereti*. Linnaeus called this plant *Plantago media*, which was sufficient to identify it. As well as being concise, the Linnaean system describes relationships between species.

Later developments

Linnaeus constantly expanded *Systema Naturae*; its 10th edition (1758) became the starting point for modern animal classification. It was he who suggested that humans were members of the primate family. Much later, aided by Charles Darwin's theory of evolution by natural selection, biologists accepted that a classification should reflect the principle of common descent, which led to the methodology known as cladistics. ▪

Whales were once thought to be fish, and were classified as such in an early edition of Linnaeus's *Systema Naturae*. Only later was it understood that they are actually mammals.

Carl Linnaeus

Born in rural southern Sweden, Linnaeus was educated at the University of Uppsala, where he began teaching botany in 1730. He spent three years in the Netherlands, and, on returning to Sweden, he divided his time between teaching, writing, and plant-collecting expeditions. At Uppsala, 17 of his students embarked on expeditions all over the world. Linnaeus was a friend of Anders Celsius, the inventor of the temperature scale. After his friend's death, Linnaeus reversed the scale so that freezing point was 0°C (32°F) and boiling point 100°C (212°F). Linnaeus has been described as the "prince of botanists", and the philosopher Rousseau said of him "I know no greater man on Earth". Linnaeus is buried in Uppsala Cathedral; his remains constitute the type specimen – the specimen that represents a species – used for *Homo sapiens*.

Key works

1735 *Systema Naturae*
1737 *Flora Lapponica*
1751 *Philosophia Botanica*
1753 *Species Plantarum*

"REPRODUCTIVELY ISOLATED" ARE THE KEY WORDS
BIOLOGICAL SPECIES CONCEPT

IN CONTEXT

KEY FIGURE
Ernst Mayr (1904–2005)

BEFORE
1686 Naturalist John Ray defines individual plant and animal species as those that derive from the same seed.

1859 Charles Darwin's *On the Origin of Species* introduces the idea that species evolve through natural selection.

AFTER
1976 *The Selfish Gene* by Richard Dawkins popularizes gene-centred evolution – natural selection at a genetic level.

1995 *The Beak of the Finch* by Jonathan Weiner follows the work of biologists Peter and Rosemary Grant on the Galapagos Islands.

2007 Massimo Pigliucci and Gerd B. Müller use the term "eco-evo-devo" to suggest how ecology is among the factors affecting evolution.

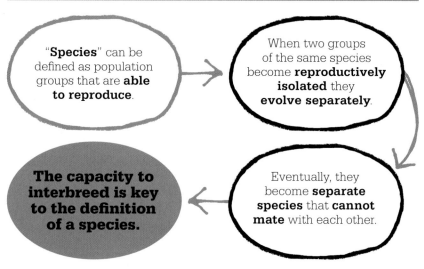

"**Species**" can be defined as population groups that are **able to reproduce**.

When two groups of the same species become **reproductively isolated** they **evolve separately**.

Eventually, they become **separate species** that **cannot mate** with each other.

The capacity to interbreed is key to the definition of a species.

By the early 20th century, it was accepted that multiple species could evolve from a common ancestor. However, it was not clear how this evolution process actually occurred. In fact, there was some debate about precisely what a "species" was. In 1942, evolutionary biologist Ernst Mayr proposed a new definition of species: groups of interbreeding natural populations that are "reproductively isolated from other such groups".

What this means is that two populations of the same species living in the same area may at some point become separated by geography, mate choice, feeding strategies, or other means, and then begin to change through natural selection or genetic drift. Over time, as a result of this initial separation, two distinct species evolve, which cannot interbreed. This type of speciation commonly occurs in small populations of creatures on remote islands.

Key differences

The biological species concept is primarily focused on the breeding potential between organisms. Two

See also: Evolution by natural selection 24–31 ▪ The role of DNA 34–37 ▪ The selfish gene 38–39
▪ Competitive exclusion principle 52–53

organisms may appear identical and live in the same place, but this does not mean that they are the same species. For example, the western meadowlark (*Sturnella neglecta*) and eastern meadowlark (*Sturnella magna*) look similar and have overlapping ranges, but they have evolved to produce different songs. This prevents them from mating with each other, making them two distinct species.

Another scenario is when members of the same species look very different, but because they can mate and reproduce they are still considered to be the same species. The most obvious instance of this is the domestic dog (*Canis familiaris*), a species in which there are great differences between individuals. However, as is also evident, different breeds are capable of reproduction with each other, and therefore belong to the same species.

Complex permutations

According to the biological species concept, the potential for inter-breeding is key to the definition

> Endless forms most beautiful and most wonderful have been, and are being, evolved.
> **Charles Darwin**

of a species. Geographical separation alone does not prevent species from reproducing if they are brought together. Evolutionary divergences – such as the different mating songs of the western and eastern meadowlark – are what prevents interbreeding.

The biological species concept is not applicable to asexual organisms, such as bacteria, or asexual creatures – for example, species of whiptail lizard.

Sometimes, too, different animal species are able to mate and produce offspring, as is the case of a female horse (*Equus ferus caballus*) and a male donkey (*Equus africanus asinus*), which together can produce a hybrid – the mule. However, mules themselves are generally incapable of reproduction, and therefore the horse and donkey remain different species. Another example is the liger, a zoo-bred hybrid of a female tiger and a male lion.

Such anomalies highlight the complexities of defining a species. The biological species concept remains the most popular, but scientists are now looking at the idea of shared genes, and using DNA sequence analysis. To date, no one has come up with a single definition that covers every known species, and it seems unlikely that anyone ever will. In the absence of better models, Ernst Mayr's biological species concept provides an extremely useful way of thinking about species and evolution. ▪

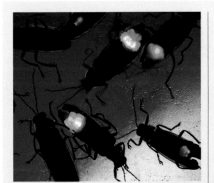

Male fireflies are an example of a typological species. They emit a pattern of flashes to attract females, who recognize their species' code and flash back – if they wish to mate.

Alternative species concepts

Although Mayr's idea about biological speciation is perhaps the most common way to define species and explain how they evolve, it is far from the only one. In fact, there are more than 20 recognized species concepts, arranged across two broad groups – typological and evolutionary concepts. Typological species concepts are based on the idea that a population of individuals of the same type – or sharing the same set of traits – are what makes up a species. The traits can be based on genetics, such as DNA or RNA base sequences, or on phenotypes, such as the size of certain body parts or particular markings, such as the arrangments of spots on insects' wings. The evolutionary species concept is based on species lineages. A species is defined as the organisms that share a lineage from the time when the species initially split off until extinction, or until an additional splitting off and creation of a new species.

ORGANISMS CLEARLY CLUSTER INTO SEVERAL PRIMARY KINGDOMS
A MODERN VIEW OF DIVERSITY

IN CONTEXT

KEY FIGURE
Carl Woese (1928–2012)

BEFORE
1758 *Systema Naturae* (10th edition) by Carl Linnaeus classifies known life into two kingdoms: animals and plants.

1937 French biologist Edouard Chatton divides life into prokaryotes (bacteria) and eukaryotes (organisms with complex cells).

1966 German biologist Willi Hennig establishes a system of classification based on clades – groups of organisms based on common ancestry.

1969 American ecologist Robert Whittaker divides life into five kingdoms: bacteria, protists, fungi, plants, and animals.

AFTER
2017 A consensus among biologists accepts a seven-kingdom classification of life.

Before biologists had the equipment and techniques needed to scrutinize the microscopic structure of living things, biological diversity was split simply into animal-like and plant-like organisms. Then, in the 20th century, better microscopes began to reveal deeper differences that could not be seen with the naked eye. By the 1960s, picking up on an idea first proposed by Edouard Chatton in the 1930s, the need for a new division of living things emerged, placed between prokaryotes (such as bacteria,

with simple nucleus-free cells), and eukaryotes (such as animals and plants with larger, more complex cells).

In the 1970s, the American biologist Carl Woese claimed that even this system failed to account for the diversity among microbes – the smallest living things. He focused on ribosomes – minuscule

Sulphur-dependent archaea
organisms thrive in the hot geothermal pools of Yellowstone National Park, Wyoming, in conditions that would kill most other organisms.

See also: Early theories of evolution 20–21 ▪ Evolution by natural selection 24–31 ▪ The role of DNA 34–35 ▪ A system for identifying all nature's organisms 86–87

Carl Woese's three-domain tree

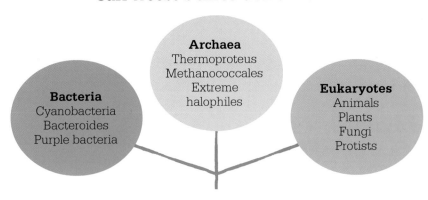

Bacteria
Cyanobacteria
Bacteroides
Purple bacteria

Archaea
Thermoproteus
Methanococcales
Extreme
halophiles

Eukaryotes
Animals
Plants
Fungi
Protists

According to Carl Woese, all organisms can be separated into three main categories or "domains". These divisions are based on similarities in the ribosome structure found in the cells of the groups of organisms within each domain.

grains that all cells need in order to make protein – and devised what he called the "three-domain system". This gave him a new perspective on the branches of Charles Darwin's evolutionary "tree of life". Woese found big differences in the chemical make-up of ribosomes among tiny microbes, with one group as far from other prokaryotes as bacteria are from humans.

Revising the tree of life

Woese's third domain of organisms, known as archaea, is superficially similar to bacteria, but has some strange properties. Many thrive in extreme habitats. Some – uniquely among living things – generate methane in oxygen-deprived places, such as deep marine sediments, or inside warm digestive cavities, such as those of belching, flatulent plant-eating mammals. Other archaea inhabit lakes that are ten times saltier than seawater, or hot acidic pools fed by geothermal heat that would kill anything else.

A decade before Woese proposed his theory, Robert H. Whittaker had recognized animals, plants, and fungi as separate eukaryotic kingdoms, with all other eukaryotes placed in the protist kingdom, and bacteria constituting a fifth kingdom. Whittaker's protist kingdom covered eukaryotic organisms such as amoebas that did not fit the other categories. Some protists were closer to animals, some closer to plants, and others not close to either. They did not match the tree of life model, in which clades – groups of organisms with a common ancestry – spring as branches from the previous fork.

Woese sought a classification system that reflected the intricacies of evolution – with main branches on the tree of life splitting into smaller ones, and even tinier twigs that end in the leaves of individual species. In the future, the complex tree of life may reveal even more evolutionary categories. ▪

Kingdom of their own

For most of the history of biology, fungi were considered to be plants. Even the great classifier of organisms Carl Linnaeus included them in his kingdom Plantae. It was only with the invention of more powerful microscopes that the differences in fungi began to be better understood. It is now known that chitin, a complex carbohydrate and component of fungus cell walls, is not found in plants. Also, fungi make their food by digesting rotted material, whereas plants make food by absorbing light energy in photosynthesis.

DNA analysis shows that fungi are far removed from plants in the evolutionary tree of life: they are, in fact, genetically closer to the branch that gives rise to animals. These same studies show that certain aquatic moulds – traditionally classified as fungi – are not related to fungi, while some disease-causing microbes are fungi that have evolved to become microscopic parasites.

Fungi, such as this bright yellow jelly fungus growing on a fallen tree, are no longer classified as plants. Fungi are genetically closer to animals.

SAVE THE BIOSPHERE AND YOU MAY SAVE THE WORLD

HUMAN ACTIVITY AND BIODIVERSITY

IN CONTEXT

KEY FIGURE
Edward O. Wilson (1929–)

BEFORE
1993 The UN proclaims
29 December as the
International Day for
Biological Diversity.

1996 *The Song of the Dodo*
by American science writer
David Quammen explores
the nature of evolution and
extinction as habitats become
more and more fragmented.

AFTER
2014 *The Sixth Extinction*
by environmental journalist
Elizabeth Kolbert shows how
humans are causing a sixth
mass extinction of species.

2016 In *Half-Earth*, Edward
Wilson proposes that Earth
can be saved by dedicating
half of it to nature.

Biodiversity is the variety of
life on Earth – in all forms
and at every level, from
genes to microbes to humans and
all other species, including those
yet to be discovered. Humans rely
on biodiversity for food and fuel,
shelter, medicine, beauty, and
pleasure. For other species, it also
provides nutrients, seed dispersal,
pollination, and reproductive
success. No living thing could
survive without biodiversity.

Ecologists have identified
growing threats to biodiversity,
many of them driven by human
actions. The current rate of species
extinction is thought to be up to
1,000 times greater than it was

See also: Biodiversity hotspots 96–97 ▪ Animal ecology 106–113 ▪ Island biogeography 144–149 ▪ Biodiversity and ecosystem function 156–157 ▪ Biomes 206–209 ▪ Mass extinctions 218–223 ▪ Deforestation 254–259 ▪ Overfishing 266–269

The effects of human activity on biodiversity

The five human activities that most seriously affect biodiversity on Earth can be represented by HIPPO, the acronym devised by Edward Wilson, with the relative severity of each reflected in the order of the letters.

1. Habitat destruction

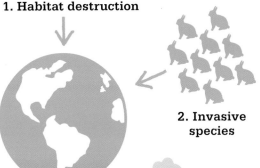

5. Overharvesting by hunting or fishing

2. Invasive species

4. Human population

3. Pollution

animals, as they may no longer be able to find places to feed or rest along their normal routes. Native species and ecosystems are also disrupted by the introduction, accidentally or deliberately, of new species. These invasive species can threaten the food supply or other resources of native species, carry disease, and become a predatory threat. The brown tree snake, for example, was brought accidentally to the island of Guam on a cargo ship, and has caused the extirpation (the extinction of a species in a particular area) of 10 of the island's 11 native bird species.

Air and water poisoning

Any kind of pollution threatens biodiversity, but air and water pollution are particularly harmful. Burning fossil fuels, for example, releases the waste gases sulphur dioxide and nitrogen oxide into the air; these return as acid rain, causing water and soil acidification and affecting ecosystem health and biodiversity. Ozone emissions at ground level can also damage cell membranes on plants, curbing their growth and development. »

before 1800, when humans began to dominate the planet. The first use of the term "biodiversity", in 1988, was by American biologist Edward O. Wilson, who became known as the "father of biodiversity". He later highlighted five key threats to biodiversity using the acronym HIPPO: habitat destruction; invasive species; pollution; human population; and overharvesting by hunting and fishing.

Habitat wreckers

The Red List of the International Union for Conservation of Nature (IUCN) covers more than 25,000 threatened species. Of these, 85 per cent are endangered by the loss of habitats that once supported particular species. This destruction can occur as a result of natural causes, such as fire or flood, or, more commonly, through the expansion of agricultural land, timber harvesting, and overgrazing by livestock. Deforestation, in particular, has contributed hugely to habitat loss, with around half of the world's original forests now cleared, mainly for agricultural use.

Some habitats are not destroyed but rather broken up or divided into more isolated units by human interventions, such as building dams or other water diversions. This habitat fragmentation is particularly dangerous for migratory

It is that range of biodiversity that we must care for – the whole thing – rather than just one or two stars.
David Attenborough
British broadcaster and naturalist

Edward O. Wilson

Born in Alabama, US, in 1929, Edward Osborne Wilson was left blind in one eye after a fishing accident aged seven, and switched interests from birdwatching to insects. He discovered the first colony of fire ants in the US when he was only 13, and later attended the universities of Alabama and Harvard. Wilson's work has focused primarily on ants but also extends to the study of isolated ecosystems, known as "island biogeography". A leading environmentalist, he has spearheaded efforts to preserve biodiversity and educate people about it. He has been awarded over 150 prizes, including the National Medal for Science, the Cosmos Prize, and two Pulitzer Prizes for non-fiction, and was named one of the century's leading environmentalists by *Time* and *Audobon* magazine.

Key works

1984 *Biophilia*
1998 *Consilience: The Unity of Knowledge*
2014 *The Meaning of Human Existence*

We should preserve every scrap of biodiversity as priceless while we learn to use it and come to understand what it means to humanity.
Edward O. Wilson

Water pollution is caused mainly by sewage or by chemicals absorbed into water as it flows off agricultural land. This pollution reduces oxygen levels in water, making survival more difficult for some species, particularly when combined with water temperatures that have risen due to climate change. Freshwater streams used by certain species of spawning fish, for example, can be made uninhabitable by pollution.

Some organisms can absorb a substance, such as an agricultural chemical, more quickly than they can excrete it, in a process known as bioaccumulation. Initial, low concentrations of chemicals may not be a problem. However, as those chemicals accumulate through the food chain – from phytoplankton to fish to mammal, for example – they can reach levels that cause birth defects and disrupt hormone levels and immune systems.

Poaching, forest clearance, and other human activities have largely contributed to the status of the African western lowland gorilla as a "critically endangered" species.

Rapid population growth has generated further damage to the environment. The world's human population has risen from less than 1 billion in 1800 to more than 7 billion, and is expected to reach nearly 10 billion by 2050. As the population grows, so do other threats to biodiversity: increasing numbers of invasive species are spread through trade and travel; urban development and resource extraction destroy habitats; more pollution is created; and land is overharvested. The impacts of human population growth will be difficult to limit, as ever more people rely on food and shelter to survive, and demand ever more goods in an increasingly global consumer society.

Upsetting the balance
Population growth also drives overharvesting, the final human-made threat to biodiversity in the HIPPO acronym. Found in forestry, livestock grazing, and commercial agriculture, overharvesting can also arise from targeted hunting, gathering, and fishing, as well as unintentional harvesting, such as fish discarded from catches.

The building of railways across the US was accompanied by hunters hired to decimate the buffalo population that had sustained Native American tribes. By the end of the 19th century, only a small number of wild buffalo survived.

When the rate of harvest exceeds the rate of replenishment through either reproduction or human activities such as tree planting, the harvest is not sustainable, and without regulation could result in the extinction or extirpation of species.

A study of the IUCN's Red List in 2016 showed that 72 per cent of species listed as threatened or near-threatened are harvested at a rate that means their numbers cannot be balanced by natural reproduction or regrowth. Some 62 per cent of species are at risk from agricultural activity alone, such as livestock farming, tree felling, and the production of crops for food, fuel, fibres, and animal fodder.

Protecting biodiversity

In reality, the five HIPPO threats identified by Wilson are interrelated, and there is generally no single reason why any particular species is endangered. Agricultural development, for example, can not only destroy a habitat, but can also releases greenhouse gases into the atmosphere, contributing to air pollution and climate change. More than 80 per cent of the species on the IUCN's Red List are affected by more than one of the five major biodiversity threats.

Biodiversity maintains the health of the ecosystems of the planet. Ecosystems are a delicate balance of living creatures, both plant and animal, as well as the soil, air, and water in which they live. Healthy ecosystems provide resources that sustain human and all other life, improve resilience against natural disasters and human-made shocks, including climate change, and provide recreational, medicinal, and biological resources.

Although the threats to biodiversity from human activity are serious, ways to protect it are being developed. Foremost is a "sustainable" approach to harvesting and agriculture that allows species – such as fish, trees, or crops – to be maintained at a stable level and even increased over time. Official protected status for areas of land, water, and ice can help to sustain threatened species, while national and international agreements and negotiations can mitigate the impact of both legal and illegal trade, such as poaching. Public education also helps people to better understand their potential impacts on biodiversity and how to protect it for future generations. ∎

Anthropogenic biomes

The biosphere – all the areas of Earth and its atmosphere that contain living things – consists of biomes, which are large ecosystems based on a specific environment, such as desert or tropical rainforest. The impact of human actions on biodiversity and the consequent reshaping of much of the planet have led ecologists to reassess biomes and suggest that a designation of anthropogenic (manmade) biomes is now necessary. Anthropogenic biomes are grouped into six main categories: dense settlements; villages; croplands; rangeland; forested; and wildlands.

Unlike other biomes, which can range across continents, anthropogenic biomes are a mosaic of pockets over Earth's surface. According to ecologists, more than 75 per cent of Earth's ice-free land has been affected by at least some form of human activity, particularly in dense settlements (urban areas), which account for over half the world's population, and villages (dense agricultural settlements).

WE ARE IN THE OPENING PHASE OF A MASS EXTINCTION

BIODIVERSITY HOTSPOTS

IN CONTEXT

KEY FIGURE
Norman Myers (1934–)

BEFORE
1950 Theodosius Dobzhansky studies plant diversity in the tropics.

AFTER
2000 Myers and collaborators re-evaluate the list of hotspots and add several new ones, bringing the total to 25.

2003 An article in *American Scientist* criticizes the concentration of conservation effort on hotspots, saying that this neglects less species-rich but still important "coldspots".

2011 A team of researchers confirm the forests of east Australia as the 35th hotspot.

2016 The North American coastal plain is recognized as meeting the criteria for a global biodiversity hotspot – and becomes the 36th.

A biodiversity hotspot is an area with an unusually high concentration of animal and plant species. The term was coined in 1988 by Norman Myers, a British conservationist, to describe areas that are both biologically rich and deeply threatened. Facing the huge

The lush hillsides and forests of Arunachal Pradesh, India, are part of the Indo-Burma biodiversity hotspot. The area contains some 40 per cent of India's animal and plant species.

and increasing challenge of mass extinctions of species caused by the destruction of premium habitats, Myers argued that priorities had to be set to establish where to concentrate resources to conserve as many lifeforms as possible.

Defining hotspots
Initially, Myers identified ten hotspots crucial for conserving plant species that were endemic (did not grow anywhere else on Earth). By 2000, he had refined the concept to focus attention on

See also: Human activity and biodiversity 92–95 ▪ The ecosystem 134–137 ▪ Deforestation 254–259 ▪ Sustainable Biosphere Initiative 322–323

> Our welfare is intimately tied up with the welfare of wildlife ... by saving the lives of wild species, we may be saving our own.
> **Norman Myers**

regions that fulfilled two criteria: the area must contain at least 1,500 vascular plants (plants with roots, stems, and leaves) that were endemic, and it must have lost at least 70 per cent of its primary vegetation (the plants that originally grew in the area). Conservation International, an environmental agency that uses Myers' concept to guide its efforts, now lists 36 such regions. Although they represent only 2.3 per cent of Earth's land surface, they are home to nearly 60 per cent of the planet's plant, amphibian, reptile, mammal, and bird species – and a high proportion of these species live only in their respective hotspot.

Most hotspots lie in the tropics or subtropics. The one facing the highest threat level is the Indo-Burma area in Southeast Asia. Only 5 per cent of the original habitat remains, but its rivers, wetlands, and forests are vital for the conservation of mammals, birds, freshwater turtles, and fish. Animals unique to this area include the saola, a forest-dwelling mammal that is related to cattle but

looks like an antelope; it was seen for the first time in 1992, in the Annamite Mountains of Vietnam. The endangered Irrawaddy dolphin is found along the coastlines of Southeast Asia and the islands of Indonesia. Other rare animals include Eld's deer, the fishing cat, and the giant ibis.

Protective measures

Conservation agencies agree on targets for every hotspot. They list species that are threatened and make plans to conserve and manage those areas with suitable habitat and viable populations of target plants and animals. Sites are ranked according to how vulnerable and irreplaceable they are.

Myers' two criteria have been criticized by those who say they do not take account of changing land use in regions where less than 70 per cent of good habitat has been destroyed. The Amazon rainforest, for example, is not within a hotspot but the forest is being cleared faster than anywhere else on Earth. ∎

> We are into the opening stages of a human-caused biotic holocaust – a wholesale elimination of species – that could leave the planet impoverished for at least five million years.
> **Norman Myers**

Norman Myers

Myers was born in 1934 and grew up in the north of England. He studied at the University of Oxford before moving to Kenya, where he worked as a government administrator and teacher. During the 1970s, Myers studied at the University of California, Berkeley, where his interest in the environment grew. He raised concerns about deforestation for cattle ranching, describing it as the "hamburger connection".

Myers raised the concept of biodiversity hotspots in the article "Threatened Biotas: 'Hotspots' in Tropical Forests", published in *The Environmentalist* in 1988. In his first book, *Ultimate Security: The Environmental Basis of Political Stability*, he argued that environmental problems lead to social and political crises. In 2007, *Time* magazine hailed Myers as a Hero of the Environment.

Key works

1988 "Threatened Biotas: Hotspots in Tropical Forests"
1993 *Ultimate Security: The Environmental Basis of Political Stability*

THE VAR
OF LIFE

ETY

Dutch lens-makers Hans and Zacharias Janssen invent the **compound microscope**.

Louis Pasteur reveals that wine's fermentation process is caused by **germs**; his discovery sparks the development of **germ theory**.

Charles Elton publishes *Animal Ecology,* which sets out many of the **fundamental principles** of animal behaviour.

1590   **1866**   **1927**

1676   **1885**

Antonie van Leeuwenhoek identifies "animalcules", opening up the field of **microbiology**.

Albert Frank coins the term "**mycorrhizae**", in reference to the **symbiotic relationship** between fungi and tree roots.

O ur understanding of the variety, behaviour, and interraction of organisms has advanced considerably since Aristotle discovered that bee colonies have a queen and workers. Huge advances in technology, field observations, and laboratory experiments have increased our knowledge, and the modern study of animal behaviour – ethology – continues to throw up surprises.

Life under the microscope

Until the microscope was invented, no one knew that bacteria even existed, let alone what they did. Bacteria were first observed by Dutch microscopist Antonie van Leeuwenhoek in 1676, using an instrument he had built himself. He called these tiny organisms "animalcules", but little was known about them for many years. In the 1860s, French chemist Louis Pasteur and German microbiologist Robert Koch developed the germ theory of disease, highlighting the harmful role played by bacteria. Subsequent research has also highlighted their positive roles: facilitating digestion; inhibiting the growth of other, pathogenetic bacteria; "fixing" or converting nitrogen into molecules that aid plant growth; and breaking down dead organic material, which releases nutrients for the food web.

Another discovery made possible by microscopy was of the mutualistic relationship between fungi and trees, published by German plant pathologist Albert Frank in 1885. Studying what he first assumed was a pathological infection, Frank discovered that trees with the fungi attached to their roots were healthier than those without. The fine filaments, or hyphae, of the fungi make the roots more efficient at obtaining nitrate and phosphate nutrients from the soil. In return, the fungi get sugar and carbon from the tree.

Connected lives

No organism lives in isolation from the rest of its ecosystem. The behavioural interactions between them are complex and much is still being discovered about them. One of the greatest contributions in this field was made by British zoologist Charles Elton, whose 1927 classic *Animal Ecology* established many important principles of animal behaviour, including food webs and food chains, prey size, and the concept of ecological niches.

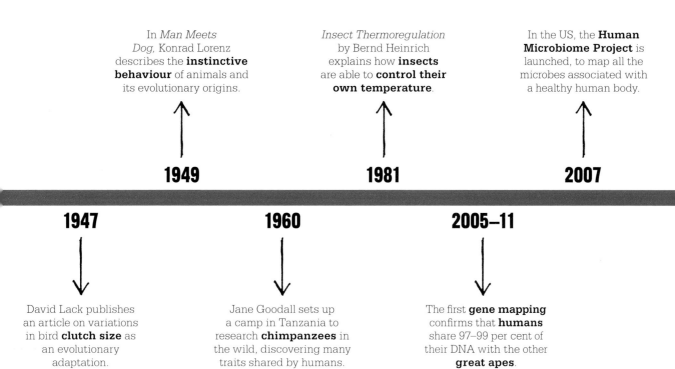

In *Man Meets Dog,* Konrad Lorenz describes the **instinctive behaviour** of animals and its evolutionary origins.

1949

Insect Thermoregulation by Bernd Heinrich explains how **insects** are able to **control their own temperature**.

1981

In the US, the **Human Microbiome Project** is launched, to map all the microbes associated with a healthy human body.

2007

1947

David Lack publishes an article on variations in bird **clutch size** as an evolutionary adaptation.

1960

Jane Goodall sets up a camp in Tanzania to research **chimpanzees** in the wild, discovering many traits shared by humans.

2005–11

The first **gene mapping** confirms that **humans** share 97–99 per cent of their DNA with the other **great apes**.

Ethology, which looks at animal behaviour and its evolutionary basis and development, is a major component in the modern study of organisms. Back in 1837, British entomologist George Newport discovered that moths and bees could raise the temperature of their thorax by quivering their muscles. From the 1970s onwards, German-American entomologist Bernd Heinrich and others uncovered more thermoregulatory adaptations that have helped insects thrive. As heterotherms, they are able to maintain different temperatures in different parts of the body.

Modern research now combines laboratory experiments, field observation, and new technology such as infrared thermography to understand insect behaviour in ever more detail.

Field observations are a key tool in ethological research. In the 1940s, British ornithologist David Lack investigated the factors controlling the number of eggs birds laid (clutch size). His food limitation hypothesis states that the number of eggs laid by a species has evolved to match the food available. Evolutionary pressure has created a correlation between clutch size and food availability.

Austrian zoologist Konrad Lorenz and Dutch biologist Nikolaas Tinbergen also studied animals in the wild to help understand their behaviour. Lorenz's 1949 work *Man Meets Dog* explains the loyalty of a pet dog to its owner in terms of canines' instinctive loyalty to their pack leader in the wild. Tinbergen's field experiments showed how gull chicks, which tap a red spot on a parent's beak when they want food, will tap coloured marks painted on a model beak.

Human traits

As well as these short term studies, British primatologist and ethologist Jane Goodall conducted field observations over a longer period, studying chimpanzees in Tanzania from 1960 to 1975. Her findings challenged the view that human behaviour is totally unique in the animal world, and indicated that chimps are behaviourally closer to people than had generally been assumed. She noted, for example, that chimps display a whole range of facial expressions and other body language to indicate their mood, are toolmakers and users, often behave cooperatively, and sometimes go into battle against rival groups. ■

IT IS THE MICROBES THAT WILL HAVE THE LAST WORD

MICROBIOLOGY

IN CONTEXT

KEY FIGURE
Louis Pasteur (1822–95)

BEFORE
1683 Dutch amateur scientist Antonie van Leeuwenhoek uses a microscope to observe bacteria and protozoa.

1796 Edward Jenner carries out the first vaccination, using the cowpox virus to protect against smallpox.

AFTER
1926 American microbiologist Thomas Rivers distinguishes between viruses and bacteria.

1928 While studying influenza, Scottish bacteriologist Alexander Fleming discovers penicillin.

2007 An inventory of all the microbes associated with a healthy human body is completed.

Microbes – bacteria, moulds, viruses, protozoa, and algae – are present in every environment, living in soil, water, and air. Some microbes cause disease but most are vital for life on Earth. Among other things, they break down organic matter so that it can be recycled back into the ecosystem.

Trillions of microbes also live on and in the human body. The most common of these are beneficial bacteria, which aid the digestion of food, produce vitamins, and help the immune system to find and attack more harmful microbes. Scientists did not understand

Microbes are the worker bees that perform most of the important functions in your body.
Dr Robynne Chutkan
Microbiome expert and author

microbes until they could see them. The first observations began in the 17th century, using the recently invented microscope. These studies revealed a previously unknown world teeming with microbiotic life. Around the same time, the word "germ", originally meaning "seed", was first used to describe these tiny organisms.

Fighting disease

Some 17th and 18th-century scientists believed that certain "germs" might cause diseases, but the prevailing view was that such maladies were the spontaneous result of inherent weakness in an organism. It was not until the painstaking laboratory work of the 19th-century French chemist Louis Pasteur that the "germ theory of disease" was proved.

Pasteur began by looking at the alcohol fermentation process. He discovered that sourness in wine was caused by external agents – microbes, or germs. A crisis in the French silk industry, caused by an epidemic among silkworms, then allowed Pasteur to isolate and identify the microorganisms that caused the particular disease. As he extended germ theory to

See also: Classification of living things 82–83 ▪ The microbiological environment 84–85 ▪ The ecosystem 134–137

> Where observation
> is concerned,
> chance favours only
> the prepared mind.
> **Louis Pasteur**

human disease, Pasteur proposed that germs invade the body and cause specific disorders. Edward Jenner, nearly 100 years before, had shown that a disease could be prevented with the application of a "vaccine" – a virus similar to that of the disease-causing microbe. Pasteur found that an attenuated, or weakened, form of a disease-causing germ, produced in a laboratory and injected into the host animal or human, was

particularly effective at enabling the body's immune system to fight off the disease. At first, Pasteur faced strong opposition and alarm at the prospect, but he was able to develop vaccines for anthrax, fowl cholera, and rabies – the latter involving his first test on a human.

Annihilating germs
The focus later shifted to finding germ-killing agents, or antibiotics, such as penicillin – discovered by Alexander Fleming. A strategy of annihilating microbes has been followed ever since. Yet this "slash and burn" approach has its drawbacks. It kills beneficial microbes as well as harmful ones, and also promotes resistance in bacteria that can ultimately render antibiotics ineffective. ▪

The bacterium *Enterococcus faecalis* is a microbe found in the gut and bowel of healthy humans. If it spreads to other areas of the body, however, it can cause serious infections.

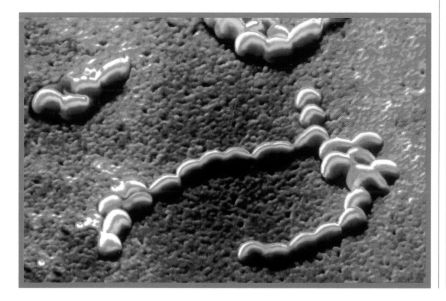

Louis Pasteur

Born in Dole, France, in 1822, Pasteur was the son of a poor tanner. He was an average student, but he worked hard, obtaining his degree in 1842 and his doctorate in science in 1847. After teaching in various universities, in 1867 he became Professor of Chemistry at the Sorbonne in Paris. His major research interest was the fermentation process. Pasteur discovered that the fermentation of wine and beer was caused by germs – microbes. He also discovered that microbes could be killed by short, mild heat treatment – a process now named after him as "pasteurization". Pasteur's "germ theory" led to the wider development of vaccines, which remain a vital method of disease control. In 1887, he established the Pasteur Insitute, which opened in 1888 and continues to help to prevent and fight diseases.

Key works

1870 *Studies on Silk Worm Disease*
1878 *Microbes: Their Role in Fermentation, Putrefaction, and the Contagion*
1886 *Treatment of Rabies*

CERTAIN TREE SPECIES HAVE A SYMBIOSIS WITH FUNGI
THE UBIQUITY OF MYCORRHIZAE

IN CONTEXT

KEY FIGURE
Albert Frank (1839–1900)

BEFORE
1840 German botanist Theodor Hartig discovers a network of filaments on the roots of pine trees.

1874 Hellmuth Bruchmann, a German biologist, notes the "Hartig net" is made of fungal filaments.

AFTER
1937 A.B. Hatch, an American botanist, shows a beneficial relationship between pine trees and mycorrhizal fungus.

1950 Swedish botanists Elias Melin and Harald Nilsson show that plant roots can extract more nutrients from the soil with the aid of mycorrhizae.

1960 Another Swedish botanist, Erik Björkman, shows that plants pass carbon into mycorrhizal fungi in exchange for phosphate and nitrate.

I n 1885, a professor of plant pathology at the Royal College of Agriculture in Berlin named Albert Frank was the first to see a connection between fungi growing on tree roots and the health of the trees. Frank realized that these were not pathological (disease-related) infections but in fact underground partnerships: far from suffering, the trees seemed to benefit from better nutrition. He invented a new term for the partnership – "mycorrhiza", from the Greek *mykes*, meaning fungus, and *rhiza*, meaning root.

Mycorrhizae in action
False truffles are an example of the fungal side of this partnership. Nineteenth-century Prussian botanists had found these fungi under spruce trees, and noticed that each tree root was drawn towards a truffle, and wrapped in a fungal husk. Although they did not know it, the botanists were witnessing a phenomenon that is vital to many ecosystems.

Fungi are typically nourished by a supply of organic matter, from which they extract food by external digestion. A deep layer of forest litter is perfect. They pour digestive chemicals onto their meal and

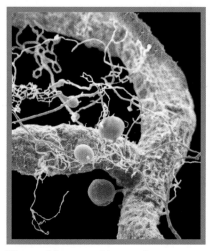

Mycorrhizae on the root of a soya bean. In arbuscular mycorrhizae, such as these, the tips of the hyphae form clusters inside the plant's root cells, optimizing the exchange of nutrients.

absorb the soluble organic compounds produced through a network of microscopic filaments – hyphae – called a mycelium.

Plants rely on root hairs to absorb water and minerals, such as nitrates and phosphates. But there is a limit to how far plant roots can grow and therefore what quantity of nutrients the root hairs can absorb. The hyphae of mycorrhizae can cover a much wider area, absorbing

See also: Evolution by natural selection 24–31 ▪ Mutualisms 56–59
▪ The ecosystem 134–137 ▪ Energy flow through ecosystems 138–139

Beneficial exchange between mycorrhizae and plant roots

Mycorrhiza

Plant

Supplies sugar from photosynthesis

Connects plants in an extensive network

Increases uptake of water and nutrients

Allows plants to share nutrients with others

Boosts protection against soil diseases

The mutualistic relationship between mycorrhizae and plants is highly evolved. As many as 90 per cent of all plant species rely on fungi for nutrients and protection. In return, plants supply the fungi with a vital food source.

a much greater amount of minerals. When the fungal hyphae attach to the plant roots, they extend the root system, causing extra nutrients to seep into the plant.

Albert Frank realized that this partnership worked both ways. It was a winning combination for both plant and fungus. In exchange for passing on a share of its minerals, the fungus receives sugar from the plant – made by photosynthesis in the leaves and transported to the roots via the plant's sap. This boosts the nutrient supply that the fungus derives from dead organic matter.

Ancient networks

Fossils of plants dating from 400 million years ago – when vegetation was first spreading across dry land – show traces of fungal threads. This suggests that the mycorrhizal partnership was key to the evolution of terrestrial life. Today, the majority of plant species continue to rely on

fungi in this way. Trees supported by mycorrhizae are more resistant to drought and disease, and can even communicate alarm signals by releasing chemicals in response to attack by herbivores. This fungal network connecting trees has been dubbed "the wood-wide web". ▪

[the fungus] performs a "wet nurse" function and performs the entire nourishment of the tree from the soil.
Albert Frank

Mycorrhizae as pollution indicators

Mycorrhizal fungi are not only good for the health of plants – they can also act as indicators of the health of the entire environment. Laboratory experiments with these fungi have shown that some grow badly in the presence of toxins, which means that they can be used to detect pollutants in the air or soil. For instance, some fungi fail to grow when exposed to heavy metals such as lead or cadmium, and because different kinds of fungi react differently to environmental change, certain species can be used to identify specific kinds of pollution.

Mycorrhizae are also useful indicators of the health of their native habitat. Many form cauliflower-like growths on tree roots, but these are smaller in polluted soil. The trees themselves may also respond to pollution with weaker shoot growth, but the mycorrhizal response is more acute and serves as a valuable early-warning sign of a habitat in decline.

Weak growth in the russet brittlegill, a mycorrhizal fungus of European and North American spruce forests, can be an early indicator of habitat air pollution.

FOOD IS THE BURNING QUESTION

ANIMAL ECOLOGY

IN CONTEXT

KEY FIGURES
Charles Elton (1900–91),
George Evelyn Hutchinson
(1903–91)

BEFORE
Ninth century Arab writer
Al-Jaziz introduces the
concept of the food chain
in *Kitab al-Hayawan* (Book
of Animals), concluding that
"every weak animal devours
those weaker than itself".

1917 American biologist
Joseph Grinnell first describes
an ecological niche in his
paper, "The niche relationships
of the California thrasher".

AFTER
1960 American ecologist and
philosopher Garrett Hardin
publishes an essay in the
magazine *Science* in which
he states that "every instance
of apparent coexistence must
be accounted for".

1973 Australian ecologist
Robert May publishes *Stability
and Complexity in Model
Ecosystems*, in which he uses
mathematical modelling to
demonstrate that complex
ecosystems do not necessarily
lead to stability.

A food web is a graphic depiction
of the feeding connections between
different species within an ecological
community. This example illustrates
the relationships within a marine
ecosystem, in which killer whales are
the apex predators and phytoplankton
are the primary producers.

The concept of food chains –
the idea that all living
things are linked through
their dependence on other species
for their food – dates back many
centuries, but it was not until the
early 20th century that scientists
developed the concept of food
chains forming a food web.

The pioneer of this thinking was
British zoologist Charles Elton,
whose book *Animal Ecology* (1927)
describes what he called the "food
cycle". He later went on to develop
theories that encompassed more
complex interactions between
animals and the environment –
insights that underpin modern
animal ecology. He likened our

knowledge of individual plant and
animal species to the cells in a
beehive – each "cell" of knowledge
is important in its own right, but
by putting them all together
something much more than the
sum of the parts is created –
the "beehive" of ecology.

Nowadays, the study of animal
ecology focuses on how animals
interact with their environment,
the roles played by different
species, why populations rise
and fall, why animal behaviour
sometimes changes, and the
impact of environmental change on
animals. The principle underlying
the work of animal ecologists is
that there is generally a balance

Food web

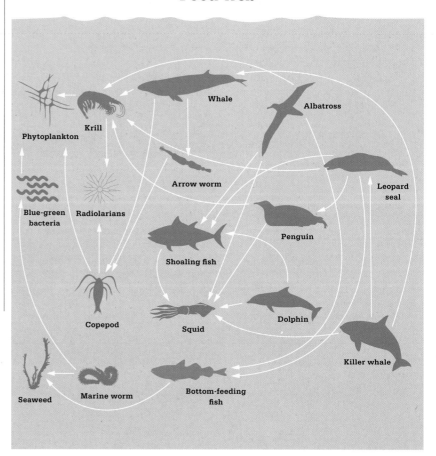

in nature, so if the population of a given species grows too large it will be regulated, most often by a lack of food. However, relationships between organisms and their environment change from place to place and through time.

Chain of dependence

In *Animal Ecology*, Elton outlined the key principles of the study of animal communities: food chains and webs, food size, and ecological niches. Each food chain and web, he asserted, is dependent on producers: plants and algae that support plant-eating consumers (herbivores). These herbivores in turn support one or more levels of meat-eating consumers (carnivores). Large carnivores generally eat smaller animals, but because small animals reproduce more quickly, their numbers are able to support the larger predators.

Competition for food is very tight near the top of a food web. Although apex (top) predators, such as big cats and large birds of prey, have no natural predators, this often means that they have to defend territories

A spider traps a damselfly, demonstrating that the principle of food size can be modified by the comparative aggression and strength of the predator and its prey.

against rivals of their own species to ensure there is enough food for themselves and their offspring.

Food size

One of Elton's most important points was the notion that food chains exist primarily because of the principle of food size. He explained that every carnivorous animal eats prey between upper and lower limits. Predators are physically unable to catch and consume other animals above a certain size because they are not large enough, strong enough, or skilful enough. That is not to say that predators cannot kill and eat larger animals than themselves; a weasel can easily kill a larger rabbit because it is more aggressive. However, an adult lioness, one of the world's top predators, is not capable of killing a healthy adult African elephant. Likewise, a dragonfly larva on the bottom of a pond may be able to prey on a small tadpole, but it would not be able to eat an adult frog. »

Every animal is closely linked with a number of other animals living around it – and these are largely food relations.
Charles Elton

Forecasting the effects of climate change

Ecologists examine changes to animal populations and distribution and apply climate change models to forecast how these will change further in the future, over the course of 5, 10, 50, or more years. In the Arctic, for example, where average temperatures are rising more rapidly than anywhere else, the sea ice is contracting. As a result, polar bears have to travel further in search of ice where they can catch seals, rest, and mate. The further they swim, the more energy they burn. As the sea ice declines, the polar bears starve. Scientists monitor their numbers and movements and compare this data with changes in sea ice.

The polar bear plays a vital part in the ecology of the Arctic. As an apex predator and keystone species, it must have access to seals, which are its almost exclusive diet. The number of seals regulates the density of polar bears, while polar bear predation in turn regulates the density and reproductive success of seals.

A lone polar bear surveys the sea for prey from a piece of floating ice in the Arctic. The shrinking area of sea ice in the region threatens this species' survival.

Snowshoe hare and lynx population cycles

In Canada's boreal forests, the favoured prey of lynx are snowshoe hares. Charles Elton examined the relationship between the populations of these two species, using data covering the period 1845–1925. When hares are numerous, lynx hunt little else. After their population reaches its peak density, the hares struggle to find enough plant food. Some starve, while others are weakened and are more easily caught by predators, including lynx, which feed very well for a time. When hare numbers continue to fall, this affects the lynx. They are forced to hunt less nutritious prey, such as mice and grouse.

As they struggle to find enough to eat, lynx produce smaller litters or even stop breeding altogether. Some starve to death. A decline in the lynx population sets in one or two years after the hare population has bottomed out, a cycle that repeats every eight to eleven years.

A Canadian lynx captures a snowshoe hare, its preferred prey. When hares are plentiful, a lynx will eat two every three days.

Animals may be capable of killing much smaller prey, but it is simply not worth the effort. Wolves hunt medium-sized or large mammals such as elk. If those mammals disappear from their environment, they find it hard to catch sufficient numbers of smaller animals such as mice to sustain them; the energy they use finding small prey is greater than the energy they gain by consuming them.

Plants cannot run away or fight back, so different considerations apply to herbivores when it comes to food size. There is a maximum size of seed that a given finch, for example, can fit in its bill, so larger finches have an advantage over smaller species. Similarly, individual species of hummingbird can drink nectar only from flowers up to a certain size, depending on the length of their bill.

Ecological niches

An animal or plant's niche is its ecological role or way of life. For American zoologist Joseph Grinnell,

The sword-billed hummingbird, a native of South America, has a long bill, which enables it to suck up nectar from the long flowers of *Passiflora mixta*, a species of passionflower. As it feeds, it spreads the plant's pollen.

working in the early decades of the 20th century, an organism's niche was defined as its habitat. He studied birds called thrashers in California and observed how they fed, nested, and hid from predators in the dense undergrowth of the chaparral shrubland. However, a niche is more complex than simply the place where an organism lives. Oxpeckers and buffalo share exactly the same habitat – open grassland – but their requirements for survival are very different: the buffalo graze on the grasses, while oxpeckers derive their food from the ticks they peck from the buffalos' hide.

Charles Elton explored the concept of ecological niches in more depth. For him, food was the primary factor in defining an animal's niche. What it ate and what it was eaten by were crucial. Depending on the habitat, a particular niche could be filled by a different animal. Elton cited the example of a niche that was filled

Observation of species in the wild convinces me that the existence and persistence of species is vitally bound up with environment.
Joseph Grinnell

> Different species press against one another, like soap bubbles, crowding and jostling, as one species acquires … some advantage over another.
> **G. Evelyn Hutchinson**

by birds of prey that hunted small ground-dwelling animals such as mice and voles. In a European oakwood, that niche would be filled by tawny owls, while on open grassland kestrels would fill the role.

Elton also argued that an animal could not only tolerate a certain set of environmental conditions, but could also change them. The tree-felling and stream-damming activity of beavers is one of the most dramatic examples, creating habitats for fish in dammed pools, woodpeckers in dead trees, and dragonflies around pool margins.

Niches and competition

British-born zoologist G. Evelyn Hutchinson, working at Yale University from the 1950s to the 1970s, examined all the physical, chemical, and geological processes at work in ecosystems and proposed that any organism's role in its niche includes how it feeds, reproduces, finds shelter, and

A true specialist, a koala bear requires 1kg (2.5lb) of eucalyptus leaves a day. This species is found in the wild only in Australia, where eucalyptus is common.

interacts with other organisms and with its environment. For example, each species of trout – and other fish – has its own range of water salinity, acidity, and temperature that it can tolerate, as well as a range of prey and river- or lake-bed conditions. This makes some better competitors than others, depending on the conditions of the habitat in which they live. Seen as the father of modern ecology, Hutchinson inspired other scientists to explore how competing animals use their environment in different ways.

An animal or plant's niche width comprises the whole range of factors it requires to allow it to thrive. Brown rats, raccoons, and starlings are examples of animals with a broad niche width in that they are able to survive in a wide variety of conditions. Such species are called generalists. Other animals have narrow requirements. For example, koalas depend almost entirely on eucalyptus leaves, and hyacinth macaws in the Pantanal region of Brazil eat virtually nothing but the hard fruits of two species of palm trees – these are specialists.

Animals rarely occupy the whole of their niche width, owing to competition between species. Part of the habitat requirement of North American bluebirds is dead trees with old woodpecker holes in which they lay their eggs and raise their young. Although suitable holes are commonplace in many forests, bluebirds cannot occupy all these holes because they are often out-competed by more aggressive starlings. Therefore their realized niche – the places they actually occupy – is not as extensive as their potential (or fundamental) niche.

Many animals share some aspects of their niche, but not others. This is called niche »

Three major types of ecological pyramid

Pyramid of numbers

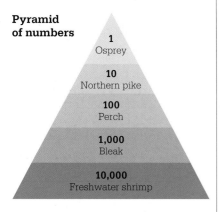

1	Osprey
10	Northern pike
100	Perch
1,000	Bleak
10,000	Freshwater shrimp

Pyramid of biomass

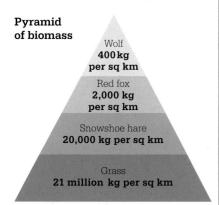

Wolf
400 kg per sq km

Red fox
2,000 kg per sq km

Snowshoe hare
20,000 kg per sq km

Grass
21 million kg per sq km

Pyramid of energy

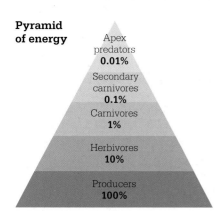

Apex predators
0.01%

Secondary carnivores
0.1%

Carnivores
1%

Herbivores
10%

Producers
100%

Ecological pyramids represent quantifiable data in an ecosystem. Numbers show the population size of individual species in a trophic level; biomass, their relative presence; and energy, who eats what and how much.

overlap. If different species live in the same habitat and have similar lifestyles, they will be in competition but they may be able to live in close proximity if some aspects of their behaviour or diet differ. This arrangement is known as niche partitioning. For example, various anole lizards on Puerto Rico successfully occupy the same areas because they select perching locations in different parts of trees.

There are limits to niche overlap. When two animals with identical niches live in the same place, one will drive the other to extinction. This concept – the competitive exclusion principle – was outlined by Joseph Grinnell in 1904 and developed in a paper published by Russian ecologist Georgy Gause in 1934, becoming known as Gause's law.

Pyramid of numbers

Charles Elton used a pyramid as a way of graphically representing the different levels in a food chain, with the producers at the bottom, the primary consumers on the level above, and so on. Often, the primary consumers – insects, in particular – will outnumber the producers, but the higher levels of consumers will become less numerous towards the top of the pyramid. This system does not take account of parasites; fleas and ticks on mammals and birds will far outnumber the total of all the vertebrates in an ecosystem.

In 1938, German-born animal ecologist Frederick Bodenheimer modified Elton's pyramid of

Microscopic organisms, including these diatoms, form a significant part of all ecological pyramids. Their huge numbers and rapid reproduction provide mass and energy for the species higher up the pyramid.

The basic process in trophic dynamics is the transfer of energy from one part of the ecosystem to another.
Raymond Lindeman

numbers to produce a pyramid of biomass that represented the amount of living matter in a given area at every level. This took into account the fact that some organisms are much larger than others, but because it showed comparative biomasses at a fixed point in time, it produced anomalies. For example, in a pond, the mass of the phytoplankton producer (microscopic organisms that are the foundation of the aquatic food web) may not be as great as the mass of the fish consumers at a particular point in time, so the pyramid will be inverted. However, phytoplankton reproduce quickly

when conditions, such as sunlight and nutrients, are right. Over time, the mass of the phytoplankton will far outweigh that of the fish.

Trophic pyramids

American ecologist Raymond Lindeman proposed a pyramid of energy, called the trophic pyramid, showing the rate at which energy is transferred from one level to the next as herbivores eat plants, and predators eat herbivores. An organism's trophic level is the position it occupies in a food chain. Plants and algae are at trophic level 1, herbivores at level 2, and the first level of predators is at 3. It is rare for there to be more than five levels. Plants convert the sun's energy into stored carbon compounds, and when a plant is eaten by a

herbivore, some of the energy transfers to the animal. When a predator eats the herbivore, it receives a smaller amount of that energy, and so on.

Published in 1942, Lindeman's Ten Percent Law explains that when organisms are consumed, only about 10 per cent of the energy transferred from them is stored as flesh at the next trophic level. The energy model creates a more realistic picture of the condition of an ecosystem. For example, if the biomass of weed and fish in a pond is the same, but the weed reproduces twice as fast as the fish, the energy of the weed would be shown to be twice as large. Also, there are no inverted pyramids – there is always more energy in the lowest trophic level than the one

Tench feed on snails, which graze on periphyton – a mixture of microbial organisms that cling to plants. By reducing the number of snails, tench increase the periphyton biomass.

above. Assessing energy transfer, however, requires a lot of information about energy intake, as well as the number and mass of organisms.

Future thinking

Relationships between organisms and their environment change from place to place and through time. Global climate change is one example of environmental factors that will increasingly affect animal communities. Some changes have already taken place, but one of the challenges of ecological thinking in the future is to forecast others. ∎

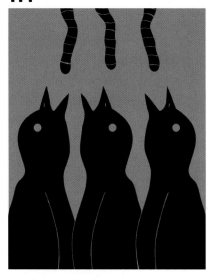

BIRDS LAY THE NUMBER OF EGGS THAT PRODUCE THE OPTIMUM NUMBER OF OFFSPRING

CLUTCH CONTROL

IN CONTEXT

KEY FIGURE
David Lack (1910–73)

BEFORE
1930 British geneticist
Ronald Fisher combines
Gregor Mendel's work
on genetics with Charles
Darwin's theory of natural
selection, and argues that the
effort spent on reproduction
must be worth the cost.

AFTER
1948 David Lack extends his
theory of optimal clutch size
in birds to include litter size
in mammals.

1954 Lack develops his food
limitation hypothesis further
in *The Natural Regulation
of Animal Numbers*, to
encompass birds, mammals,
and some insect species.

1982 Tore Slagsvold proposes
the nest predation hypothesis,
which states that clutch size
is related to the likelihood
of the nest being attacked.

Why do some birds lay more eggs than others? For example, on average blue tits lay nine eggs, robins six, and blackbirds four. In the 1940s, British ornithologist and evolutionary ecologist David Lack proposed an explanation that rapidly gained support. He argued that the clutch size (number of eggs laid) was not controlled by the female's ability to lay eggs, since birds can lay many more eggs than they typically do. This fact can be demonstrated by replacement experiments, in which eggs are removed from a nest; the bird will then re-lay repeatedly to compensate for the loss.

Instead, Lack said, the number of eggs laid by any species has evolved to fit with the food supply available. In other words, nature favours clutch sizes that correspond to the maximum number of young

Blue-tit nests contain an average of nine eggs, although the females can lay many more. David Lack proposed that the clutch size is determined by the likely amount of available food.

See also: Animal ecology 106–113 ▪ Animal behaviour 116–117 ▪ The food chain 132–133 ▪ The ecosystem 134–137 ▪ Ecological resilience 150–151

that the parents are likely to be able to sustain. So, if a pair of birds can only find enough food to feed six chicks, but the female has laid 12 eggs, those young will be hungry and may starve. If she has laid just one egg, although the chick will be raised successfully, most of the available food will have been unused. So neither the 12-egg nor the one-egg scenarios are good reproductive strategies; instead, laying six eggs offers the best chance of raising the most offspring.

This theory became known as the food limitation hypothesis, or Lack's principle, and it was later generalized by him and others to cover litter size in mammals and clutch size in fish and invertebrates.

The "latitude trend"

Lack's hypothesis also suggested an answer to another puzzle: why most bird species have bigger clutches at higher latitudes. On average, birds near the equator lay about half the number of eggs laid by the same species in the far north. This "latitude trend" could be explained by a greater availability

Clutch size increases with increasing latitude and day length because … a longer day enables the parents to find more food.
David Lack

Laying a clutch which will result in a smaller brood than … could be fed and reared successfully … confers advantages.
Tore Slagsvold

of food during the long day-length of summer compared with the shorter day-length in the tropics.

However, other factors may also apply. Higher mortality rates in high latitudes – where winters are harsh – may have led to the evolution of large clutch sizes. This is because the chances of survival until the next breeding season are low, and the reduced population results in more food being available for the survivors next season.

In 1982, Tore Slagsvold, a Norwegian evolutionary ecologist, advanced the nest predation hypothesis, which proposes that high rates of nest predation result in smaller clutches. If a nest with many chicks is found by a predator, more work by the parent birds will have been wasted than if the nest contained fewer chicks. Also, parents raising a large clutch are more likely to be seen by predators, because of the extra activity. Some ecologists have argued that the relative abundance of predators in the tropics has been more important than food supply in the evolution of small clutch sizes at low latitudes. ▪

Siblicide and the blue-footed booby

Blue-footed boobies are seabirds native to the Pacific Ocean. They get their food from the ocean, but come to coastal cliffs to breed. The female lays two eggs, roughly five days apart, so that by the time the second chick hatches, the first one has already grown considerably. When food is plentiful, the parents can find enough to feed both offspring until they fly the nest (fledge). However, when food is scarce, the larger chick will peck its junior sibling to death. The older chick can then get more food, and is more likely to fledge. If it does not murder its sibling when food is scarce, both chicks may starve.

This behaviour, based exclusively on the availability of food, is called "facultative siblicide". In contrast, masked boobies practise "obligate siblicide" – the first-hatched chick nearly always kills its brother or sister, regardless of how much food is available.

Blue-footed boobies are driven to siblicide by genetic factors. The murder of a sibling can benefit the perpetrator while also ensuring the survival of the entire species.

THE BOND WITH A TRUE DOG IS AS LASTING AS THE TIES OF THIS EARTH CAN EVER BE

ANIMAL BEHAVIOUR

IN CONTEXT

KEY FIGURES
Konrad Lorenz (1903–89),
Nikolaas Tinbergen
(1907–88)

BEFORE
1872 Charles Darwin's *The Expression of the Emotions in Man and Animals* posits that behaviour is instinctive and has a genetic basis.

1951 Nikolaas Tinbergen's *The Study of Instinct* lays down the foundations and theory behind ethology, the study of animal behaviour.

AFTER
1967 Desmond Morris, a British zoologist, brings ethology to bear on human behaviour in his popular book *The Naked Ape*.

1976 British evolutionary biologist Richard Dawkins publishes *The Selfish Gene*, describing how most of an animal's behaviour is designed to pass on its genes.

Any dog owner will describe the companionable and loyal relationship they enjoy with their pet. The Austrian zoologist Konrad Lorenz set out to explain this behaviour in *Man Meets Dog* (1949). He described the behaviour of dogs and other pets as substantially innate, "instinctive activity", as opposed to behaviour learned through conditioning. Lorenz proposed that such hard-wired behaviour helped the animal survive as a species. For example, a domestic dog's loyalty to its human master originates in the natural behaviour of its wild ancestors, which were loyal to the pack leader because this had benefits in terms of hunting success and safety.

Field experiments
Lorenz was not alone in his theories. Other biologists working in the field included fellow Austrian Karl von Frisch and Dutch biologist Nikolaas Tinbergen, who studied animals in their natural environments. Until then, most animal behaviour studies had taken place in laboratories or artificial settings, so the behaviour witnessed was not entirely natural. Studying animals in the wild had its own challenges, particularly

Ducklings imprinting is an example of instinctive behaviour that can be manipulated – to make them imprint on humans or even inanimate objects.

when devising rigorous field experiments that could be repeated, so that the findings could be recognized as facts, not anecdotes.

The term "ethology" was coined by American entomologist William Morton Wheeler in 1902 to describe the scientific study of animal behaviour. Ethologists study animals in their natural habitats, combining laboratory studies and fieldwork in order to describe an animal's behaviour in relation to its ecology, evolution, and genetics.

Ethologists found that in certain situations, an animal will have a predictable behavioural response. They called this a "fixed action

See also: The selfish gene 38–39 ▪ Field experiments 54–55 ▪ Keystone species 60–65 ▪ Animal ecology 106–113 ▪ Clutch control 114–115 ▪ Using animal models to understand human behaviour 118–125 ▪ Thermoregulation in insects 126–127

pattern" (FAP). A FAP has set characteristics. It is species-specific; it is repeated in the same way every time and is not affected by experience. The triggers for the behaviour ("sign stimuli") are highly specific and may involve a colour, pattern, or sound. For example, male sticklebacks respond aggressively when another male enters their streambed patch. Ethologists suggest this is triggered by seeing the male's red underbelly.

Nikolaas Tinbergen found that some an artificial sign stimuli work better than the real thing. He investigated the begging behaviour of herring gull chicks, which peck at a red spot on the parent gull's beak to make it regurgitate food. He found that chicks will also peck at a model of the gull's beak, yet when they were offered a narrow red pencil with three white lines at the end, the chicks pecked at this even more enthusiastically. Tinbergen called this a "supernormal stimulus", showing that instinctive animal behaviour can be manipulated artificially. ▪

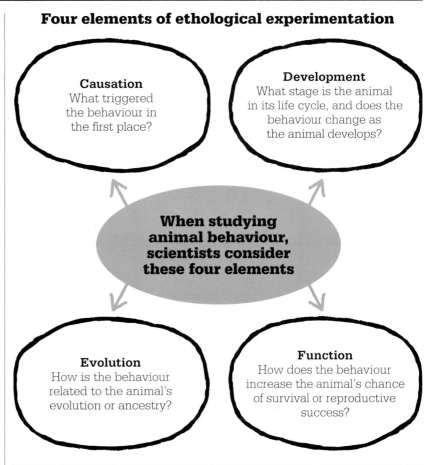

Four elements of ethological experimentation

Causation
What triggered the behaviour in the first place?

Development
What stage is the animal in its life cycle, and does the behaviour change as the animal develops?

When studying animal behaviour, scientists consider these four elements

Evolution
How is the behaviour related to the animal's evolution or ancestry?

Function
How does the behaviour increase the animal's chance of survival or reproductive success?

Konrad Lorenz

Born in Vienna, Austria, Lorenz was enthralled by animals from an early age and kept fish, birds, cats, and dogs. The son of an orthopaedic surgeon, he studied medicine at Vienna University, graduating in 1928, and gained his PhD in Zoology in 1933. His numerous pets became the first subjects of his studies. Lorenz is perhaps best known for describing the phenomenon known as "imprinting". This is when a newly hatched chick bonds with the first thing it sees (usually its parent) and will follow it around. The behaviour, seen in ducks

and other birds, as well as mammals, is instinctive and occurs shortly after birth. Lorenz demonstrated the theory by quacking like a duck at newly hatched ducklings. He soon had a tribe of ducklings that followed him everywhere.

Key works

1952 *King Solomon's Ring: New Light on Animal Ways*
1949 *Man Meets Dog*
1963 *On Aggression*
1981 *The Foundations for Ethology*

REDEFINE "TOOL", REDEFINE "MAN", OR ACCEPT CHIMPANZEES AS HUMANS

USING ANIMAL MODELS TO UNDERSTAND HUMAN BEHAVIOUR

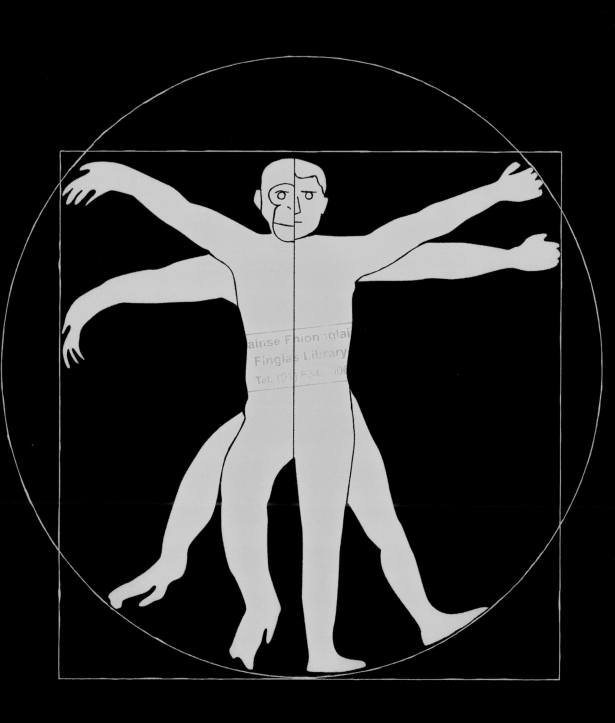

IN CONTEXT

KEY FIGURE
Jane Goodall (1934–)

BEFORE
1758 Carl Linnaeus, the father of taxonomy, dares to classify humans within the rest of nature, calling us *Homo sapiens* ("wise man").

1859 Charles Darwin's theory of evolution further challenges the established view that man is different from the animal kingdom.

AFTER
1963 Konrad Lorenz publishes *On Aggression*, proposing that warlike behaviour in humans is innate.

1967 Desmond Morris, a British zoologist and ethologist, publishes *The Naked Ape: A Zoologist's Study of the Human Animal*, a major study that describes human behaviour in the context of the animal kingdom.

In reality, we are Pan narrans, the storytelling chimpanzee.
Terry Pratchett
British fantasy author

The Primates Tree

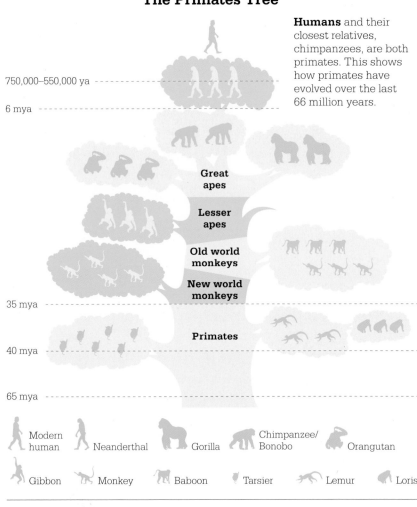

Humans and their closest relatives, chimpanzees, are both primates. This shows how primates have evolved over the last 66 million years.

750,000–550,000 ya

6 mya

Great apes

Lesser apes

Old world monkeys

New world monkeys

35 mya

Primates

40 mya

65 mya

Modern human — Neanderthal — Gorilla — Chimpanzee/Bonobo — Orangutan

Gibbon — Monkey — Baboon — Tarsier — Lemur — Loris

Modern molecular studies mapping the genomes of humans and other animals have confirmed a theory that was first suggested by Charles Darwin in the mid-19th century – that we share a common ancestor with the great apes. Today few scientists would dispute that the common chimpanzee (*Pan troglodytes*) and the bonobo or pygmy chimpanzee (*Pan paniscus*) are our closest living relatives. The study of these animals therefore offers us a unique chance to learn about ourselves and the origins of our behaviour. Yet for many years the scientific community remained convinced that humankind was different from the rest of nature.

It was largely the work of British primatologist Jane Goodall that opened our eyes to the similarities between chimps and man. In 1961, in an excited communication to her mentor, Louis Leakey, Goodall announced an observation that would shake the scientific establishment: she had seen a chimp using a tool. It was the first time this behaviour had been documented and it would challenge perceived ideas of what it means to be human.

See also: Evolution by natural selection 24–31 ▪ A system for identifying all nature's organisms 86–87 ▪ Animal ecology 106–113 ▪ Animal behaviour 116–117

Goodall's knowledge of natural history had impressed Leakey on their first meeting in 1957 and he offered her a job studying the behaviour of chimpanzees. As an anthropologist and palaeontologist, Leakey believed in evolutionary theory, which proposed that humans and the great apes – chimpanzees, bonobos, gorillas, and orangutans – in the family Hominidae (Great Apes), share a common ancestor.

Making connections

Leakey's fieldwork focused on looking for the "missing link" – fossils of transitional forms between that common ancestor and humans. Chimpanzees had not been studied seriously in the wild and such a study, he reasoned, could throw light on the evolution of early humans. Goodall, a keen observer and free of academic ties, was the ideal choice for the work. As Leakey had hoped, she provided a fresh perspective on the theory and was brave enough to say that chimps and humans were more alike than had been imagined.

Until this point, the scientific and popular consensus was that the ability to devise and make tools marked humans out as superior to the rest of the animal kingdom. Goodall's findings forced scientists to think again.

Goodall's camp was in Gombe Stream National Park, Tanzania, where she studied a chimp community on the eastern shore of Lake Tanganyika. In choosing to live among chimps to witness their

A chimp uses a twig stripped of its leaves – a modified "tool" – to catch termites for consumption. Goodall first recorded the ability of chimpanzees to invent simple technologies in Gombe.

true unfettered behaviour, Goodall was one of the first people to work in the field of ethology, whereby biologists monitor animals in their natural environments and try to understand their natural behaviours. In her first few months at the »

Jane Goodall

Born in London in 1934, Jane Goodall's first meeting with a chimp was a stuffed animal that her father named Jubilee. She was interested in animal behaviour from an early age – once, she hid in a henhouse for hours so that she could watch a chicken lay an egg. She left school at 18 and worked in various jobs, before going to Kenya in 1957 and meeting palaeoanthropologist Louis Leakey. With his support, in 1960 Goodall set up a research base in Gombe, Tanzania, where she was to study chimpanzees until 1975. Her work radically

transformed our understanding of chimpanzees and challenged perceived ideas of our own place in the natural world. In 1965, she earned a PhD in ethology from Cambridge University. Her many awards include France's Legion of Honour, given to her in 2006.

Key works

1969 *My Friends the Wild Chimpanzees*
1986 *The Chimpanzees of Gombe: Patterns of Behaviour*
2009 *Hope for Animals and Their World*

camp, the chimps fled from her, but they then began to forget she was there.

Goodall sat for many hours observing the chimps, keeping her distance and quietly making field notes. One morning in November 1961, she noticed a chimpanzee she called David Greybeard sitting over a termite mound. He was poking blades of grass into the mound, pulling them out, and then putting them into his mouth. She watched for some time before the chimp moved off. On reaching the spot where the chimp had been sitting, Goodall saw discarded grass stems lying on the ground. Picking one up and poking it into the mound, she found that the agitated termites bit onto the stem. She realized the chimp had been "fishing" for termites with the grass stems, and transferring them into his mouth.

From talks with Leakey, Goodall knew this was a major discovery. She also saw chimps modifying thin twigs by stripping them of leaves and then using them in termite mounds; the chimps were not only using tools but making them.

Chimp technology
Goodall went on to witness nine different tools being used by chimps in the Gombe community. At the time, scientists questioned Goodall's methods and ridiculed her for giving the chimps names

I viewed my
fellow man not
as a fallen angel,
but as a risen ape.
Desmond Morris
British Zoologist

Jane Goodall working, notebook in hand, at Gombe National Park in 2006. The pioneering primatologist continues her lifelong commitment to protect endangered chimpanzees.

instead of numbers, suggesting that her fieldwork was less than rigorous. Since then, however, many other studies around the world have corroborated her findings: chimps in the Congo have been observed stripping twigs to use in termite mounds; chimps in Gabon have been seen heading into the forest with a five-piece "toolkit" that included a heavy stick for opening bee hives and pieces of bark for scooping up the honey. In Senegal, hunting parties of chimps have been observed travelling with sticks that they chew to a sharp point and use like spears to kill bush babies.

More alike than different
Ethologists take behaviours studied across several species to formulate generalizations that apply

> We admit that we are like apes, but we seldom realize that we are apes.
> **Richard Dawkins**
> *British evolutionary biologist*

to many species. The idea that animal behaviour could be a model for human behaviour took root in the work of ethologists in the 1950s and '60s, such as Konrad Lorenz, Nikolaas Tinbergen, and Karl von Frisch. Studying animals in their natural habitats, they saw how complex the lives of animals were. They began to understand social interactions arising from instinct as well as learned behaviours. The animal studies held a mirror up to human behaviours.

The persistent belief that humans are totally different from other species was firmly rebutted with the advent of gene mapping. When the chimpanzee genome was mapped in 2005 – followed by the other great apes – and compared with the human genome, the results were clear. Humans share 98.8 per cent of their DNA with chimps, 98.4 per cent with gorillas, and 97 per cent with orangutans. Humans and great apes are more alike than they are different. Yet

An alpha male's body language says "keep away" to begging chimps wanting a share of his prey in Gombe National Park. The main source of prey for chimps is the colobus monkey.

it is worth noting that these percentages are based on genes that instruct the body how to make proteins, which make up a very small part of the human genome (about 2 per cent). It is likely that the things that make humans different from chimpanzees can be found in the regions of DNA called "junk DNA" because they were previously thought to be redundant. It is now understood that this junk DNA holds vital information about how and when genes are expressed. Still, the similarities between the DNA of humans and the great apes are striking.

Meat-eating hunters

During her studies, Goodall also witnessed chimps eating meat and hunting. As with tool-making, the idea that chimpanzees were carnivorous predators went against all received knowledge. At first, scientists claimed it was aberrant behaviour, but as the research continued and more sightings were made, it became established fact. Meat-eating has been reported in just about every area where chimps have been studied, from Gombe and Mahale Mountains National Park, Tanzania, to Tai National Park, in Côte d'Ivoire. »

Chromosomal evidence

A strong piece of evidence in favour of a shared common ancestor is seen by comparing chromosomes. Chimpanzees (and gorillas) have 24 pairs of chromosomes. Humans have only 23. Evolutionary scientists believe that when we diverged from a common ancestor, two chromosomes in humans fused and this is why we have one less pair than other apes.

On the ends of every chromosome, there are genetic markers – or sequences of DNA – called telomeres. In the middle of each chromosome there is a different sequence, known as a centromere. If two chromosomes have fused, it should be possible to see telomere-like regions in the middle of the chromosome as well as at each end. Also, the fused chromosome would have two centromeres. Scientists looked and found just this. Human chromosome 2 appears to be the fusion of chimp chromosomes 2a and 2b. It is almost beyond doubt that we share a common ancestor with chimps, bonobos, and gorillas.

Orphaned chimps – their mothers killed for bushmeat – walk along a mud track with their keeper at a conservation centre in West Africa.

Conservation of chimpanzees

According to the Jane Goodall Institute in Tanzania, the number of chimpanzees living in the wild has plummeted over the last century. In 1900, there were an estimated 1 million chimpanzees in Africa; today, there are fewer than 300,000. Habitat loss due to a rising human population in need of more space has had a huge impact, as have industries such as logging and mining, which destroy habitat and fragment chimp communities when roads are built through their territories. Roads also encourage another damaging activity – hunting for bushmeat, a highly valued meat in Africa that includes great apes. Roads enable hunters from towns to travel directly into the bush. The protection of chimps focuses on land conservation and on raising awareness both locally and across the globe.

Such behaviour has implications for human evolution. Science has long questioned why and when humans first began eating meat. From prehistoric stone tools and marks on bones, palaeontologists know that the early hominids were using stone tools to cut meat from animals bones 2.5 million years ago, but it is not known what they were eating between then and 7 million years ago, when the common ancestor of chimpanzees and humans is thought to have lived.

It is likely that these early hominids hunted prey. Although they did not have large canine teeth like chimpanzees, these are not necessary for hunting and killing small prey. Biologists have observed that chimps hunting colobus monkeys grab them from the trees and then kill them by repeatedly thumping the bodies on the ground; early hominids could have hunted and killed in a similar fashion long before the earliest known tools.

Cooperative behaviour

Another aspect of chimps' hunting behaviour that is similar to that of humans is the social element.

Although chimps sometimes hunt alone, hunting tends to be a group activity. Chimps rampage through the forest, coordinating their positions and surrounding their prey. After the hunt, the food is shared. This shows how early ancestors of humans may have developed cooperative behaviour, a factor that may have contributed to their evolutionary success.

Chimp warfare

A shocking revelation that came out of the Gombe camp was that chimps are capable of violence, murder, and in particular warfare – once believed to be the preserve of humans. Between 1974 and 1978, Jane Goodall watched as her peaceful community of chimps fractured into two rival groups that then waged savage war on each other. Goodall was deeply upset about the chimps' activity, which included ambushes, kidnappings, and bloody murder. The trigger

Chimps may fight over territory in order to acquire more resources or mates, but some primatologists maintains that such aggression is unnatural and provoked by human impact on their habitat.

for the war was unclear; some researchers blamed the feeding stations Goodall had set up in the area, which may have encouraged unnatural congregations of chimps. The answer to the mystery came in March 2018, when a research team at Duke and Arizona State Universities, US, digitized Goodall's meticulous check sheets and field notes from 1967 to 1972 and fed them into a computer in order to analyse the social networks and alliances of all the male chimps. Their findings revealed that the fracture in the community occurred two years before the war broke out, when an alpha male Goodall called Humphrey took over the troupe, alienating two other high-

> I'm determined my
> great grandchildren
> will be able to
> go to Africa and
> find wild great apes.
> **Jane Goodall**

ranking males called Charlie and Hugh and causing them to split off with some other chimps to the south. The two groups became more and more separate, feeding in different parts of the forest. At first there was the odd aggressive skirmish and then war broke out. Over four years, Humphrey and his cohorts killed every male in the southern group and took over their territory, as well as three surviving females. It is thought that the full-blown war may have been due to a lack of mature females in the northern group. Power struggles and fighting over a female all sound very human.

Fights over resources

The long-running war witnessed by Goodall is the only sustained conflict among chimpanzees to have been fully documented, but violence within chimp communities has been recorded many times. Chimps have been observed stealing and killing baby chimps and rounding on a disliked alpha male. In communities studied in Uganda, males routinely beat the females they mate with. It is thought that this violent streak running through chimps may be

associated with food resources and meat eating. When food is limited, the chimps become more violent in order to obtain the resources they need. Chimps are known to eat more meat when fruit is scarce.

Kissing cousins

The link between food scarcity and aggression in the common chimpanzee may explain why our other evolutionary cousin in the primate world, the bonobo (pygmy chimp), is so peace-loving. These small, placid chimps are omnivores but live in an environment where fruit is plentiful most of the time. They forage in groups, and tend to use sex to relieve tensions in social situations. Conflict is rare in bonobo societies, which are also matriarchal, unlike the male-dominated chimp communities.

An experiment carried out by researchers at Duke University, North Carolina, in 2017 showed that bonobos are also altruistic. Two bonobos (unknown to each other) were put in adjacent rooms (A and B) with a fence between them and a

piece of fruit hanging over one room (B). The bonobo in room A could release the fruit but not get to it himself. The researchers found that this bonobo would consistently release the fruit, so that the other one could reach it, helping a stranger, with no reward for himself.

Researchers also observed how the sight of an unknown bonobo yawning in a film would trigger a yawning response in bonobos watching the film, suggesting a capacity for empathy. Other studies have shown how bonobos comfort each other when in distress. Unlike the "negative" behaviour that humans share with chimps, these traits mirror more laudable human characteristics, such as compassion. Understanding such behaviour in bonobos could shed light on how our human social behaviour developed. ■

Bonobos are very social primates. Their capacity for empathy makes them less aggressive and may align them more closely with their human cousins than the common chimpanzee.

ALL BODILY ACTIVITY DEPENDS ON TEMPERATURE
THERMOREGULATION IN INSECTS

IN CONTEXT

KEY FIGURE
Bernd Heinrich (1940–)

BEFORE
1837 In the UK, George
Newport observes that flying
insects are capable of raising
their body temperature above
the ambient temperature.

1941 Danish researchers
August Krogh and Eric
Zeuthen conclude that the
temperature of an insect's
flight muscles just before
take-off determine the muscles'
rate of work during flight.

AFTER
1991 German biologist Harald
Esch describes how muscle
"warm-up" plays a role in brood
incubation and colony defence
as well as flight preparation.

2012 Using infrared
thermography, Spanish
zoologist Jose R. Verdu shows
how some dung beetle species
heat or cool their thorax to
improve flight performance.

Insects are usually described as "cold-blooded", or ectotherms. Unlike mammals and other "warm-blooded" endotherms, animals that maintain their body temperature at a more or less constant level, insects have a variable body temperature that changes with their environment.

In the early 19th century, however, British entomologist George Newport discovered that some moths and bees raise the temperature of their thorax (the central part of the body, to which wings and limbs attach) above that of the surrounding air by rapidly flexing their muscles. It is now

In insects… the active flight muscles… are, metabolically, the most active tissues known.
Bernd Heinrich

known that many insects are heterotherms, maintaining different temperatures in different parts of the body, and are sometimes far warmer than the ambient temperature.

The right temperature
The main challenge facing insects is how to get warm enough to fly but cool enough not to overheat. German–American entomologist Bernd Heinrich explained in 1974 how moths, bees, and beetles could continue to function by controlling their own temperature. He realized that insects' thermal adaptations do not differ as much from those of vertebrates as had been thought.

Most flying insects have higher metabolic rates than other animals but their small body size means they lose heat rapidly, so they cannot keep their temperature constant at all times. The minimum temperature that allows an insect to fly varies from species to species, but the maximum temperature falls within 40–45°C (104–113°F). To prevent overheating, insects can transfer heat from the thorax to the abdomen.

Many larger flying insects would remain grounded if they were not able to increase the temperature of their flight muscles. These insects

See also: Evolution by natural selection 24–31 ▪ Ecophysiology 72–73
▪ Animal ecology 106–113 ▪ Organisms and their environment 166

Heat regulation

Honeybees are renowned for controlling the temperature of their hive. When it gets too hot, they ventilate it by using their wings to fan the hot air out of the nest. When it gets too cold, the bees generate metabolic heat by rapidly contracting and relaxing their flight muscles. They also use heat as a defence mechanism. Japanese giant hornets are fierce predators of honeybees. Capable of killing large numbers quickly, they pose a serious threat to bees' nests. Hornets begin their attacks by picking off single honeybees at the entrance to the hive. However, Japanese honeybees defend themselves with self-generated heat. If a hornet attacks, they swarm around it, vibrating their wings to raise their collective temperature. Since the hornet cannot tolerate a temperature above 46°C (114.8°F) whereas the bees can survive at almost 48°C (118.4°F), the attacker eventually dies.

"quiver" the muscles that control the upbeat and downbeat of the wings to generate heat before taking off. Once flying, the muscles use large amounts of chemical energy but only some of it is used to beat the wings; the rest becomes more heat. This, combined with the warmth of direct sunlight, means a flying insect risks overheating.

To solve this problem, many species have a heat-exchange mechanism that shifts excess heat from the thorax to the abdomen, allowing the insect to maintain a steady temperature in its thorax.

Range of techniques

By changing the angle of their wings, butterflies control their body temperature. When they are trying to warm up, holding their wings wide open maximizes the amount of sunshine falling on them. When they are trying to cool down, they move into shade or angle their wings upwards so that less direct sunlight shines on their surface.

A tortoiseshell butterfly feeds on a dandelion. Most butterflies can angle their wings upwards in an attempt to cool down, in a process called behavioural thermoregulation.

Other insects use even more remarkable methods to regulate their body temperature. When a mosquito drinks the warm blood of a mammal, this raises its body temperature. To compensate, it produces droplets of fluid that are kept at the end of the abdomen; evaporative cooling of these droplets lowers the insect's temperature. Dung beetles construct balls of dung in which females lay their eggs. Some dung beetles are able to raise the temperature of their thorax so they can roll heavier balls.

The range of thermoregulation techniques shows how life forms evolve to better fit their environment. They can also inspire technology: arrays of solar panels angled to track the Sun capture maximum amounts of solar radiation – just like butterfly wings. ▪

This Japanese giant hornet is raiding nursery cells in a bees' nest in Hase Valley, Japan. Hornets seek to devour the bee larvae inside the cells.

ECOSYST

EMS

Richard Bradley describes how plants, pollinating insects, and insectivores rely on one another in a **food chain**.

Charles Elton develops the idea of the food web in *Animal Ecology* and introduces the concept of the **ecological niche**.

The journal *Ecology* posthumously publishes Raymond Lindeman's article "The **trophic-dynamic aspect** of ecology".

Nelson Hairston, Frederick Smith, and Lawrence Slobodkin's "**green world hypothesis**" argues that the **predator–prey balance** is key to flourishing ecosystems.

1718   **1927**   **1942**   **1960**

1859   **1935**   **1957**

Charles Darwin describes **food webs** in his *On the Origin of Species*.

Arthur G. Tansley coins the term **ecosystem**, arguing that an environment and all its living organisms have to be seen as a **single, interactive whole**.

G. Evelyn Hutchinson establishes the concept of **niche breadth** at the Cold Spring Harbor Symposia on Quantitative Biology.

When Aristotle wrote about plant and animal species existing for the sake of others, he showed a basic understanding of food chains – as have countless observers of the natural world since ancient Greek times. Arab scholar Al-Jahiz described a three-level food chain in the 9th century, as did the Dutch microscopist Antonie van Leeuwenhoek in 1717. British naturalist Richard Bradley published more detailed findings on food chains in 1718, and in 1859, Charles Darwin described a "web of complex relations" in the natural environment in his book *On the Origin of Species*. The concept of a food web, with many predator–prey interactions, was then further developed by Charles Elton in his classic *Animal Ecology* (1927).

The concept of the ecosystem ("a recognizable self-contained entity") followed soon after, when in 1935, British botanist Arthur Tansley wrote that organisms and their environment should be considered one physical system. In his PhD thesis, American ecologist Raymond Lindeman expanded on Tansley's work, positing that ecosystems are composed of physical, chemical, and biological processes "active within a space–time unit of any magnitude".

Lindeman also conceived the idea of feeding levels, or trophic levels, each of which is dependent on the preceding one for its survival. In 1960, the American team of Nelson Hairston, Frederick Smith, and Lawrence Slobodkin published findings on the factors controlling animals on different trophic levels.

They identified both the top-down pressures exerted by predators and the bottom-up pressures exerted by limitations on food supply. Twenty years later, American ecologist Robert Paine wrote of the trophic cascade effect – the way a system is changed by the removal of a key species within it. He described changes to the food web after the experimental removal of the ochre starfish from an intertidal zone. This predatory starfish was shown to be a keystone species, playing a crucial role within its ecosystem.

Island isolation

Habitat fragmentation is now a major problem in most terrestrial environments because it leaves specialist organisms isolated. For that reason, research into the biogeography of islands – those

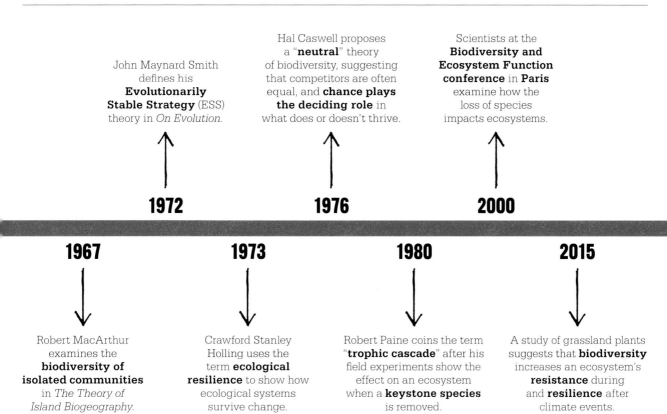

John Maynard Smith defines his **Evolutionarily Stable Strategy** (ESS) theory in *On Evolution*.

Hal Caswell proposes a "**neutral**" theory of biodiversity, suggesting that competitors are often equal, and **chance plays the deciding role** in what does or doesn't thrive.

Scientists at the **Biodiversity and Ecosystem Function conference** in **Paris** examine how the loss of species impacts ecosystems.

1972

1976

2000

1967

1973

1980

2015

Robert MacArthur examines the **biodiversity of isolated communities** in *The Theory of Island Biogeography*.

Crawford Stanley Holling uses the term **ecological resilience** to show how ecological systems survive change.

Robert Paine coins the term "**trophic cascade**" after his field experiments show the effect on an ecosystem when a **keystone species** is removed.

A study of grassland plants suggests that **biodiversity** increases an ecosystem's **resistance** during and **resilience** after climate events.

surrounded by ocean but also "islands" of distinct habitat surrounded by a very different environment – is so important in ecology. In the US in the 1960s, Edward O. Wilson and Robert MacArthur discovered key factors determining species diversity, immigration, and extinction on islands. James Brown later did similar work on animal populations in isolated patches of forest ridge in California. Such work has showed how to identify species most at risk of extinction due to isolation.

Stability and resilience

One major contribution to the understanding of ecosystem dynamics was the concept of the evolutionarily stable state. In the 1970s, British biologist John Maynard Smith used the term

evolutionarily stable strategy (ESS) to describe the best behavioural strategy for an animal competing with others living in its vicinity. This strategy depends on how the other animals behave and is determined by the animal's genetic success – if it makes the wrong decisions, it will not live long and cannot pass on its genes. The overall balance between the evolutionarily stable strategies of all the animals in an ecosystem is called the evolutionarily stable state.

Canadian ecologist Crawford Stanley Holling introduced the idea of resilience – how ecosystems persist in the wake of disruptive changes such as fire, flood, or deforestation. A system's resilience can be seen in its capacity to absorb disturbance, or the time it takes to return to a state of equilibrium

after a trauma. Ecologists now understand that ecosystems can have more than one stable state, and that resilient systems are not always good for biodiversity.

When the populations of many species are declining, or becoming locally extinct, ecologists are once more focusing their attention on ecosystem resilience. Many, including French ecologist Michel Loreau, believe that if diversity in an ecosystem is reduced, the whole system will be less likely to resist major impacts such as the effects of climate change. Today, Loreau and others are working towards finding a new general theory that can explain the relationship between ecosystem biodiversity and resilience in order to understand and combat the effects of today's environmental challenges. ∎

EVERY DISTINCT PART OF NATURE'S WORKS IS NECESSARY FOR THE SUPPORT OF THE REST

THE FOOD CHAIN

IN CONTEXT

KEY FIGURE
Richard Bradley (1688–1732)

BEFORE
9th century Arab scholar
Al-Jahiz describes a three-
level food chain of plant matter,
rats, snakes, and birds.

1717 Dutch scientist Antonie
van Leeuwenhoek observes
how haddock eat shrimp and
cod eat haddock.

AFTER
1749 Swedish taxonomist Carl
Linnaeus introduces the idea
of competition.

1768 John Bruckner, a Dutch
naturalist, introduces the idea
of food webs.

1859 Charles Darwin writes
about food webs in *On the
Origin of Species*.

1927 British zoologist Charles
Elton's *Animal Ecology* outlines
principles of animal behaviour,
including food chains.

The food chain

Apex predator

↑

Larger predator
(tertiary consumer)

↑

Carnivore
(secondary consumer)

↑

Herbivore
(primary consumer)

↑

Producer (autotroph)

All animals must eat other living things in order to receive the nutrients they need to grow and function. A food chain shows the feeding hierarchy of different animals in a habitat. For example, the chain would show that foxes eat rabbits but rabbits never eat foxes. Although there were earlier notions of a hierarchy of animals linked to each other in a food chain, British naturalist Richard Bradley brought more detail to this idea in his book *New Improvements in Planting and Gardening* (1718). He noted that each plant had its own particular set of insects that lived off it and proposed that the insects in turn received the attentions of other organisms of "lesser rank" that fed on them. In this way, he believed that all animals relied upon each other in a self-perpetuating chain.

Producers and consumers

The modern concept of a food chain explains that some organisms produce their own food. These are known as producers, or autotrophs. Plants and most algae fall into this category, normally using the energy of sunlight to convert water and carbon dioxide into glucose, at the

See also: Predator–prey equations 44–49 ▪ Mutualisms 56–59 ▪ Keystone species 60–65 ▪ Optimal foraging theory 66–67 ▪ Animal ecology 106–113 ▪ The ecosystem 134–137 ▪ Trophic cascades 140–143 ▪ Ecological resilience 150–151

> Each species has a specific place in nature, in geographic location and in the food chain.
> **Carl Linnaeus**

same time as releasing oxygen. This process, photosynthesis, is the first step towards creating food. In places where there is no sunlight, organisms producing their own food are called chemotrophs. Those in the deep ocean, for example, get the energy they need from chemicals released by hydrothermal vents.

Animals that eat producers and creatures that eat other animals are called consumers, or heterotrophs. There may be two, three, or more levels of these in any particular part

of the food chain, but there is always a producer at the bottom, and all levels above it are consumers. Animals that only eat plants are herbivores, or primary consumers, and they include cattle, rabbits, butterflies, and elephants. Those that eat only other animals are carnivores, or secondary consumers; these include thrushes, dragonflies, and hedgehogs. In turn, secondary consumers may be eaten by larger predators, or tertiary consumers, such as foxes, small cats, and birds

An apex predator, such as the bronze whaler shark, has no natural predators. In the temperate waters of the ocean off South Africa the shark can find vast quantities of sardines to eat.

of prey. The animals at the top of their food chain are apex predators. They include consumers such as tigers, killer whales, and golden eagles that are not preyed upon by other animals.

The food chain does not break when plants and animals die. Detritus feeders (detritivores) prey on the remains, recycling nutrients and energy for the next generation of producers to use.

Food webs

Observers after Bradley suggested that animals were not simply part of a food chain, but a larger and more complex "food web" that comprises all the food chains in a location. This idea was put forward by Dutch naturalist John Bruckner in 1768, and later taken up by Charles Darwin, who called the variety of connected feeding relationships between species a "web of complex relations". ▪

Richard Bradley

Born around 1688, noted British botanist Richard Bradley gained patrons after writing a *Treatise of Succulent Plants* at the age of 22. With no university education, he was nonetheless elected a Fellow of the Royal Society and later became the first professor of botany at Cambridge.

Bradley's research interests were wide-ranging, including fungal spore germination and plant pollination. In some cases, Bradley was ahead of his time; he argued, for example that

infections were caused by tiny organisms, visible only with a microscope. His investigations into the productivity of rabbit warrens and fish lakes led to his theories about predator–prey relations. Bradley died in 1732.

Key works

1716–27 *The History of Succulent Plants*
1718 *New Improvements in Planting and Gardening*
1721 *A Philosophical Account of the Works of Nature*

ALL ORGANISMS ARE POTENTIAL SOURCES OF FOOD FOR OTHER ORGANISMS

THE ECOSYSTEM

IN CONTEXT

KEY FIGURE
Arthur Tansley
(1871–1955)

BEFORE
1864 George Perkins Marsh, an American conservationist, publishes *Man and Nature*, which hints at the concept of ecosystems.

1875 Austrian geologist Eduard Suess proposes the term "biosphere".

AFTER
1953 American ecologists Howard and Eugene Odum develop a "systems approach" to studying the flow of energy through ecosystems.

1956 American ecologist Paul Sears highlights the role of ecosystems in recycling nutrients.

1970 Paul Ehrlich and Rosa Weigert warn of potentially destructive human interference in ecosystems.

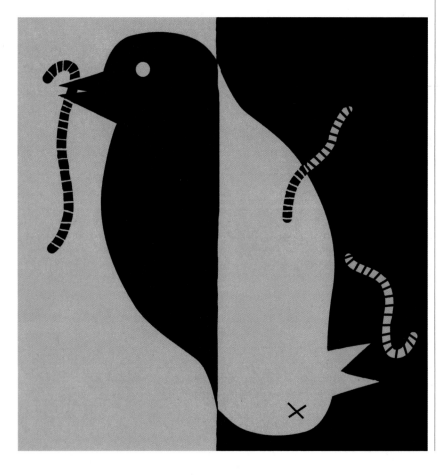

British biologist Arthur Tansley was the first to insist that communities of organisms in a particular area had to be seen in a wider context, including the non-living elements of that area. Tansley argued that in a given region, all the living organisms and their geophysical environment together form a single, interactive entity. Borrowing a concept from engineering, he saw the network of interactions as a dynamic, physical system. On the suggestion of his colleague Arthur Clapham, he coined the word "ecosystem" to describe it.

See also: Animal ecology 106–113 ▪ The food chain 132–133 ▪ Energy flow through ecosystems 138–139 ▪ The biosphere 204–205 ▪ The Gaia hypothesis 214–217 ▪ Environmental feedback loops 224–225 ▪ Ecosystem services 328–329

Tropical coral reefs are some of the most diverse ecosystems of all, full of fish, sea turtles, crustaceans, molluscs, and sponges, as well as corals.

This idea had been developing long before Arthur Tansley published his influential paper on the subject in 1935. As early as 1864, conservationist George Perkins Marsh, in his book *Man and Nature*, had identified "the woods", "the waters", and "the sands" as different types of habitat. He examined how the relationship between them and the animals and plants that lived in them could be upset by human activity.

Interconnected systems

By the 20th century, the idea had taken hold that these and other environments could be understood as discrete entities, with distinctive interactions between the living and inert elements within them. In 1916, American ecologist Frederic Clements built on this idea in his work on plant succession, referring to a "community" of vegetation as a single unit, and using the term "biome" to describe the whole complex of organisms inhabiting a given region.

Tansley envisaged ecosystems as being made up of biotic (living) elements and abiotic (nonliving) elements such as energy, water, nitrogen, and soil minerals, which are essential to the functioning of the systems as a whole. The biotic components within an ecosystem not only interact with one another, but also with the abiotic parts. Thus, within any given ecosystem, the organisms adapt to both the »

Arthur G. Tansley

A free-thinking Fabian socialist and atheist, Arthur Tansley was one of the most influential ecologists of the 20th century. Born in London in 1871, he studied biology at University College London, where he later taught. In 1902, he founded the journal *New Phytologist* and he later established the British Ecological Society, becoming founding editor of its *Journal of Ecology*. In 1923, he took a break from teaching to study psychology with Sigmund Freud in Vienna. He was later Sherardian Professor of Botany at the University of Oxford.

He retired in 1937, but maintained a special interest in conservation. Tansley was appointed the first Chairman of the UK's Nature Conservancy in 1950, five years before his death.

Key works

1922 *Types of British Vegetation*
1922 *Elements of Plant Ecology*
1923 *Practical Plant Ecology*
1935 "The use and abuse of vegetational terms and concepts", *Ecology*
1939 *The British Islands and Their Vegetation*

biological and physical elements of the environment. The different types of ecosystem can be defined by their physical environments. There are four categories of ecosystem: terrestrial, freshwater, marine, and atmospheric. These can be further subdivided into various types according to different physical environments and the biodiversity within them. Terrestrial ecosystems, for example, can be subdivided into deserts, forests, grasslands, taigas, and tundras.

Dynamic feedback

Tansley's most important insight was that these discrete communities of living and nonliving components form dynamic systems. In a terrestrial ecosystem, for example, the organisms interact to recycle

matter: plants absorb carbon dioxide (CO_2) from the atmosphere and nutrients from the soil to grow. These plants release life-sustaining oxygen into the atmosphere by respiration, and provide food for animals. Animal excreta and dead matter also release carbon, and provide material to be decomposed by bacteria and fungi, in turn providing soil nutrients for plants.

Arthur Tansley also argued that these internal processes within an ecosystem conform to what he described as "the great universal law of equilibrium". Self-regulating, these processes have a natural tendency towards stability. The cycles within an ecosystem contain feedback loops that correct any fluctuations from a state of equilibrium.

A small glacial lake, or tarn, in the English Lake District. Each tarn has an ecosystem that varies according to many factors, including the degree of nutrient enrichment in the water.

Each ecosystem is located in a particular area, with characteristics unique to its environment, and behaves as a self-contained and self-regulating system. Together, the patchwork of ecosystems across the globe form what Austrian scientist Eduard Suess called the biosphere – the sum total of all ecosystems.

External factors

Various external factors, such as climate and the geological make-up of the surrounding environment, can affect an ecosystem. One constant external force that affects all ecosystems is the Sun. The supply of energy that it provides enables photosynthesis and the capture of CO_2 from the atmosphere; some of this energy is distributed through the ecosystem and through the food chain. In the process, some of this energy dissipates as heat.

Other external factors, however, can arise unexpectedly to create pressures on ecosystems. All ecosystems are subjected to external disturbances from time to time, and must then go through a process of recovery. These disturbances include storms,

The dynamic transfer of energy

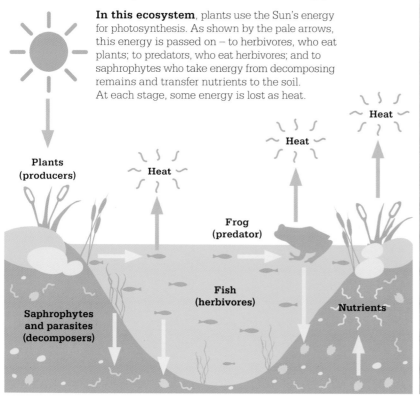

In this ecosystem, plants use the Sun's energy for photosynthesis. As shown by the pale arrows, this energy is passed on – to herbivores, who eat plants; to predators, who eat herbivores; and to saprophytes who take energy from decomposing remains and transfer nutrients to the soil. At each stage, some energy is lost as heat.

Plants (producers)

Heat

Heat

Heat

Frog (predator)

Fish (herbivores)

Nutrients

Saphrophytes and parasites (decomposers)

> There is no waste in functioning … ecosystems. All organisms, dead or alive, are potential sources of food for other organisms.
> **G. Tyler Miller**
> *Science writer*

earthquakes, floods, droughts, and other natural phenomena, but are increasingly the result of human activity – through the destruction of natural habitats by deforestation, urbanization, pollution, and the cumulative effects of anthropogenic (human-induced) climate change. Humans can also be responsible for the introduction of invasive species. Without these external factors, an ecosystem would maintain its state of equilibrium, and retain a stable identity.

Resistance and resilience

Ecosystems are often strong enough to withstand some natural external disturbances and retain their equilibrium. Some are more resistant to disturbance than others, and have adapted to the particular disturbances normally associated with their environment. In some forest ecosystems, for example, the periodic fires caused by electrical storms cause only a minor imbalance in the ecosystem.

Even when severely disrupted by external disturbances, some ecosystems have a resilience that enables them to recover. However, other ecosystems are more fragile, and when they are disturbed may never be able to regain their equilibrium.

The resistance and resilience of an ecosystem is generally thought to be related to its biodiversity. If, for example, there is only one species of plant performing a particular function in the system, and that species is not frost-resistant, an abnormally severe winter could deplete the species enough to have a major impact on the system as a whole. In contrast, if there are several species with that role in the system, it is more likely that one will be resistant to the disturbance.

The human factor

Some disturbances can be severe enough to be catastrophic for an ecosystem, damaging it beyond the point of recovery and so causing a permanent change in its identity, or even its demise. The fear is that much of the disturbance caused by human activity has the potential to cause such permanent damage, particularly when it involves the wholesale destruction of habitat and the consequent depletion of its biodiversity. In addition, some have suggested that human influence has created a new category of ecological systems, dubbed "techno-ecosystems". For example, "cooling ponds" are man-made ponds, built to cool down nuclear power plants, but they have become ecosystems for aquatic organisms.

The relationship between humans and natural ecosystems is not all negative. In recent years, scientific data has fuelled public awareness of the benefits that ecosystems afford humankind, including the provision of food, water, nutrients, and clean air, as well as the management of disease and even climate. There is now a growing commitment from many governments across the world to use these benefits both responsibly and sustainably. ∎

The Eden Project, in Cornwall, UK, simulates a rainforest ecosystem in one of its giant dome greenhouses. The domes' panels are slanted to absorb plenty of light and thermal energy.

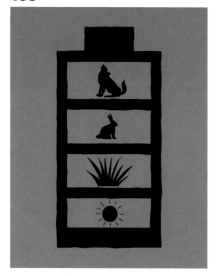

LIFE IS SUPPORTED BY A VAST NETWORK OF PROCESSES

ENERGY FLOW THROUGH ECOSYSTEMS

IN CONTEXT

KEY FIGURE
Raymond Lindeman
(1915–42)

BEFORE
1913 American zoologist Victor Shelford produces the first illustrated food webs.

1920 Frederic Clements describes how groups of plant species are associated in communities.

1926 Russian geochemist Vladimir Vernadsky sees that chemicals are recycled between living and non-living things.

1935 Arthur Tansley develops the concept of the ecosystem.

AFTER
1957 American ecologist Eugene Odum uses radioactive elements to map food chains.

1962 Rachel Carson draws attention to the accumulation of pesticides in food chains, in her book *Silent Spring*.

In 1941, an American student called Raymond Lindeman submitted the final chapter of his PhD thesis for publication in the prestigious journal *Ecology*. Called "The Trophic-dynamic Aspect of Ecology", it was about the relationship between food chains and the changes over time in a community of species.

Lindeman had spent five years studying the life forms in an ageing lake at Cedar Creek Bog, Minnesota, and was especially interested in the changes in the lake as, year by year, aquatic habitat gradually gave way to land. He received his PhD, but his paper was initially rejected by *Ecology*, for being too theoretical.

Lindeman had painstakingly sampled everything in the lake, from the aquatic plants and microscopic algal plankton to the worms, insects, crustaceans, and fish that fed upon one another and depended on each other for their existence. He stressed that the community of organisms could not be properly understood on its own; instead, it must be examined in the context of its wider surroundings. The living (biotic) organisms and

Producers (plants and algae) depend on **energy** gathered from the **Sun** and nutrients from **decomposed organic matter**.

→

Primary consumers are dependent on an **abundance of plants** and **algae** to eat.

↓

Secondary consumers rely on an abundance of **primary consumers** as their food source.

←

Life is supported by a vast network of processes.

See also: Ecological niches 50–51 ▪ Non-consumptive effects of predators on their prey 76–77 ▪ The food chain 132–133 ▪ The ecosystem 134–137

the non-living (abiotic) components (air, water, soil minerals) were linked together by nutrient cycles and energy flows. This entire system – the ecosystem – was the central ecological unit.

Producers and consumers

Lindeman's research showed how an ecosystem is powered by a stream of energy from one organism to another. The organisms can be grouped into discrete "trophic levels" (feeding levels) – from producers (plants and algae), which absorb energy in the form of sunlight to make food, to consumers (animals). "Primary consumers" are the herbivores that eat the plants; "secondary consumers" are animals that eat the herbivores. Each trophic level depends on the preceding one for its survival. At the same time, dead material accumulating from each stage is broken down by decomposers, such as bacteria and fungi, and materials in the form of nutrients are recycled back to feed plants and algae.

Boneworms are deep-sea creatures that feed on the remains of animals such as whales. They grow "roots" to break down the bones, thereby recycling nutrients from the dead material.

Lindeman also demonstrated how some of the energy at each trophic level is lost as waste, or converted into heat when organisms respire. By combining the results of his own study with data from a wide range of other sources, he was able to build up a picture of this system as it worked in Cedar Creek Bog.

British ecologist G. Evelyn Hutchinson, considered to be one of the founding fathers of modern ecology, was Lindeman's mentor at Yale University. He recognized the importance of his student's work to the future development of ecology, and he lobbied for Lindeman's paper to be accepted. Lindeman, who had always suffered from ill health, died in 1942 from cirrhosis of the liver at the tragically young age of 27, just four months before his trophic-dynamic paper – now seen as a classic in its field – was finally published. ▪

… biological communities could be expressed as networks or channels through which energy is flowing and being dissipated…
G. Evelyn Hutchinson

Measuring productivity

Lindeman's trophic-dynamic theory helped to clarify the idea of ecosystem productivity, which ecologists had previously defined in rather vague terms. The productivity of a plant or animal is measured by its growth in organic material, or biomass. This is never equal to the organism's energy input, because the conversion of solar energy into leaf in the case of plants, or the conversion of food into flesh in the case of an animal, is never 100 per cent efficient. Some energy is released as heat, most of which is lost via respiration – an essential aspect of metabolism in all living things.

Warm-blooded animals lose a lot of heat when their body temperature is much higher than that of their surroundings. All animals also lose energy when they excrete urine. In addition, not all the material in an animal's food can be digested in its gut, and the material that is expelled as faeces represents unused chemical energy.

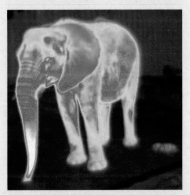

This thermal image of an elephant shows how some of the animal's heat is lost. Both its body temperature and its manure are warmer than the surroundings.

THE WORLD IS GREEN

TROPHIC CASCADES

IN CONTEXT

KEY FIGURE
Nelson Hairston (1917–2008)

BEFORE
1949 Aldo Leopold publishes
A Sand County Almanac,
drawing attention to the
ecological impact of hunting
wolves on mountain plant life.

AFTER
1961 Lawrence Slobodkin,
an American marine
ecologist, publishes *The
Growth and Regulation of
Animal Populations*, a key
ecology textbook.

1980 Robert Paine describes
the "trophic cascade effect"
when predators are removed
from an intertidal ecosystem.

1995 The reintroduction of
grey wolves to Yellowstone
National Park sets in motion
a series of ecosystem changes.

S oon after the end of World
War II, Aldo Leopold, an
ecologist and one of the top
wildlife management experts in the
United States, challenged the view
that wolves should be eradicated
because they threatened livestock.
In *A Sand County Almanac*, he
wrote of the destructive effect that
removing this top predator would
have on the rest of the ecosystem.
In particular, he said, it would lead
to overgrazing of mountainsides by
deer. Leopold's view was an early
expression of the idea of trophic
cascades, although he himself did
not use that term.

Predators help to keep a balance
in a food web by regulating the
populations of other animals. When

See also: Predator–prey equations 44–49 ▪ The food chain 132–133 ▪ The ecosystem 134–137 ▪ Energy flow through ecosystems 138–139 ▪ Evolutionarily stable state 154–155 ▪ Biodiversity and ecosystem function 156–157

Ochre starfish prey on sea creatures such as mussels and limpets. In a famous experiment, Robert Paine took them out of their rock pools to observe the effect on the rest of the food web.

they attack and eat prey, they affect the number and behaviour of that prey – since prey move away when predators are present. The impact of a predator can extend down to the next feeding level (trophic level), affecting the population of the prey's own food source. In essence, by controlling the population density and behaviour of their prey, predators indirectly benefit and increase the abundance of their prey's prey.

Indirect interaction that occurs across feeding levels is described by ecologists as a trophic cascade. By definition, trophic cascades must cross at least three feeding levels. Four- and five-level trophic cascades are also known, although these are less common.

Controlling factors

In 1960, the American ecologist Nelson Hairston and his colleagues Frederick Smith and Lawrence Slobodkin published a key paper entitled "Community Structure, Population Control, and Competition", which examined the factors that control populations of animals on different trophic levels. They concluded that populations of producers, carnivores, and decomposers are limited by their respective resources. Competition occurs between species on each of these three trophic levels. They also found that herbivore populations are seldom limited by the supply of plants, but are limited by predators, so they are unlikely to compete with other herbivores for common resources. The paper highlighted the important role of top-down forces (predation) in ecosystems, and bottom-up forces (food supply).

American ecologist Robert Paine was the first to use the term "trophic cascade" when, in 1980, he described changes in food webs that were brought about by the experimental removal of predatory starfish from the intertidal zone in Washington State. The concept of trophic cascades is now generally accepted, although debate continues as to how widespread they are.

Top-down cascades

This type of cascade is clearly demonstrated when a food chain is interrupted by the removal of a top predator. The ecosystem may continue to function despite the shift in species composition; alternatively, the removal of one species may lead to the »

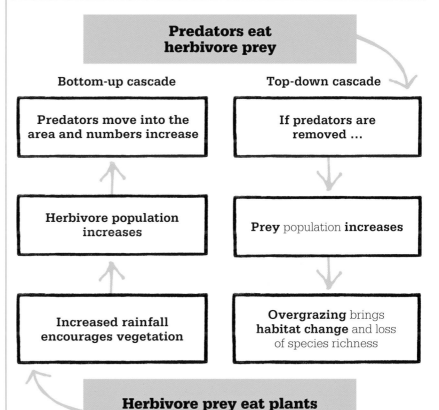

Predators eat herbivore prey

Bottom-up cascade

Predators move into the area and numbers increase

↑

Herbivore population increases

↑

Increased rainfall encourages vegetation

Top-down cascade

If predators are removed ...

↓

Prey population **increases**

↓

Overgrazing brings **habitat change** and loss of species richness

Herbivore prey eat plants

ecosystem's breakdown. In the US, a trophic cascade on the coast of southern New England is believed to be responsible for the die-off of saltmarsh habitat. Recreational anglers have reduced the number of predatory fish to such an extent that the number of herbivorous crabs has expanded dramatically. The resulting increase in the consumption of marsh vegetation has had a knock-on effect on other species that depend on it.

Trophic cascades can also be disturbed by the introduction and spread of a non-native species, as happened when the omnivorous mud crab indigenous to waters on the east coast of North America and Mexico became common in the Baltic Sea in the 1990s. Crabs, which are keystone species in many coastal food webs, feed on benthic (sea floor) communities – bivalves, gastropods, and other small invertebrates – with devastating efficiency, creating a strong top-down cascade. The increase in the number of mud crabs in the Baltic, where there are no equivalent predators, resulted in a dramatic decline in the mix of benthic invertebrate species. This in turn led to an increase in floating nutrients, which ultimately boosted phytoplankton rather than species on the sea floor. The net effect of the crabs' arrival was to transfer nutrients from the sea floor to the water column – the water between the sediment and the surface – and to degrade the ecosystem.

Bottom-up cascades

If a plant – a primary producer – is removed from an ecosystem, a bottom-up cascade may result. For example, if a fungal disease causes grass to die-off, a rabbit population that depends on it will crash. In turn, the predators that eat rabbits will starve or be forced to move away, and the entire ecosystem could break down. Conversely, if planting or conservation efforts boost the mix of plant life, more herbivores (including the pollinators that help plants to reproduce and spread) will be attracted, and with them more predators.

In the bottom-up model, the responses of herbivores and their predators to increased plant variety follow in the same direction: more

> Just as a deer herd lives in mortal fear of its wolves, so does a mountain live in mortal fear of its deer.
> **Aldo Leopold**

plants support more herbivores and more predators. This is in contrast to top-down cascades, in which more predators lead to fewer herbivorous prey and a greater mix of plants.

Beetles, ants, and moths

Investigating trophic cascades in four-tier systems is more difficult, because predators at the top feeding level may eat predators at the level below and also herbivores below them, so the relationships become very complex. In 1999, researchers studying trophic cascades in tropical rainforest in Costa Rica got around this problem by studying a system of three trophic levels of invertebrates, in which the top predator – a clerid beetle – ate the predatory ants in the level below it, but not the herbivores in the level below that. When the number of predatory beetles in the study area was increased, the population of predatory ants fell dramatically. This reduced the pressure on dozens of species of herbivorous

Californian yellow bush lupines are fast growing and invasive. The plant can upset the ecosystem by causing elevated nitrogen levels that attract non-native species.

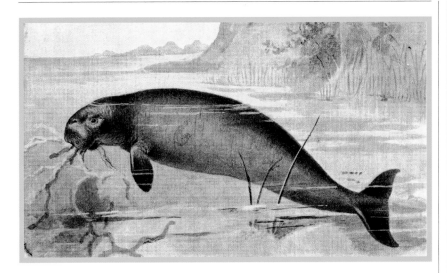

Steller's sea cow was a giant sirenian discovered by the naturalist Georg Steller in 1741. Its extinction is the cause of debate: was it hunted to death, or did its food source disappear?

invertebrates, which therefore ate more vegetation. The leaf area of the plants in the study was consequently reduced by half.

Not all the "players" in trophic cascades are obvious or visible. Some are tiny and live underground. For example, yellow bush lupines – plants that live on the Californian coast – are consumed by the caterpillars of ghost moths, which eat the lupines' roots. In turn, worm-like invertebrates called nematodes parasitize the caterpillars. If these nematodes are present in the soil, they will limit the population of caterpillars, and fewer of the lupines' roots will be affected.

Extinction events

In extreme cases, a trophic cascade can lead to species extinction – as in the case of Steller's sea cows, marine mammals that once lived in the Bering Strait but were declared extinct in 1768. It has recently been argued that this extinction was caused by a calamitous trophic

cascade, triggered by the hunting to virtual extinction of sea otters for the fur trade. The over-exploitation of sea otters allowed the population of sea urchins, their usual prey, to rise above a critical threshold. Sea urchins eat kelp, so the growth of their population led to a collapse in the extent of kelp forests – the sea cows' food source. Although the sea cows themselves were not being hunted, they soon became extinct. Understanding how such interventions, and the introduction of non-indigenous species, can damage trophic cascades is vital in shaping conservation measures today. ■

Herbivores are usually expected to be well fed and carnivores are usually expected to be hungry.
Lawrence Slobodkin

Early humans and megafauna

In the last 60,000 years, which includes the end of the last ice age, about 51 genera of large mammals became extinct in North America. Most were herbivores, including ground sloths, mastodons, and large armadillos, but many were carnivores, such as American lions and cheetahs, scimitar cats, and short-faced bears.

Many of the extinctions occurred between 11,500 and 10,000 years ago, shortly after the arrival and spread of the Clovis people, who were hunters. One of the most convincing theories about these developments is the "second-order predation hypothesis", which suggests that the humans triggered a trophic cascade. The people killed the large carnivores, which competed with them for prey. As a result, predator numbers were reduced and prey populations rose disproportionately, resulting in overgrazing. The vegetation could no longer support the herbivores, with the result that many herbivores starved.

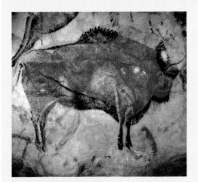

Cave paintings in Altamira, Spain, show the importance of bison to early humans. The wild population became extinct in 1927, but captive herds have since been reintroduced.

ISLANDS ARE ECOLOGICAL SYSTEMS

ISLAND BIOGEOGRAPHY

IN CONTEXT

KEY FIGURES
Robert H. MacArthur
(1930–72),
Edward O. Wilson (1929–)

BEFORE
1948 Canadian lepidopterist
Eugene Munroe suggests
a correlation between island
size and butterfly diversity
in the Caribbean.

AFTER
1971–78 In the US, biologist
James H. Brown studies
mammals and bird species
variety on forest "islands" in
the Great Basin of California
and Utah.

2006 Canadian biologists
Attila Kalmar and David Currie
study bird populations on
346 oceanic islands and
discover that species variety
depends on climate as well
as area and isolation.

Unless we preserve
the rest of life, as a
sacred duty, we will be
endangering ourselves
by destroying the home
in which we evolved.
Edward O. Wilson

Mangrove-fringed islands in the
Florida Keys – now protected for their
diverse range of marine and terrestrial
life – were the focus of research to test
the island biogeography theory.

Island, or insular, biogeography
examines the factors that
affect the species richness of
isolated natural communities.
Charles Darwin, Alfred Russel
Wallace, and other naturalists had
written about island flora and fauna
in the 19th century. Their studies
were conducted on actual islands
in the ocean, but the same methods
can be used to look at any patch
of suitable habitat surrounded by
unfavourable environment that
limits the dispersal of individuals.
Examples include oases in the
desert, cave systems, city parks in
an urban environment, freshwater
pools within a dry landscape,
or fragments of mountain forest
between non-forested valleys.

In the mid-20th century,
ecologists began more intensive
studies into species distribution
on different islands, and how and
why they varied. In the United
States, biologists Edward Wilson
and Robert MacArthur constructed
the first mathematical model of the

See also: Evolution by natural selection 24–31 ▪ Predator-prey equations 44–49 ▪ Field experiments 54–55 ▪ The ecosystem 134–137

Random dispersal of organisms to islands

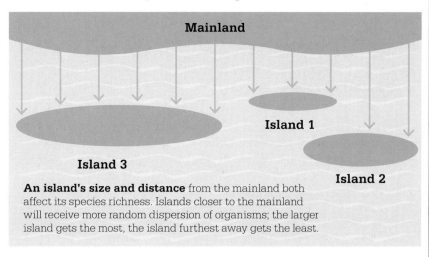

An island's size and distance from the mainland both affect its species richness. Islands closer to the mainland will receive more random dispersion of organisms; the larger island gets the most, the island furthest away gets the least.

Robert H. MacArthur

Born in Toronto, Canada in 1930, and later relocating to Vermont in the US, Robert MacArthur originally studied mathematics. In 1957, he received his PhD from Yale University for his thesis exploring ecological niches occupied by warbler species in conifer forests. MacArthur's emphasis on the importance of testing hypotheses helped to transform ecology from an exclusively observational field to one that employed experimental models as well. This methodology is reflected in *The Theory of Island Biogeography*, which he co-authored with Edward O. Wilson. MacArthur received awards throughout his career, and was elected to the National Academy of Sciences in 1969. In 1972, he died of renal cancer. The Ecological Society of America awards a biennial prize in his name.

Key works

1967 *The Theory of Island Biogeography*
1971 *Geographical Ecology: Patterns in the Distribution of Species*

factors at play in island ecosystems and, in 1967, they outlined a new theory of island biogeography.

Their theory proposed that each island reflected a balance between the rate of new species arriving there and the rate at which existing species become extinct. For example, a habitable but relatively empty island would have a low extinction rate since there are fewer species to become extinct. When more species arrive, competition for limited resources increases. At a certain point, smaller populations will be outcompeted, and the rate of species extinction will rise. An equilibrium point occurs when the species immigration rate and the rate of those becoming extinct are equal; this may remain constant until a change occurs in either rate.

The theory also proposes that the rate of immigration depends on the distance from the mainland, or another island, and declines with increased distance. The area of an island is a further factor. The larger it is, the lower its rate of extinction,

because if native species are pushed out of prime habitat by new immigrants, they have a better chance of finding an alternative, albeit imperfect ("suboptimal") habitat. Larger islands are also likely to have a greater variety of habitats or microhabitats in which to accommodate new immigrants. A combination of variety and lower rates of extinction produces a greater species mix than on a small island – the "species-area effect". The actual species in the mix will change over time, as a result of colonization and extinction, but will remain relatively diverse.

Monitoring mangroves

In 1969, Wilson and his student Daniel Simberloff conducted a field experiment that tested the theory on six small mangrove islands in the Florida Keys in the US. They recorded the species living there, then fumigated the mangroves to remove all the arthropods, such as insects, spiders, and crustaceans. In each of the next two years, »

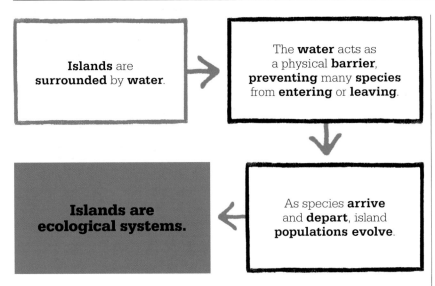

Islands are **surrounded** by **water**.

The **water** acts as a physical **barrier**, **preventing** many **species** from **entering** or **leaving**.

As species **arrive** and **depart**, island **populations evolve**.

Islands are ecological systems.

they counted returning species to observe their recolonization. The Florida Keys experiment showed that distance did indeed play an important role: the further an island was from the mainland, the fewer invertebrates returned to recolonize the area.

New waves of immigration can, however, save even faraway island species from extinction. This is more likely to happen with certain bird species – which can travel long distances quickly – than with, for example, small mammals. There is also the so-called target effect, where some islands are more favoured destinations because of the habitat they provide. Given the choice of a treeless island and one with woodland, a tree-nesting bird will naturally opt for trees.

Human impact
The key factors influencing the species mix on an oceanic island are its degree of isolation, how long it has been isolated, its size, the suitability of its habitat, its location relative to ocean currents, and chance arrivals (for example, organisms washed up on mats of floating vegetation). Most of these factors apply to any similar, isolated habitats, not just actual islands.

The impact of humans – who probably began visiting isolated islands in the Pacific at least 3,000 years ago – has sometimes been dramatic. In recent centuries, people took dogs, cats, goats, and pigs with them when they colonized islands in the Pacific and elsewhere; inadvertently, they also carried rats on their boats. On many islands, rats ate the eggs of breeding seabirds and the seeds of endemic plants, some of which grew nowhere else. On the Galapagos Islands, dogs ate tortoise eggs, native iguanas, and even penguins. Goats competed with Galapagos tortoises for food and wiped out up to five species of plant on the Santiago Islands.

The arrival of humans, however, has not always reduced species richness on islands. Researchers discovered the important role of ship-assisted colonization of islands in the Caribbean. Despite its relatively small size, Trinidad, for instance, has more species of anole lizards than the much larger island of Cuba, because economic sanctions since the 1960s have meant that fewer boats (and their lizard stowaways) dock in Cuba.

"Island" habitats
In the early 1970s, American biologist James H. Brown applied the Wilson–MacArthur model to "islands" of coniferous forest on 19 mountain ridges in the Great Basin of California and Utah. The ridges are separated from each other by a vast sagebrush desert. Brown discovered that the diversity and distribution of small mammals (excluding bats) in the isolated forests could not be explained in terms of an equilibrium between colonization and extinction. Some species had become extinct, but no new species had arrived for millions of years, so Brown dubbed the mammals "relicts". A few years later, his analysis of resident bird populations on the ridges revealed that new bird species had arrived from larger, similar forests in the Rocky Mountains to the east and in the Sierra Nevada to the west. Brown concluded that certain species groups – especially those that fly – are more likely to be successful immigrants than others.

Destroying rainforest for economic gain is like burning a Renaissance painting to cook a meal.
Edward O. Wilson

The rebirth of Krakatau

In 1883, volcanic eruptions devastated the Indonesian island of Krakatau, wiping out flora and fauna on the island and nearby Sertung and Panjang. By 1886, mosses, algae, flowering plants, and ferns had returned to Krakatau, borne either on the wind or as seeds on the surf. The first young trees emerged in 1887; various insect species, and a single lizard, were discovered in 1889. Recent research shows that the level of immigration to Krakatau and its neighbours peaked during the period of forest formation, from 1908 to 1921, but extinctions were at their height when the dense tree canopy prevented sunlight from reaching the forest floor, between 1921–33. Although the immigration of land birds and reptiles has almost stopped, new species of land mollusc and many insect groups are still arriving from Sumatra and Java, both just under 45 km (28 miles) away.

Ecologists also studied the diversity of beetles and flies in nine parks of different sizes in Cincinnati, Ohio. Area was the best predictor of species richness, but when the ecologists coupled their findings with data on population sizes, they calculated that an increased size of parkland acts primarily to reduce extinction rates rather than to provide habitats for new species.

Conservation practices
Soon after island biogeography theory was developed, ecologists began to apply it to conservation. Nature reserves and national parks were seen as "islands" in landscapes altered by human activity. When first creating protected areas, ecologists debated the optimum size: was one big reserve better than several smaller ones? As the island theory shows, biodiversity depends on a number of factors, and different species benefit in different settings. A sizeable mammal will not survive in a small reserve, but many small organisms will thrive there. In places under pressure from human

Central Park in Manhattan, New York City, is an "island" in an urban setting. Its checklist includes 134 bird species, 197 insect, 9 mammal, 5 reptile, 59 fungi, and 441 plant species.

activity, the island theory has also encouraged the creation of wildlife corridors. These link areas of suitable habitat, which helps to maintain ecological processes – for example, allowing animal movement and enabling viable populations to survive – without requiring a great expansion of protected areas. ■

I will argue that every scrap of biological diversity is priceless …
Edward O. Wilson

Krakatau's deadly eruption sent up an ash cloud 80 km (50 miles) high that altered global weather patterns and caused a temperature drop of 1.2°C (2.2°F) for five years.

IT IS THE CONSTANCY OF NUMBERS THAT MATTERS
ECOLOGICAL RESILIENCE

IN CONTEXT

KEY FIGURE
Crawford Stanley Holling (1930–)

BEFORE
1859 Charles Darwin describes the interdependence between species as an "entangled bank".

1955 In the US, Robert MacArthur proposes a measure of ecosystem stability that increases as the number of interactions between species multiplies.

1972 In contrast with MacArthur, Australian ecologist Robert May argues that more diverse communities with more complex relationships may be less able to maintain a stable balance between species.

AFTER
2003 Australian ecologist Brian Walker works with Crawford Holling to refine the definition of resilience.

The capacity for ecosystems to recover following a disturbance – such as a large fire, flood, hurricane, severe pollution, deforestation, or the introduction of an "exotic" new species – is known as ecological resilience. Any of these impacts can upset food webs, often dramatically, and human activity is responsible for an increasing number of them.

Staying resilient
Canadian ecologist Crawford Stanley Holling first proposed the idea of ecological resilience to describe the persistence of natural systems in the face of disruptive changes. Holling argued that natural systems require stability and resilience, but – contrary to what previous ecologists had assumed – these are not always the same qualities.

A stable system resists change in order to maintain the status quo, but resilience involves innovation and adaptation. Holling wrote that natural, undisturbed systems are likely to be continually in a transient state, with populations of some species increasing and others decreasing. However, these population changes are not as important as whether the whole system is being fundamentally altered. The resilience of a system can be described either by the time it takes to return to equilibrium after a big shock or by its capacity to absorb disturbance.

One example that Holling studied was the fisheries of the Great Lakes in North America. A large tonnage of sturgeon, herring, and other fish was harvested in the early decades of the 20th century, but overfishing dramatically reduced the catches. Despite subsequent controls on fishing, populations in the Great Lakes did not recover. Holling

Ecosystems are dynamic – constantly changing and inherently uncertain, with potential multiple futures …
Crawford Stanley Holling

See also: The food chain 132–133 ▪ The ecosystem 134–137 ▪ Energy flow through ecosystems 138–139 ▪ Trophic cascades 140–143

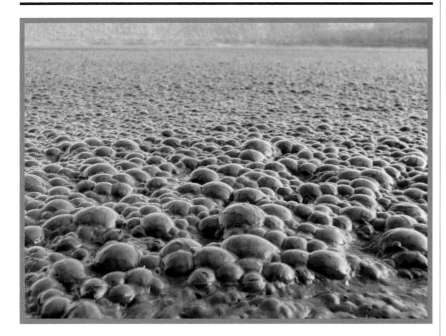

A thick green scum of algae covers parts of Lonar Lake, in Maharashtra, India. Algae thrive in high-nutrient conditions, but decomposing algae consume oxygen, and depleted levels of oxygen lead to fewer fish surviving.

suggested that the intense fishing had progressively reduced the resilience of the ecosystem.

Holling argued that ecological resilience is not always positive. If a freshwater lake experiences a large input of nutrients from agricultural fertilizers, for example, it will become eutrophic: algae will thrive, depleting the lake's oxygen and making it unsuitable for fish. Such a lake may be resilient, but it will become less biodiverse. Holling claimed that three critical factors determine resilience: the most a system can be changed before crossing a threshold that makes total recovery impossible; the ease or difficulty in making a big change to the system; and how close to the threshold a system is currently.

Changing states

According to Holling's view, resilience at the ecosystem level is enhanced by its populations not being too rigid – meaning that the components of the ecosystem can change. One example is the disappearance of most American chestnuts from forests in eastern North America, which was largely compensated for by the expansion of oaks and hickories. For Holling, this counted as resilience, because although the exact mix of tree species had changed, broad-leaved forest still remained.

Ecologists now understand that ecosystems can have more than one stable state. In Australia, for example, woodlands dominated by mulga trees can exist in a grass-rich environment that supports sheep-farming, or in a shrub-dominated environment that is totally unsuitable for sheep. ▪

The role of budworm

Spruce budworm caterpillars have devastated balsam fir forests in eastern North America six times since the 18th century. Holling described this process as having two very different states: one with young, fast-growing trees and few budworms; and one with mature trees and very large numbers of budworms.

Between outbreaks of budworm, young balsam fir grow alongside spruce and white birch trees. Eventually, the fir becomes dominant. A combination of this dominance and a sequence of very dry years stimulates a huge increase in the budworm population. The mature fir is destroyed, giving the spruce and birch an opportunity to regenerate. By keeping the balsam fir in check, the budworm also maintains the spruce and birch. Without it, the fir trees would crowd out the others. So the system is unstable but at the same time resilient.

Spruce budworm larvae in Quebec, Canada, feed voraciously on fir and spruce before they pupate. Moths emerge about a month later, ready to mate.

POPULATIONS ARE SUBJECTED TO UNPREDICTABLE FORCES
THE NEUTRAL THEORY OF BIODIVERSITY

IN CONTEXT

KEY FIGURES
Hal Caswell (1949–),
Stephen P. Hubbell (1942–)

BEFORE
1920 Frederic Clements
describes how species
of plants are associated with
each other in communities.

1926 Henry Gleason proposes
that ecological communities
are organized more randomly.

1967 Richard Root introduces
the concept of the ecological
guild – a group of species
exploiting resources in
similar ways.

AFTER
2018 A review headed
by Dutch ecologist Marten
Scheffer suggests that,
although species that use
the same resources may
be competitively equivalent,
they may also differ according
to their response to stress-
inducing factors, such as
drought or disease.

Biodiversity is shaped globally by new species appearing and others becoming extinct. Community ecology has traditionally held that interactions between species play a vital role in determining this process. If two species compete for similar resources, for example, either the stronger pushes the weaker to extinction, or each is driven into a narrower niche of specialism.

In 1976, however, American ecologist Hal Caswell proposed a "neutral" theory of biodiversity. It maintained that ecologically similar species are competitively equal, and whether species become common or rare is down to chance processes.

The "null" model
In the early 2000s, American ecologist Stephen P. Hubbell developed a mathematical model known as the "null" hypothesis, published in *The Unified Theory of Biodiversity and Geography* (2001), that supported Caswell's theory. He tested his model by studying real communities.

Caswell made a bold attempt to create a neutral theory of community organization.
Stephen P. Hubbell

Neutral theories of biodiversity have dominated community ecology in recent years. However, an Australian study of coral reefs, published in 2014, focusing on once-dominant species that have been almost lost to overfishing, did not support the theory. According to Hubbell, species are interchangeable, so others should have increased to take their place. The fact that this did not happen in this case suggests that the neutral theory is flawed. The question of what maintains diversity remains an open one. ∎

See also: Human activity and biodiversity 92–95 ▪ Island biogeography 144–149 ▪ Climax community 172–173 ▪ Open community theory 174–175

ONLY A COMMUNITY OF RESEARCHERS HAS A CHANCE OF REVEALING THE COMPLEX WHOLE
BIG ECOLOGY

IN CONTEXT

KEY ORGANIZATION
National Science Foundation (1950–)

BEFORE
1926 Russian geochemist and mineralogist Vladimir Vernadsky formulates the theory of the biosphere in which everything on Earth lives.

1935 Pioneering British ecologist Arthur Tansley defines an ecosystem as encompassing all the interactions between a group of living creatures and their environment.

AFTER
1992 At the Earth Summit in Rio de Janeiro, there is international consensus on the importance of protecting the biosphere.

1997 The Kyoto Protocol to reduce greenhouse gas emissions is signed by 192 countries.

An in-depth understanding of ecosystems requires long-term study. In 1980, the US National Science Foundation set up six Long Term Ecological Research (LTER) sites to study long-term, large-scale ecological phenomena. There are currently 28 sites, five of which have been running since 1980. Ecologists are amassing datasets that will enable in-depth knowledge to be shared.

A forest ecosystem
One of the original six research sites is Andrews Forest in Oregon. It provides a good example of a temperate rainforest, enjoying mild, wet winters and cool, dry summers. With 40 per cent being old-growth conifer forest, there is a high degree of biodiversity across its forest, stream, and meadow ecosystems. Ecologists have recorded thousands of species of insects, 83 bird species, 19 conifer species, and 9 species of fish. Projects aim to observe how land-use (such as forestry) and natural phenomena (fires, floods, climate) affect hydrology, biodiversity, and carbon dynamics – the way carbon and nutrients move through the ecosystem. There are many other long-term research sites worldwide with researchers logging data on ecosystems. With free access to the information, the research can be easily disseminated globally. ∎

Log decomposition is being studied over a 200-year period at six old-growth forest sites in Andrews Forest, Oregon. The experiment began in 1985.

See also: The ecosystem 134–137 ▪ The biosphere 204–205 ▪ Sustainable Biosphere Initiative 322–323 ▪ Ecosystem services 328–329

THE BEST STRATEGY DEPENDS ON WHAT OTHERS ARE DOING
EVOLUTIONARILY STABLE STATE

IN CONTEXT

KEY FIGURE
John Maynard Smith
(1920–2004)

BEFORE
1944 Mathematician John von Neumann and economist Oskar Morgenstern use a theory of games of strategy to devise a mathematical theory of economic and social organization.

1964 British biologist W.D. Hamilton applies game theory to the evolution of social behaviour in animals.

1965 Hamilton uses game theory to describe the ecological consequences of natural selection.

1976 Richard Dawkins popularizes the idea of evolutionarily stable strategies.

AFTER
1982 John Maynard Smith applies the theory to evolution, sexual biology, and life cycles.

Animals come into **conflict** with each other over **food, territory,** and **mate selection**.

They have evolved to **react** to the behaviour of other animals in certain **pre-programmed ways**.

The best strategy depends on what others are doing.

The field of behavioural ecology seeks to explain how the behaviour of animals – what they eat, how they socialize, and so on – has evolved to suit their particular environment. The driving force is natural selection, as the environment favours individuals with certain genes – some genes are "better" for certain situations and not for others – which are then passed on to offspring. Because the behaviour of animals is influenced by genes, behaviour must be influenced by natural selection as well.

Adaptive behaviour

In 1972, British evolutionary biologist John Maynard Smith introduced a theory known as the evolutionarily stable strategy (ESS), that helped to explain how behavioural strategies appear by natural selection. Just as factors such as food and temperature can affect animals, so can the behaviour of other species. Maynard Smith suggested that an ESS adapts to the behaviour of other animals, and cannot be beaten by competing strategies, thus giving animals the best chance to pass on their genes. He argued that only natural

See also: Evolution by natural selection 24–31 ▪ The selfish gene 38–39 ▪ Predator–prey equations 44–49 ▪ Ecological niches 50–51 ▪ Trophic cascades 140–143 ▪ Biodiversity and ecosystem function 156–157

Behaviour arising from conflicts over space and territory might emerge as evolutionarily stable strategies. Fruit bats jostle for the best spots in the trees, with alpha males driving weaker bats down to lower branches.

selection could upset this balance – hence why an ESS is "stable" – and that these behaviour patterns are genetically pre-programmed.

ESS has its roots in game theory: a mathematical way of working out the best strategy in a game. Many examples of how animals behave emerge as being evolutionarily stable strategies, such as territorial behaviour and hierarchies. For example, the genetically pre-programmed "rules" of "if resident, fight and defend" or "if visiting, give in and retreat", which would help animals retain territory, combine to make territorial behaviour an ESS.

Balancing strategies
The payoff that an individual animal gains – or the price it risks paying – by displaying a particular

behaviour can be quantified, so biologists can work out which strategies are likely to be most stable by using mathematical models (see box). If the model does not match the behaviour of animals in the real world, then it suggests that stability has not evolved.

In real rather than hypothetical ecosystems, it is not a single strategy that is stable, but the

balance between two or more strategies within the system as a whole. The overall balance is therefore better called an evolutionarily stable state, and not a strategy. Such a balance emerges when all individuals have equal fitness: they pass on their genes to the same extent. The state remains stable, even when there are minor changes in the animal's environment. ▪

The hawk-dove "game"

The simplest demonstration of John Maynard Smith's evolutionarily stable strategy (ESS) concerns a hypothetical response to aggression known as the hawk-dove "game". In this, individuals can either be hawkish and fight until badly injured, or dovish and posture, but then retreat. Hawks will outmatch doves, but could be seriously harmed in a fight with another hawk. Doves routinely escape injury, but waste time

in posturing. Which strategy would be better for passing on genes? Maynard Smith and his collaborators devised a mathematical model to provide the answer, and – in this instance – being more hawkish than dovish emerged as the ESS. It predicts a ratio of seven hawks for every five doves, which is equivalent to any one individual being hawkish seven-twelfths of the time, and dovish five-twelfths of the time.

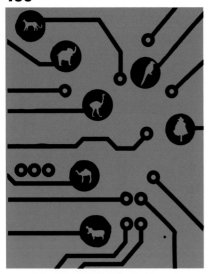

SPECIES MAINTAIN THE FUNCTIONING AND STABILITY OF ECOSYSTEMS
BIODIVERSITY AND ECOSYSTEM FUNCTION

IN CONTEXT

KEY FIGURE
Michel Loreau (1954–)

BEFORE
1949 At the California Institute of Technology in the US, the first phytotron (research greenhouse) is built to study how an artificial ecosystem can be manipulated.

1991 In the UK, an Ecotron, a set of experimental ecosystems in computer-controlled units, is created at Imperial College, London.

AFTER
2014 Leading ecologists in the US say that the effect of diversity loss on ecosystems is at least as great as – or even greater than – that of fire, drought, or other drivers of environmental change.

2015 A paper published in *Nature* provides evidence that biodiversity increases an ecosystem's resilience in a broad range of climate events.

In an age when human activities are rapidly eroding the complex mix of species in different habitats, ecologists have increasingly focused on how biodiversity loss affects the way ecosystems work. If species are replaced or lost altogether, can an ecosystem remain intact – or does this damage ecosystem function?

Such questions were the focus of the Biodiversity and Ecosystem Function (BEF) conference held in Paris in 2000. More than 60 leading international ecologists, including

A phytotron built in 1968 in North Carolina, US, now includes 60 growth chambers, four greenhouses, and a controlled-environment facility for studying plant diseases and insects.

Michel Loreau, director of the Centre for Biodiversity Theory and Modelling in Moulis, France, outlined diverse research; some looked more closely at species, others at what makes an ecosystem work. Loreau maintains that a new unified ecological theory is necessary to combat extreme

> Biodiversity loss… is likely to decrease the ability of ecosystems to resist the effects of climate change.
> **Michel Loreau**

environmental challenges. That, he says, requires the integration of community ecology (the study of how species interact in ecosystems) with ecosystem ecology (research into the physical, chemical, and biological processes that connect organisms and their environment).

Complex cycles

Scientists of both disciplines firmly believe that biodiversity, especially species and genetic diversity, is an important driver of ecosystem functioning. Ecosystems are powered by an input of energy and recycling of nutrients: plants and animals grow, die, and decompose, returning nutrients to the soil and restarting the cycle. These processes depend on the species within the ecosystems, which in turn depend upon one another as they interact – as predators and prey, for example. Many ecologists argue that a large variety of complementary species are needed to keep an ecosystem working and make it resilient to change. Others say that a few key species may be more important to stop ecosystems from collapsing.

When researching such issues, ecologists have tended to use both traditional observational fieldwork and also sophisticated mathematical models. More recently, research has begun to incorporate the manipulation of ecosystems in a more controlled way, on plots of land, for example, or within closed systems housed in giant greenhouse-like facilities called phytotrons. The experiments help to establish what factors – such as numbers of species, or species type and dominance – affect ecosystems in the long term. Their findings show that the effects of biodiversity on ecosystem functions are complex. While the most diverse ecosystems tend to be the most productive, their success also depends on climate and soil fertility.

There is more to be learned about how plant diversity affects soil processes, the role of microbe biodiversity in the soil, and the effects of mutualistic species, such as flowering plants and pollinating insects. Much has been achieved, but questions remain, and the unifying theory that Loreau is seeking has still to be devised. ∎

> One of the distinctive and fascinating features of ecological systems is their extraordinary complexity.
> **Michel Loreau**

Habitat fragmentation

Barro Colorado Island in the Panama Canal of Central America was formed in 1914, when tropical rainforest was flooded by damming, creating an isolated fragment of forest surrounded by water. Since 1946, the area has been studied in detail by biologists of the Smithsonian Institution and elsewhere to determine the effects of this habitat fragmentation: species diversity on the island has declined, and top predators are among the most vulnerable species. In the US, studies of habitat fragmentation and its effects on diversity in the Florida Keys led to Robert MacArthur and E.O. Wilson's seminal *Theory of Island Biogeography* (1967).

From such environments, planners have learned important lessons about how to conserve species in isolated patches of habitat – sometimes in the midst of cities – that are set aside as reserves. Barro Colorado, and places like it, have also provided vital opportunities for study, where ecologists can explore how changing species diversity affects the functioning of an ecosystem at every level.

ORGANIS
IN A CHAN
ENVIRON

MS
GING
MENT

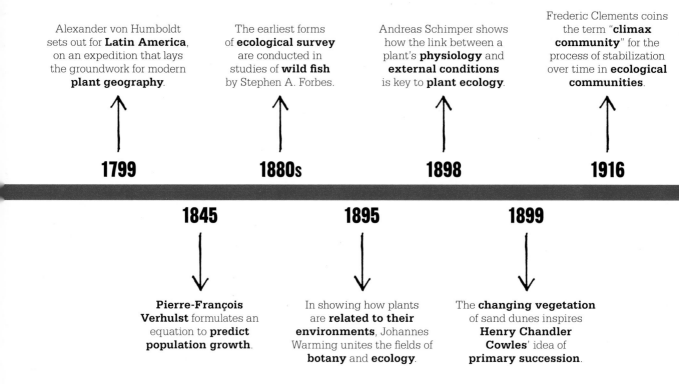

Alexander von Humboldt sets out for **Latin America**, on an expedition that lays the groundwork for modern **plant geography**.

The earliest forms of **ecological survey** are conducted in studies of **wild fish** by Stephen A. Forbes.

Andreas Schimper shows how the link between a plant's **physiology** and **external conditions** is key to **plant ecology**.

Frederic Clements coins the term "**climax community**" for the process of stabilization over time in **ecological communities**.

1799 **1880s** **1898** **1916**

1845 **1895** **1899**

Pierre-François Verhulst formulates an equation to **predict population growth**.

In showing how plants are **related to their environments**, Johannes Warming unites the fields of **botany** and **ecology**.

The **changing vegetation** of sand dunes inspires **Henry Chandler Cowles**' idea of **primary succession**.

The distribution of organisms through space and time is a fundamental interest of ecology. Early in the 19th century, Prussian explorer Alexander von Humboldt, a founding father of ecology, made detailed studies of plant geography in Latin America. Philip Sclater described the global distribution of bird species, and Alfred Russel Wallace did the same for other vertebrates, proposing six zoogeographic regions that are largely still in use today.

Communities

Early fieldwork concentrated on the distribution and abundance of organisms, but later in the 19th century scientists increasingly recognized that survey data could also throw light on interactions between species. In a sense, this represented the true birth of the field of ecology. Pioneers included American naturalist Stephen A. Forbes, who studied wild fish populations in the 1880s, and Danish botanist Johannes Warming, who examined the interaction between plants and their environment and introduced the idea of plant communities.

The link between climate and a region's dominant vegetation type was set out by German botanist Andreas Schimper, who produced a worldwide classification of vegetation zones in 1898. In the early years of the 20th century, ecologists devoted more attention to the interrelatedness of all organisms within an ecosystem, exemplified by Russian scientist Vladimir Vernadsky's concept of the biosphere.

While studying the vegetation growing on sand dunes along the shore of Lake Michigan in the 1890s, American botanist Henry Chandler Cowles realized that there was a succession of plant species, with "pioneer" plants being replaced by others, which were in turn themselves supplanted. Fellow American Frederic Clements used the term "climax community" to describe the endpoint of this succession. In 1916, he proposed that global vegetation patterns could be thought of as "formations", or large communities of plants – and the organisms that depended on them – which reflected the regional climate. In relatively wet, temperate regions, for example, deciduous forest may dominate, but grassland tends to dominate in drier, more temperate areas.

The concept of the "**ecological guild**" is introduced in Richard B. Root's thesis on the blue-grey gnatcatcher.

Robert May applies **chaos theory** to predicting rates of growth and decline in **animal population dynamics**.

James H. Brown and Robert Maurer devise the concept of **macroecology**, in which **ecological patterns** are analysed across large areas.

Mathew Leibold's "**metacommunities**" concept looks at how populations of a single species **disperse and interact**.

1967   **1976**   **1989**   **2004**

1957   **1975**   **1988**   **1991**

The first **satellite** goes into space, heralding new technologies in **wildlife tracking**.

Citizen science enables Fred and Norah Urquhart to discover where **monarch butterflies** go in winter.

John Odling-Smee suggests that "**niche constructors**" actively change their environment.

Ilkka Hanski outlines his **metapopulation theory** for species in fragmented habitats.

Clements argued that these climax communities were bound together and could be thought of as single, complex organisms.

Clements was soon challenged by American botanist Henry Gleason, who agreed that plant communities could be mapped, but argued that since individual plant species have no common purpose, the idea of integrated communities was invalid. His view found support in the 1950s, in the field studies of Robert Whittaker and the numerical research of John Curtis.

In 1967, American ecologist Richard Root proposed the idea of the "guild", a group of organisms – closely related or otherwise – that exploit the same resources. Later, ecologists James MacMahon and Charles Hawkins refined the definition of a guild to species that "exploit the same class of environmental resources", regardless of how they do it.

New ideas

Many new ideas enriched the study of ecology in the late 20th and early 21st centuries. The metapopulation concept was advanced by the Finn Ilkka Hanski, who argued that a population of a species is made up of differing, dynamic components. One part of a population may become extinct, while another thrives. The thriving element may subsequently help re-establish the population that has died out.

In the process, British ecologist John Odling-Smee argued, so-called "niche-constructor" species create a more favourable environment for themselves – as seen in countless examples, from ancient oxygen-producing cyanobacteria that altered the composition of the atmosphere in prehistoric times, to beavers creating wetlands.

Modern methods

Traditionally, the task of monitoring environmental change has been the responsibility of academics and professional ecologists, but millions of interested amateurs now provide enormous amounts of raw data on everything from flowering dates to butterfly numbers, and from the state of coral reefs to the breeding populations of birds. With computer power to process vast amounts of data at speed, and with Earth's ecology changing faster than ever, this "citizen science" looks set to become an invaluable resource for ecology. ■

THE PHILOSOPHICAL STUDY OF NATURE CONNECTS THE PRESENT WITH THE PAST

THE DISTRIBUTION OF SPECIES OVER SPACE AND TIME

IN CONTEXT

KEY FIGURE
Alexander von Humboldt
(1769–1859)

BEFORE
1750 Carl Linnaeus explains that the distribution of plants is determined by climate.

AFTER
1831–36 Charles Darwin makes various observations on the voyage of HMS *Beagle*, confirming that many animals living in one area are not found in similar habitats elsewhere.

1874 British zoologist Philip Sclater produces a description of the zoogeography (the geographical distribution of animals) of the world's birds.

1876 Alfred Russel Wallace publishes his two-volume book *The Geographical Distribution of Animals*, which becomes the definitive biogeography text for the next 80 years.

Species are **distributed** throughout the **world**.

Plants and **animals** move over **time** as **Earth** and its habitats **change**.

Scientists study **where** and **how species** live **now** but also where they were **before**, and what has **changed**.

The philosophical study of nature connects the present to the past.

The distribution, or range, of biological communities and species varies according to many factors – including latitude, climate, elevation, habitat, isolation, and the species' characteristics. The study of species distribution is called biogeography. Biogeography is also concerned with how and why the patterns of distribution change over time.

Early zoologists and botanists such as Carl Linnaeus were well aware of geographical variations in species' distributions, but the first to make detailed studies of this aspect of zoology was the Prussian polymath Alexander von Humboldt, who travelled to Latin America with French botanist Aime Bonpland in 1799. Their five-year expedition laid the basis of plant geography. Humboldt believed observation *in situ* to be paramount, and used sophisticated instruments to make meticulous records of both plant and animal species, noting all the factors that could influence the

See also: Modern view of diversity 90–91 ▪ Animal ecology 106–113 ▪ Island biogeography 144–149 ▪ Big ecology 153 ▪ Climate and vegetation 168–169

The unity of nature means the interrelationship of all physical sciences.
Alexander von Humboldt

data. This holistic approach is best illustrated in his highly detailed map and cross-section of Chimborazo mountain in Ecuador.

Wallace's contribution

Many 19th-century naturalists contributed to biogeographical knowledge, but one of the most significant was British naturalist Alfred Russel Wallace. After reading Philip Sclater's account of the global distribution of bird species, Wallace set out to do the same for other animals. He examined all the factors known at the time to be relevant, including changes in land bridges and the effects of glaciations. He produced maps to demonstrate how vegetation influenced animal ranges, and he summarized the distribution of all known families of vertebrates.

Wallace then proposed six zoogeographic regions, which are still largely in use today: the Nearctic (North America), Neotropics (South America), Palearctic (Europe, north Africa, and most of Asia), Afrotropics (south of the Sahara), Indomalaya (South and Southeast Asia), and Australasia (Australia, New Guinea,

and New Zealand). The dividing line between these last two regions, which runs through Indonesia, is still known as "Wallace's Line".

Plate tectonics

Wallace also made some remarkable discoveries from the fossil record. For example, he worked out that early rodents had evolved in the northern hemisphere, moving via Eurasia into South America. Later, in 1915, German geologist Alfred Wegener proposed the radical idea that the continents of South America and Africa were once connected, which allowed the spread of tapirs and other species.

Wegener understood that the distribution of species was in part a record of geological history. Species colonize new areas as conditions change, and over time have become separated by barriers such as new oceans or mountain ranges. Today, as human-made changes to climate and the environment gather pace – creating new barriers – this understanding has taken on a new and vital importance. ∎

Tapirs evolved in North America at least 50 million years ago. They spread to and now live in South and Central America, as well as southeast Asia, but died out in North America.

Alexander von Humboldt

Known as the "founder of plant geography", Humboldt also made valuable contributions to geology, meteorology, and zoology. Born in Berlin in 1769, he started collecting plants, shells, and insects at an early age. His expedition to Latin America in 1799–1804 encompassed Mexico, Cuba, Venezuela, Colombia, and Ecuador, and his team broke the world altitude record when they climbed to 5,878m (19,285ft) on Chimborazo.

Humboldt also speculated that volcanoes result from deep subterranean fissures, investigated the decrease in temperature with altitude, and discovered that the strength of Earth's magnetic field decreases away from the poles. The 23-volume work detailing his expedition set a new standard for scientific writing, cementing his fame.

Key works

1807 *Essay on the Geography of Plants*
1805–1829 *Personal Narrative of Travels to the Equinoctial Regions of the New Continent During Years 1799–1804*

THE VIRTUAL INCREASE OF THE POPULATION IS LIMITED BY THE FERTILITY OF THE COUNTRY

THE VERHULST EQUATION

IN CONTEXT

KEY FIGURES
Thomas Malthus
(1766–1834), **Pierre-François Verhulst** (1804–49)

BEFORE
1798 Thomas Malthus argues that populations increase exponentially, based on a common ratio, whereas food supplies grow more slowly at a constant rate, leading to potential food shortages.

1835 Belgian statistician Adolphe Quetelet suggests that population growth tends to slow down as population density increases.

AFTER
1911 Anderson McKendrick working as an army physician, applies the Verhulst equation to bacteria populations.

1920 American biologist Raymond Pearl proposes the Verhulst equation as a "law" of population growth.

Pierre-François Verhulst was a Belgian mathematician who, after reading Thomas Malthus's *An Essay on the Principle of Population*, became fascinated by human population growth. In 1845, he published his own model for population dynamics, which was later named the Verhulst equation.

Although influenced by the ideas of Malthus, Verhulst realized that there was a major flaw in his predictions. Malthus had claimed that human population tends to increase geometrically, doubling at regular time intervals. Verhulst thought this to be too simplistic, reasoning that the Malthus model did not take account of a larger population's difficulty in finding food. He argued instead that "the population gets closer and closer to a steady state", in which the rate of reproduction is proportionate to both the existing population and the amount of available food. In Verhulst's model, after the point of maximum population growth – the "point of inflection" – the growth rate becomes progressively slower, gradually levelling off to reach the "carrying capacity" of an area – the number of individuals it can sustain. When visualized, Verhulst's model produces an S-shaped curve, which was later called a logistic curve.

Practical demonstrations

Verhulst's model was ignored for several decades, partly because he himself was not entirely convinced. However, in 1911, Scottish army physician and epidemiologist Anderson McKendrick used the logistic equation to forecast growth in populations of bacteria. Then, in 1920, Verhulst's equation was adopted and promoted in America by Raymond Pearl.

Pearl conducted experiments with fruit flies and hens. He gave a constant quantity of food to fruit

The hypothesis of geometric progression can hold only in very special circumstances.
Pierre-François Verhulst

See also: Distribution of species over time and space 162–163 ▪ Metapopulations 186–187 ▪ Metacommunities 190–193 ▪ Overpopulation 250–251

> Biologists are at the present time in no way likely to suffer ostracism if they venture to study human problems.
> **Raymond Pearl**

flies kept in a bottle. Initially, their fertility rate increased. However, as the population density grew, competition for resources increased, and eventually reached a bottleneck. After this, the flies' fertility rate dropped; their numbers continued to increase, but slowly, and generally the population level stabilized.

Similarly, Pearl found that when the number of hens in a pen increased, the birds struggled to find enough food. As the space between them reduced, the hens laid fewer eggs and, as their fertility rate declined, the rate of population growth slowly stabilized.

Variable strategies

The two key variables in Verhulst's equation are the maximum capacity of a species to reproduce (r), and the carrying capacity of the area (K). Organisms are either r-strategists or K-strategists. R-strategists, such as bacteria, mice, and small birds, reproduce rapidly, mature early, and have a relatively short life. K-strategists, such as humans, elephants, and giant redwood trees, have a slower reproduction rate, take longer to mature, and tend to live longer. Ecologists study r-strategists, which are often found in unstable environments, to assess risks to their necessary high reproduction levels, and study K-strategists in more predictable environments to ensure long-term species survival. ▪

Fruit flies are small, common flies that are attracted to ripe fruit and vegetables. They are popular for laboratory studies because they reproduce so quickly and are easy to cultivate.

Thomas Malthus

Malthus was born in Surrey, UK, in 1766, the seventh child of a prosperous family. After studying languages and mathematics at the University of Cambridge, he took a post as curate of a rural church. In 1798, he published an essay arguing that the rate of increase in human populations outstrips much steadier rises in food production, leading to inevitable starvation. Malthus went on to publish six further editions of the essay, and he made a number of visits to Europe to gather population data. In 1805, he was appointed Professor of History and Political Economy at the East India Company College in Hertfordshire. He became increasingly involved in debate about economic policy, and criticized the Poor Laws for causing inflation and failing to improve life for the poor. Malthus died in 1834.

Key works

1798 *An Essay on the Principle of Population*
1820 *Principles of Political Economy*
1827 *Definitions in Political Economy*

THE FIRST REQUISITE IS A THOROUGH KNOWLEDGE OF THE NATURAL ORDER
ORGANISMS AND THEIR ENVIRONMENT

IN CONTEXT

KEY FIGURE
Stephen A. Forbes
(1844–1930)

BEFORE
1799–1804 Alexander von Humboldt pioneers the field of biogeography in his travels in Latin America.

1866 German naturalist Ernst Haeckel coins the term "ecology" to describe the study of organisms in relation to their environments.

1876 After travelling extensively, British naturalist Alfred Russel Wallace publishes *The Geographical Distribution of Animals*.

AFTER
1890s Frederic Clements proposes the notion of ecological communities.

1895 In *Ecology of Plants* Johannes Warming describes the impact of the environment on the distribution of plants.

The notion of a naturalist – someone who studies organisms in the natural world – dates back to ancient Greece. Aristotle made copious observations of wildlife, and his work laid the foundations for later naturalists. It was not until the 19th century, however, that the potential of such surveys was really understood.

The new study of ecology

As naturalists undertook longer field trips, the global distribution of species became more apparent, and the concept of ecology as a science gained traction.

One of the first scientists to employ ecological methods was US biologist Stephen A. Forbes. In the 1880s, while studying fish in a Wisconsin lake, he realized that survey data could be interpreted to give a picture of interactions between different species – not just their abundance. Forbes extended the scope of the conventional survey, combining practical fieldwork with theoretical analysis and experiments. These rounded ecological surveys created a picture of the natural order within an environment. By shedding light on the interrelated effects of its plant and animal life, they could also help to explain the distribution of species and variations over time. ∎

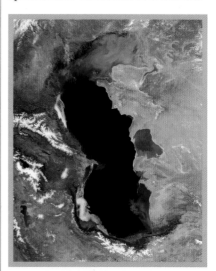

Satellite images enable ecologists to observe large-scale changes easily. The green areas in this image of the Caspian Sea are evidence of algal growth – the product of nutrient enrichment.

See also: Classification of living things 82–83 ▪ Animal ecology 106–113 ▪ Biodiversity and ecosystem function 156–157

PLANTS LIVE ON A DIFFERENT TIMESCALE
THE FOUNDATIONS OF PLANT ECOLOGY

IN CONTEXT

KEY FIGURE
Johannes Eugenius Warming (1841–1924)

BEFORE
1859 Charles Darwin's detailed descriptions of plants and animals in their natural environment mark the start of an appreciation of what is later termed "ecology".

AFTER
1935 British botanist Arthur Tansley publishes an article in *Ecology* in which he defines the term "ecosystem".

1938 American botanists John Weaver and Frederic Clements further develop the concepts of plant communities and succession.

1995 David Attenborough's television documentary "The Private Life of Plants" depicts plants as dynamic influencers of their environment.

Plant ecology examines how plants interact with one another and with their environments. Danish botanist Johannes Eugenius Warming first brought the sciences of botany and ecology together in his book *The Ecology of Plants* in 1895. He described how plants react to their surroundings, and how their life cycles and structures relate to where they grow. The book introduced the concept of plant communities, and outlined how a group of species interact and develop in reponse to the same local conditions.

Plants and ecosystems

For many years, plant ecology and animal ecology were studied separately, but in the early 20th century a more connected perspective emerged. Important theories on plant communities and succession – the process by which an ecological community changes over time – were established during this time period. In 1926, Russian geochemist Vladimir Vernadsky introduced the idea of Earth's

That land is a community is the basic concept of ecology.
Aldo Leopold
American ecologist

biosphere, the parts of its surface and atmosphere where all living organisms exist and interact.

Plants are sensitive barometers of change within an environment. The study of their anatomy, physiology, distribution, and abundance, as well as their interactions with other organisms and their response to environmental factors, such as soil conditions, hydrology, and pollution, can provide invaluable information about the entire ecosystem. ∎

See also: Climate and vegetation 168–169 ▪ Ecological succession 170–171 ▪ The biosphere 204–205 ▪ Endangered habitats 236–239 ▪ Deforestation 254–259

THE CAUSES OF DIFFERENCES AMONG PLANTS
CLIMATE AND VEGETATION

IN CONTEXT

KEY FIGURE
Andreas Schimper
(1856–1901)

BEFORE
1737 Carl Linnaeus's *Flora Lapponica* includes details of the geographical distribution of Lapland plants.

1807 Alexander von Humboldt publishes his seminal *Essay on the Geography of Plants*.

AFTER
1916 In *Plant Succession: an Analysis of the Development of Vegetation*, Frederic Clements describes how communities of species are indicators of the climate in which they have matured.

1968 "The Role of Climate in the Distribution of Vegetation", by American geographers John Mather and Gary Yoshioka, explains how temperature and rainfall alone are not enough to define plant distributions.

That different plants grow in different climates was likely common knowledge for as long as agriculture has existed; many cultures have traded plants for thousands of years. However, the clear link between a region's dominant vegetation type and climate was not categorically spelled out until German botanist Andreas Schimper published his ideas on plant geography in 1898.

Botanists such as Carl Linnaeus and Alexander von Humboldt had written about plant distributions in the 18th and early 19th centuries. The widely travelled Humboldt understood that climate was one of the key factors governing where plants did and did not grow. Schimper went one step further than Humboldt by explaining that similar vegetation types arise under similar climatic conditions in different parts of the world. He then produced a global classification of vegetation zones that reflected this observation.

Schimper's 1898 book *Plant-geography upon a Physiological Basis* ran to 870 pages and is one of the largest ecology monographs written by a single author. A synthesis of plant geography and

"Flowering stones" (*Lithops*) are native to southern Africa, their thick, fleshy leaves well suited to dry, rocky conditions. Related species also occur in similar arid habitats in Australia.

plant physiology (the functioning of plants), it became the foundation of the study of plant ecology. Schimper explained that the connection between the structures of plants and the external conditions they faced in different places was the key to what he described as "oecological plant-geography". Vegetation was divided into broad tropical, temperate, arctic, mountain, and aquatic zones, then subdivided further, according to

See also: Evolution by natural selection 24–31 ▪ Ecophysiology 72–73 ▪ The ecosystem 134–137 ▪ The foundations of plant ecology 167 ▪ Biogeography 200–201 ▪ Biomes 206–209

… the time is not far distant when all species of plants and their geographical distribution will be well known.
Andreas Schimper

the prevailing climate. For example, tropical vegetation was divided into savanna, thorn-forest, woodland, tropical rainforest, or woodland with a pronounced dry season, according to whether the climate was wet all year round, seasonally wet, or mostly dry.

Adaptations for extremes

Schimper made a close study of plant physiology – the structures of plants and how they had adapted to varying temperature and moisture conditions. He was particularly interested in plants growing in extreme climatic conditions. Salty environments, for example, require plants to survive high levels of soil and water salinity. Schimper found that vegetation growing on the coastal mangroves of Brazil, on Caribbean and Sri Lankan beaches, and in sulphur-emitting volcanic craters in Java, were similarly tolerant to salt.

Schimper also studied how plants coped in the challenging conditions of arid environments. He found that plants growing in hot, dry places had evolved "varied contrivances for regulating the passage of water". To illustrate this, he chose a type of vegetation with tough leaves, short internodes (the distances between the leaves along a stem), and leaf orientation parallel or oblique to direct sunlight. This type grew in various parts of the world, where arid conditions meant that water was scarce. The name Schimper gave to these plants – sclerophyll, from the Greek words *skleros* ("hard") and *phullon* ("leaf") – is still used today.

Epiphytes, plants that grow on the surface of other plants and derive their moisture and nutrients from the air or rain, also fascinated Schimper. He observed epiphytes such as Spanish moss growing in the southern US and the Caribbean islands and similar species in South America, South Asia, and southeast Asia. He found that they were linked by warm temperatures and year-round moisture – traits of what he called a tropical rainforest.

Although the broad geographic divisions devised by Schimper still hold true, there is now a better understanding of how vegetation develops in response to many different stimuli beyond simple climatic differences. For example, measures of potential water evaporation into the atmosphere, water surplus, and water deficit, which can be combined in a moisture index, are more useful determinants of plant distribution than simple temperature and rainfall figures. ▪

Like other epiphytes, Spanish moss lives on other species but draws water and nutrients from the air rather than from its host. It thrives in tropical and subtropical environments.

I HAVE GREAT FAITH IN A SEED

ECOLOGICAL SUCCESSION

IN CONTEXT

KEY FIGURE
Henry Chandler Cowles
(1869–1939)

BEFORE
1825 Adolphe Dureau de
la Malle coins the term
"succession" when describing
new growth in forest cuttings.

1863 Austrian botanist
Anton Kerner publishes
a study of plant succession
in the Danube river basin.

AFTER
1916 Frederic Clements
suggests that communities
settle into a climax, or stable
equilibrium, at the end of a
succession period.

1977 Ecologists Joe Connell
and Ralph Slatyer argue that
succession occurs in diverse
ways, highlighting facilitation
(preparing the way for later
species), tolerance (of lower
resources), and inhibition
(resisting competitors).

The Indiana Dunes comprise a windswept section of shifting sand along the southern shore of Lake Michigan, US. In 1896, American botanist Henry Chandler Cowles saw these dunes for the first time, and so began his career in the emerging field of ecology. Dunes are among some of the planet's least stable landforms, and therefore changes to their ecology happen relatively quickly. As Cowles walked among the dunes, he noticed that when certain plants died off, their

15,000 years ago, there would only have been bare sand around Lake Michigan's shore. Vegetation developed in a physical gradient, with sand nearest the water and forests furthest back.

decomposing matter created favourable conditions for other plants. As these new plants died, even more plants could grow.

Based on his observations, Cowles developed the idea of ecological succession, although groundwork for the concept had been laid by earlier naturalists.

See also: Field experiments 54–55 ▪ The ecosystem 134–137 ▪ Climax community 172–173 ▪ Open community theory 174–175 ▪ Biomes 206–209 ▪ Romanticism, conservation, and ecology 298

Primary succession

The process of primary succession begins in barren environments such as bare rock. Hardy species, usually lichens, appear first and then give way to a stable climax community of more complex and diverse life forms over hundreds of years.

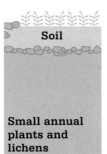

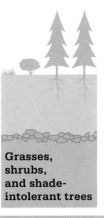
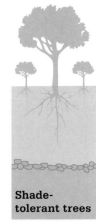

Soil

Bare rock **Lichens** **Small annual plants and lichens** **Grasses and perennials** **Grasses, shrubs, and shade-intolerant trees** **Shade-tolerant trees**

Hundreds of years

Pioneer species **Intermediate species** **Climax community**

In an 1860 address to members of the Middlesex Agricultural Society, Massachusetts, Henry David Thoreau had stated: "Though I do not believe that a plant will spring up where no seed has been, I have great faith in a seed."

Growth of an ecosystem

French geographer Adolphe Dureau de la Malle is regarded as the first person to use the term "succession" with reference to ecology when he witnessed the progression of plant communities after all the trees were removed from a forest. Cowles provided a more formal articulation of his ecological succession theory, in *The Ecological Relations of the Vegetation on the Sand Dunes of Lake Michigan*, published in 1899. In this seminal paper, he proposed the idea of primary succession – the gradual growth of an ecosystem originally largely devoid of plant life. The stages of primary succession include pioneer plants (often lichens and mosses), followed by grassy plants, small shrubs, and trees.

Life after disturbance

Secondary succession occurs after a disturbance that destroys plant life, such as a flood or a fire. The plant life re-establishes itself and develops into an ecosystem similar to the one that existed before the disturbance. The stages of secondary succession are similar to those of primary succession, although the ecosystem may start at different points in the process, depending on the level of damage caused by the trigger.

A common example of secondary succession occurs after a wildfire in oak and hickory forests. Nutrients from burned plants and animals provide the right conditions for growth of annual plants. Pioneer grasses soon follow. After several years, due at least in part to the environmental and soil changes resulting from the pioneer species, shrubs and oak, pine, and hickory trees will begin to grow. As the trees grow higher, shading out more of the underbrush, the grasses are replaced by plants able to survive with low sunlight, and, after around 150 years, the forest once more resembles the pre-fire community. ▪

> I … found indisputable evidence (a) that forests succeeded prairie, and (b) that prairie had succeeded forest.
> **Henry Allan Gleason**
> *American ecologist*

THE COMMUNITY ARISES, GROWS, MATURES, AND DIES
CLIMAX COMMUNITY

IN CONTEXT

KEY FIGURE
Frederic Clements
(1874–1945)

BEFORE
1872 German botanist
August Grisebach classifies
the world's vegetation patterns
in relation to climate.

1874 British philosopher
Herbert Spencer suggests that
the human population can be
thought of as a giant organism.

1899 In the US, Henry Cowles
proposes that vegetation
develops in stages, a process
called succession.

AFTER
1926 US ecologist Henry
Gleason argues that a climax
community is a coincidental
collection of individuals.

1939 British botanist Arthur
Tansley suggests there is not
a single climax community but
"polyclimaxes" responding to
various factors.

In every region, **plants grow and develop** through a **series of successions**.

At each stage, they become **bigger**, more **complex,** and **interconnected**.

Eventually the **vegetation takes on** the most complicated **interconnected form** the **climate** will allow.

Once a community reaches this "climax", vegetation stops changing.

The term "climax community" was first proposed in 1916 by American botanist Frederic Clements. He used it to describe an enduring ecological community that has reached a steady state, such as a naturally stable forest of old-growth trees that has not undergone or been subjected to any unnatural changes, such as logging.

Regional communities
In the 19th century, German botanists August Grisebach and Oscar Drude were among those who recognized that patterns of vegetation around the world reflect factors such as climate variations. It was clear, for example, that the typical vegetation in a wet, tropical climate was very different to that in a dry, temperate climate. Then in a landmark paper in 1899, American botanist Henry Cowles described how plants colonized sand dunes around Lake Michigan in stages – or "successions" – of increasing size and complexity.

In an influential book, *Plant Succession* (1916), Frederic Clements developed Cowles's idea, which he combined with the biogeographic thinking of the two

See also: The ecosystem 134–137 ▪ The distribution of species over space and time 162–163 ▪ Ecological succession 170–171 ▪ Open community theory 174–175 ▪ The ecological guild 176–177 ▪ Biomes 206–209

The Sonora Desert is often seen as an example of a climax community. It has both winter and summer rains, so its unique plants, which include the tall saguaro cactus, are unusually lush.

German botanists to produce a theory of the development of natural communities.

Clements suggested that the way to understand patterns of vegetation across the world is to think in terms of "formations". A formation is a large, natural community of plants dominated by a range of life forms that reflects the regional climate. In each region, plants go through stages or successions until they reach the most complex, highly developed form of vegetation possible. Once it finally reaches this climax, the community stabilizes, in what was later termed a "steady state", and stops changing.

Clements then proposed that climax communities are bound together. Although an ecological community is made up of a multitude of plants at different

For Clements, climates are like genomes, and vegetation is like an organism whose characteristics its genome determines.
Christopher Elliott
Philosopher of science

stages of growth, he argued that it can be considered as a single complex organism. A community grows towards a climax in the same way that an individual develops through life stages. Clements expanded the idea to embrace all organisms in a "biome" that comprised "all the species of plants and animals at home in a particular habitat". From this, the idea of the ecosystem as a "super-organism" later developed.

A fluctuating process
Clements's ideas were challenged from the start, although the idea of a "steady state" proved influential and dominated thinking about ecosystems up until the 1960s. However, scientists realized that communities change constantly in response to conditions, and it is almost impossible to observe a true climax community. A $10,000 prize for identifying such a community, offered by US botanist Frank Egler in the 1950s, was never claimed. Despite the difficulties, ecologists continued to use the theory of a climax community to decide how to respond to invasive species that threatened to disturb an established native community, and in recent decades Clements's ideas have regained support.

Succession remains a core principle of ecology. In general, early succession phases consist of fast-growing and well-dispersed species that are replaced by more competitive species. Initially, ecologists thought that ecological succession ended in what they described as the climax phase, when the ecosystem reached a stable equilibrium. However, it is now accepted that ecological succession is a dynamic process that is constantly in flux. ▪

AN ASSOCIATION IS NOT AN ORGANISM BUT A COINCIDENCE

OPEN COMMUNITY THEORY

Plants **grow** according to
their **individual needs**

→

There is **no evidence of
integrated development**
between plants

↓

They **grow randomly**,
influenced only by the
environmental conditions

←

**An ecological
community is not
an organism**

W hen American plant
ecologist Frederic
Clements proposed the
idea of climax communities in 1916,
he envisioned the community as a
superorganism in which all plants
and animals interact to develop the
community. A year later, American
plant ecologist Henry Gleason
dismissed the idea; he argued that
plant species have no common
purpose but merely pursue their
own individual needs. Gleason's
hypothesis became known as the
"open community" theory. The
dispute initiated a debate that still
rages in ecological circles today.

Gleason did not deny that plant
communities could be mapped and
their interactions identified, but he
could see none of the integration
proposed by Clements. Instead,
Gleason believed that groups
of plants were random growths
of individuals and species,
responding to local conditions.

Individual needs
Gleason maintained that the
changes that occur during plant
succession, as the composition
of a community evolves, are
not integrated stages, as in the
development of a single organism.
Rather, they are a combination of
responses from individual species
as they seek to meet their own
needs within a locality. "Every
species of plant," Gleason argued,
"is a law unto itself". Gleason also

See also: The ecosystem 134–137 ▪ The distribution of species over space and time 162–163 ▪ Ecological succession 170–171 ▪ Climax community theory 172–173 ▪ The ecological guild 176–177 ▪ Biomes 206–209

Diseases such as American chestnut blight challenge the idea of a fully integrated climax community, as the loss of the dominant tree species should cause the entire ecosystem to collapse.

denied that there is any endpoint or climax community; he believed that communities are always changing.

Changing opinions

Gleason's argument with Clements caused quite a stir at the time. Clements seemed to be creating an overview in which natural patterns of vegetation were determined by clear rules, just as in Newtonian science the movement of the planets is dictated by incontrovertible laws. Clements and his supporters were able to look at the bigger picture, while Gleason was viewed as a reductionist, myopically intent on the details and challenging the entire idea of ecology as a science controlled by laws.

Gleason appeared to be saying that there are no patterns in nature: it is all random. Worse still, he was accused by some of justifying exploitative farming, since his ideas seemed to imply that man

need not worry too much about disturbing the balance of the natural environment – because there is no balance. Gleason's ideas were therefore forgotten in the enthusiasm for developing ecology as a science. He became so frustrated that he gave up ecology

during the 1930s as holism became progressively supported by the idea of the interactive "ecosystem".

Nonetheless, as ecologists continued to study the world, they found more and more flaws in Clements's theory. In the 1950s, the work of American plant ecologists Robert H. Whittaker and John Curtis showed how impossible it was to identify communities as neat units of holistic theory, and that the real world was more nuanced and complex. When it comes to studying ecosystems in the field, Gleason's ideas seem to provide a better fit.

In the ensuing decades, while environmentalists continue to champion holistic ideas, ecologists have also increasingly incorporated Gleason's concepts into their work. He is now considered to be one of the most significant figures in 20th-century ecology. ▪

Henry Allan Gleason

Born in 1882, Henry Gleason studied biology at the University of Illinois. He held faculty posts and conducted acclaimed early ecological research in Sand Ridge State Forest, Illinois. In the 1920s, Gleason's theory of individualistic – rather than holistic – plant communities was not accepted by ecologists. This rejection led Gleason to abandon ecology in the 1930s. He had long held posts at the New York Botanical Garden and became famed for his work on plant

classification. With botanist Arthur Cronquist, he co-wrote a definitive guide to the plants of the northeastern US. He retired in 1950 but continued to write and study. He died in 1975.

Key works

1922 "On the Relation between Species and Area"
1926 "The Individualistic Concept of the Plant Association"

A GROUP OF SPECIES THAT EXPLOIT THEIR ENVIRONMENT IN A SIMILAR WAY

THE ECOLOGICAL GUILD

IN CONTEXT

KEY FIGURE
Richard B. Root (1936–2013)

BEFORE
1793 Alexander von Humboldt uses the word "association" to describe the mix of plant types within a particular habitat.

1917 In the US, Joseph Grinnell coins the term "niche" to describe how a species fits into its environment.

1935 British botanist Arthur Tansley identifies ecosystems – integrated biotic communities – as fundamental units of ecology.

AFTER
1989 In the US, James MacMahon suggests that it does not matter how ecological guild members use resources.

2001 Argentinian ecologists Sandra Diaz and Marcelo Cabido propose grouping species that have a similar effect on their environment.

Ecologists have long sought to understand how species in a community interact to exploit resources. A key concept in the explanation of this interplay is the idea of guilds, first developed by American biologist and ecologist Richard B. Root in 1967.

Root had researched the way the blue-grey gnatcatcher exploits its ecological niche for his doctoral thesis. The concept of ecological niches dates back to 1917, when American biologist Joseph Grinnell used the term to describe how a bird called the California thrasher fitted into its dry, scrubby chaparral environment. The thrasher's "niche" describes the aspects of its habitat for which it is suitably adapted.

Root observed that the blue-grey gnatcatcher feeds on insects that live on oak leaves. By diligently analysing stomach contents, he showed that several other birds

The blue-grey gnatcatcher is a member of a guild of small birds that eat insects living on oak trees. Other members of the guild include Hutton's vireo and the oak titmouse.

See also: Evolution by natural selection 24–31 ▪ Predator–prey equations 44–49 ▪ Optimal foraging theory 66–67 ▪ Animal ecology 106–113 ▪ Open community theory 174–175 ▪ Niche construction 188–189 ▪ Metacommunities 190–193

also consume oak-leaf insects and proposed that these oak-leaf feeding birds could be grouped into a "guild" – the "oak-foliage gleaners guild" – because they exploited the same resource.

Shared resources

Root defined a guild as a group of species that "exploit the same class of environmental resources in a similar way". It does not matter whether species in a guild are related or not – all that matters is how they use their environment. They do not even have to occupy the same niche; they just have to use the same resource.

Guilds are typically identified by the food resource they have in common, although it could be any other resource that they share. Sharing the same resource means that guild members often compete with one another, but they are not necessarily in constant competition. For example, although they may compete for the same food, on other occasions they might cooperate to deal with predators.

…does it matter that a particular insect species is captured by a silken spider web as opposed to a bird's beak?
Charles Hawkins and James MacMahon

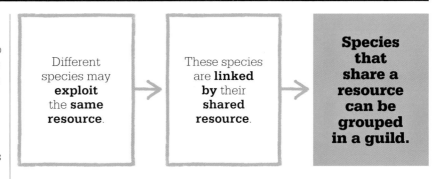

Different species may **exploit** the **same resource**.

These species are **linked by** their **shared resource**.

Species that share a resource can be grouped in a guild.

The guild concept was a major breakthrough in thinking about connections between organisms in ecosystems. The theory implied that the entire functioning of an ecosystem could be understood by identifying all the guilds within it. Although that was potentially a huge undertaking, ecologists have now managed to identify many more guilds that confirm links between species. For example, the birds of North America can be grouped into guilds of gleaners, excavators, hawkers, aerial chasers, and scavengers.

Broad associations

In the rush to identify guilds, there was some confusion over just what the term meant. By the 1980s, the American ecologists Charles Hawkins and James MacMahon felt the need to redefine the term. They argued that the words "in a similar way" should be dropped from Root's original definition. It does not matter, they maintained, whether an organism removes a tree leaf to build a nest or for food. It is the resource of the tree leaf that matters rather than the way it is utilized. Either way, the leaf-users belong to a common guild because they are exploiting the same resource. ▪

Richard B. Root

American biologist and ecologist Richard Root was born in Dearborn, Michigan, in 1936. He grew up on a farm, exploring nature and longing to know "how the woods worked". By the time he completed his doctorate at the University of Michigan, Root was already a knowledgeable ecologist. His 1967 thesis on the blue-grey gnatcatcher, in which he introduced the key concept of the guild, cemented his reputation. Root was invited to join the staff of Cornell University, where he taught biology and ecology. While there, he researched the relationship between arthropods (a large group of invertebrates including insects and arachnids) and goldenrod flowers. Root received many awards during his career, including the Ecological Society of America's Eminent Ecologist award in 2003 and its Odum award in 2004.

Key works

1967 "The niche exploitation pattern of the blue-gray gnatcatcher"

THE CITIZEN NETWORK DEPENDS ON VOLUNTEERS

CITIZEN SCIENCE

IN CONTEXT

KEY FIGURES
Fred Urquhart (1911–2002),
Norah Urquhart (1918–2009)

BEFORE
1883 The Bird Migration
and Distribution recording
programme starts in the US.

1966 The North American
Breeding Birds Survey,
conducted by volunteers,
begins in Maryland.

AFTER
2007 The Global Biodiversity
Information Facility (GBIF)
launches a global online portal
for collecting data on plants
and animals from citizen
scientists and professionals.

2010 The eBird online project,
created in the US in 2002 by
the Cornell Laboratory of
Ornithology for volunteers to
report real-time bird sightings,
becomes a global survey.

Butterflies – millions
upon millions … carpeted
the ground in their
flaming myriads on this
Mexican mountainside.
Fred Urquhart

North American migration pathways

KEY

Pacific flyway

Central flyway

Mississippi flyway

Atlantic flyway

Migratory birds
in North America
use paths that can
be divided into four
north–south zones,
called flyways – Pacific,
Central, Mississippi,
and Atlantic. Citizen
scientists can play a
key role in recording
the birds as they stop
to feed or rest along
the way, during their
flights north in spring
and south in autumn.

Citizen science is research
and observation carried
out by nonprofessional
individuals, teams, or networks
of volunteers, often in partnership
with professional scientists. It is
based on an appreciation that the
scientific community should be
responsive to the environmental
concerns of society as a whole, and
an understanding that citizens can
produce reliable scientific evidence
that leads to greater scientific
knowledge. The involvement of
ordinary people allows research
bodies to accomplish projects that
would be far too expensive or time-
consuming to run otherwise.

Early enthusiasts
While the term "citizen science"
is relatively new, dating from the
1980s, the concept and practice
of using the public to observe the
natural world and record data has
a long pedigree. In the 1870s,
small groups of ornithologists
in Germany and Scotland began
to collect reports on the autumn

migration of birds, the Scottish
enthusiasts using lighthouses
around the coast as observation
posts. Then, in the early 1880s, the
idea of collective observation was
extended onto a national scale by
American ornithologist Wells
Cooke, who began a project to show
arrival dates for migratory North
American birds and provide
evidence for migration pathways.
Cooke's project ran until World War
II, gathering 6 million data cards
on more than 800 bird species and
utilizing 3,000 volunteers at its
peak. In 2009, the North American
Bird Phenology Program began
to digitize the data from the cards,
which has provided valuable
evidence of changed bird migration
dates and routes resulting from
global climate change.

The world's longest-running
citizen science survey is the
Christmas Bird Count (CBC) held
each year in the US. Christmas
"side hunts" of birds were a popular
pastime in many rural districts
of the US in the 19th century,

See also: A system for identifying all nature's organisms 86–87
■ Big ecology 153 ■ The distribution of species over space and time 162–163

regardless of whether the birds were suitable for eating. In 1900, Frank Chapman, an officer of the Audubon Society – named after American ornithologist and painter John James Audubon – proposed counting birds, rather than shooting them. He encouraged 27 birdwatchers to participate in the first event, and the counts then grew every year. In 2016–17, 73,153 observers submitted counts from 2,536 different locations in North and Latin America, the Pacific, and the Caribbean. The data on the distribution and number of birds has provided a huge data set for ecologists, allowing comparison over time and between habitats.

In search of the monarch

Perhaps the most celebrated act of citizen science was one that set out to solve the mystery of where the migrating monarch butterfly went in winter. In 1952, a Canadian couple, zoologists Fred and Norah Urquhart, who had long been fascinated by the butterfly, set up a tagging scheme in an attempt to find where the insect ended its journey after setting out from southern Canada and the northern states of the US in autumn. They enlisted the help of a small group of "citizen scientists" to help to tag the wings of the butterflies and report sightings. From a dozen or so helpers, their Insect Migration Association, as it became known, grew to hundreds of volunteers who persisted for years, tagging hundreds of thousands of monarchs with the message "Send to Zoology, University of Toronto".

Despite the Urquharts' best efforts, the trail went cold in Texas. Finally, on 2 January 1975, two amateur naturalists, Ken Brugger and Catalina Aguado, discovered the butterflies' wintering site in montane forest north of Mexico »

Observations of birds made and recorded by "citizen scientists" in parks and gardens can provide ecologists with vital data on many species, such as the European goldfinch.

Fred and Norah Urquhart

Born in 1911, Fred Urquhart grew up near a railway line on the edge of Toronto, Canada, and became intrigued by the monarch butterflies that laid their eggs close to the track. After graduating in 1937 from the University of Toronto with a degree and a masters in biology, Urquhart began to research the butterfly. Having taught meteorology to pilots during World War II, he returned to the university to lecture in the department of zoology and married Norah Roden Patterson, another Toronto graduate, who joined his quest to find the monarch's winter home. Fred Urquhart also worked as Curator of Insects and Director of Zoology and Palaeontology at the Royal Ontario Museum. In 1998, Fred and Norah Urquhart were awarded their nation's highest civilian award, the Order of Canada.

Key works

1960 *The Monarch Butterfly*
1987 *The Monarch Butterfly: International Traveller*

Monarch butterflies form a cluster to stay warm during migration. Tagging by volunteers revealed the monarch's migratory routes, and continues with the annual "Monarch Watch".

Science should be dominated by amateurship instead of money-biased technical bureaucrats.
Erwin Chargaff
Austro-Hungarian biochemist

City. No tagged monarchs were found, however, and it was not until the following January that the Urquharts found one – tagged by two schoolboys in Minnesota the previous August. Citizen science had provided the hard evidence that the butterflies migrated from North America to Mexico. Now it is known where millions of monarchs spend the winter, the emphasis has changed to tracking their movements each spring and autumn. Thousands of people in Mexico, the US, and Canada are helping to build an ever clearer picture of what routes the monarch follows and how it deals with changing weather patterns.

Citizens march on

More volunteer-based projects were launched during the 1960s and 1970s, including the North American Breeding Bird Survey, the British Nest Records Card project, and a survey of sea turtle egg laying in Japan. In 1979, the Royal Society for the Protection of Birds (RSPB) launched the Big Garden Birdwatch in the UK, which did not even require people to leave their own homes, but simply to record what they saw in their gardens, back yards, or streets. By 2018, more than 500,000 people were participating, recording 7 million birds. The vast amount of data gathered can now be compared for every year back to 1979. Without public help, this would simply not be possible.

In 1989, the term "citizen science" first appeared in print, in the journal *American Birds*. It was used to describe a volunteer project sponsored by the Audubon Society that sampled rain for acidity. The aim of the project was to raise awareness of the acidification of rivers and lakes that was killing fish and invertebrates, and, indirectly, the birds that preyed on them. It was also designed to put pressure on the US government, which soon after introduced the 1990 Clean Air Act.

Citizen science has also proved its worth for marine conservation. In the Bahamas, a report in 2012 on declining numbers of the queen conch, a large sea snail, led to the formation of "Conchservation", a campaign that encourages locals to tag conches. Another project, set up in the US in 2010, at the

University of Georgia, uses an app, the Marine Debris Tracker, to record sightings of debris in the ocean. Understanding patterns of rubbish build-up in the world's seas helps scientists to track how it is transported by currents and where to concentrate removal efforts for maximum effect.

The advent of new technology has led to a proliferation of citizen science projects. Online recording systems mean people can log sightings of anything from stag beetles to wildflowers or migrating birds. In the UK, for example, the Greenspace Information for Greater London (GiGL) website, created by the National Biodiversity Network, allows people to submit records online or by phone, adding to a database used by scientists working to conserve species and habitats.

Limitations and potential

Some ecology research projects are beyond the reach of untrained amateurs because they require too high a degree of skill, or technology that is too complex or expensive. People unfamiliar with scientific methods may also

Young volunteers at Siyeh Pass, in the US state of Montana, record their sightings of mountain goats for the high country citizen science project in the Glacier National Park.

introduce bias into recordings, such as by the omission of a species that cannot be identified.

Most simple citizen science tasks, though, require no training, and some other, more complex, procedures can be tackled after basic tuition. People are often attracted to citizen science precisely because they gain new skills in the process. Increasing pressure on Earth's natural environments and resources creates an ever greater need for data that records presence, absence, and change in species, their habitats, and the wider ecosystems. Projects such as Zooniverse, the world's largest citizen science platform, help to fill this need, accumulating data from around 1.7 million volunteers worldwide. Such projects will be an invaluable resource for conservation organizations, research institutions, non-government agencies, and governments for years to come. ∎

Painting the complete picture

Citizen scientists are now the biggest global providers of data on the occurrence of living organisms. Data is easier than ever to submit and artificial intelligence (AI) algorithms can process data in minutes where once it would have taken weeks. For example, if a person records sightings of birds coming to a garden feeder and sends a report from a phone to Cornell University's eBird website, the information is compared with previous data on factors such as population numbers and migration routes. More than 390,000 people have submitted millions of bird sightings to eBird from nearly 5 million locations around the world. This data is fed into the Global Biodiversity Information Facility (GBIF, coordinated in Denmark), which collects information on plants, animals, fungi, and bacteria. GBIF now contains more than 1 billion observations, and the number is growing daily.

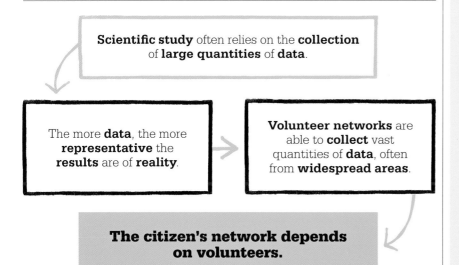

Scientific study often relies on the **collection** of **large quantities** of **data**.

The more **data**, the more **representative** the **results** are of **reality**.

Volunteer networks are able to **collect** vast quantities of **data**, often from **widespread areas**.

The citizen's network depends on volunteers.

POPULATION DYNAMICS BECOME CHAOTIC WHEN THE RATE OF REPRODUCTION SOARS
CHAOTIC POPULATION CHANGE

IN CONTEXT

KEY FIGURE
Robert May (1936–)

BEFORE
1798 Thomas Malthus argues that human populations will increase at an ever-faster rate, inevitably causing suffering.

1845 Belgian demographist Pierre-François Verhulst argues that checks to population growth will increase in line with population growth itself.

AFTER
1987 Per Bak, Chao Tang, and Kurt Wiesenfeld, a research team in New York, US, describe "self-organized criticality" – elements within a system interacting spontaneously to produce change.

2014 Japanese ecologist George Sugihari uses a chaos theory approach called empirical dynamic modelling to produce a more accurate estimate of salmon numbers in Canada's Fraser River.

Chaos theory – the idea that predictions are limited by time and the nonlinear nature of behaviour – took hold in the 1960s. American meteorologist Edward Lorenz observed the effect in weather patterns, and described it in 1961. Since then, the theory has been applied to many sciences, including population dynamics.

Chaotic populations
In the 1970s, Australian scientist Robert May became interested in animal population dynamics, and worked on a model to forecast growth or decline over time. This led him to the logistic equation. Devised by Belgian mathematician Pierre-François Verhulst, this equation produces an S-shaped curve on a graph – showing population growing slowly at first, then rapidly, before tapering off into a state of equilibrium.

May experimented with Verhulst's formula to create the "logistic map", which showed the population trends on a graph. Although it created predictable

Chaos: when the present determines the future, but the approximate present does not approximately determine the future.
Edward Lorenz

patterns at the lowest rates of growth, May found that the logistic equation produced erratic results when the growth rate was equal to or above 3.9. Instead of producing repeating patterns, the map plotted trajectories that appeared completely random. May's work showed how a simple, constant equation could produce chaotic behaviour. His logistic map is now used by demographers to track and predict population growth. ∎

See also: Predator–prey equations 44–49 ▪ Non-consumptive effects of predators on their prey 76–77 ▪ The Verhulst equation 164–165 ▪ Metapopulations 186–187

TO VISUALIZE THE BIG PICTURE, TAKE A DISTANT VIEW
MACROECOLOGY

IN CONTEXT

KEY FIGURE
James H. Brown (1942–)

BEFORE
1920 Swedish ecologist
Olof Arrhenius produces
a mathematical formula for
the relationship between
area and species diversity.

1964 British entomologist
C.B. Williams documents
patterns of species abundance,
distribution, and diversity in
his book *Patterns in the
Balance of Nature*.

AFTER
2002 British ecologists Tim
Blackburn and Kevin Gaston
argue – contrary to some –
that macroecology should be
treated as a discipline distinct
from biogeography.

2018 A team of scientists uses
practical macroecological
methods to show that bird
species living on islands have
relatively larger brains than
their mainland relatives.

Scientists seeking faster ways to analyse and counter the many threats to plant and animal populations increasingly turn to macroecology. The term, coined by American ecologists James Brown and Brian Maurer in 1989, describes studies that examine relationships between organisms and their environment across large areas to explain patterns of abundance, diversity, distribution, and change.

Brown had tried and tested this methodology in the 1970s while studying the potential effects of global warming on species in cool, moist forest and meadow habitats on 19 isolated ridges of the Great Basin, in California and Utah. He realized it would take years of fresh fieldwork to collect enough data. Instead, he used existing findings to draw new conclusions. First, he predicted how much shrinkage would occur in the area of ridge-top habitat with an assumed increase in temperature. Using known data on the minimum area required to support a population of each small mammal species, Brown was able to work out the extinction risk on each ridge as temperatures rose, and suggest conservation priorities.

Enhancing fieldwork
Macroecology often supplements fieldwork and can lead to surprising discoveries. In Madagascar, satellite data was used to develop models for chameleon species and predict them in areas beyond their known ranges. As a result, scientists investigating these areas found several new sister species. ∎

By comparing community studies made in deserts around the world, macroecologists can determine the greatest threats to a desert species such as this banner-tailed kangaroo rat.

See also: Field experiments 54–55 ▪ Animal ecology 106–113 ▪ Island biogeography 144–149 ▪ Big ecology 153 ▪ Endangered habitats 236–239

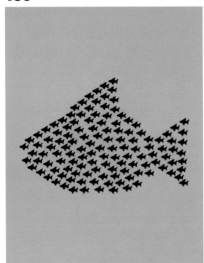

A POPULATION OF POPULATIONS
METAPOPULATIONS

A **species becomes extinct** in one habitat patch.	A **species colonizes** an **empty** habitat patch.

Extinction and **colonization** are **dynamic processes**.

A local extinction does not signal the extinction of the species.

IN CONTEXT

KEY FIGURE
Ilkka Hanski (1953–2016)

BEFORE
1931 In the US, geneticist Sewall Wright explores the influence of genetic factors on species populations.

1933 In Australia, ecologist Alexander Nicholson and physicist Victor Bailey develop their model of population dynamics to describe the host–parasite relationship.

1954 In *The Distribution and Abundance of Animals*, Australian ecologists Herbert Andrewartha and Charles Birch challenge the idea that species populations are controlled by density alone.

AFTER
2007 American ecologist James Petranka links metapopulation theory to the metamorphosis stages of amphibians.

A metapopulation is a combination of separate, local populations of the same species. The term was coined by American ecologist Richard Levins in 1969 to describe how insect pest populations rise and fall on farm fields. Since then, its use has expanded to cover any species broken up into local populations in fragmented habitats, both on land and in the oceans.

A particular species of bird, for instance, may be found in separate populations in a lowland forest, in mountain woodlands, and various other places. The species is like a family whose members have moved to different cities yet are still related. The combined effect of many populations may boost the long-term survival of the species.

Apart but together
A crucial aspect of metapopulation theory is the level of interaction between the separate local populations. If the level is high, it is not considered to be a metapopulation – all the local groups are part of one big population. In a metapopulation

See also: Animal ecology 106–113 ▪ Clutch control 114–115 ▪ Island biogeography 144–149 ▪ Metacommunities 190–193

contact between the various local groups is limited, and they remain partly cut off in their own local habitat or "patch". Yet there has to be at least some interaction. It may be just a single brave or outcast member of one group that enters another patch and mates with the local population there. Isolation for too long pushes local populations apart to the point where they can no longer mate with one another, and in time they become separate species or subspecies.

In the 1990s, Finnish ecologist Ilkka Hanski showed that at the core of metapopulation theory is the notion that local populations are unstable. The metapopulation as a whole may well be stable, but the local populations are likely to rise and fall in their individual patches in response to inside and outside influences. Some patch members may emigrate and join a much reduced population in danger of extinction, giving it renewed strength – a metapopulation feature known as the "rescue effect". Other groups may completely vanish, leaving vacant patches for another population to recolonize. Hansk argued that there is persistent balance between "deaths" (local extinctions) and "births" (the establishment of new populations at unoccupied sites). He likened this balance to the spread of disease, with the susceptible and the infected representing in turn empty and occupied "patches" for disease-carrying parasites.

Ecologists see the concept of metapopulations as increasingly important in understanding how species will survive, particularly in the face of human influence on habitats. The theory helps them to analyse the way populations rise and fall, using mathematical models to play out interactions, and enables them to predict how much habitat fragmentation a species can endure before it is driven to extinction. ▪

The Glanville fritillary butterfly metapopulation, in its fragmented habitats on Finland's Åland Islands, provided the ideal subject for Ilkka Hanski's studies into species patches.

Ilkka Hanski

Widely seen as the father of metapopulation theory, Ilkka Hanski was born in Lempäälä, Finland, in 1953. As a child, he collected butterflies, and after finding a rare species, he devoted his life to ecology, studying at the universities of Helsinki and Oxford.

Ecologists at the time paid little attention to the distribution of local species populations, but Hanski realized this was crucial, and spent much of his career testing his metapopulation theory by mapping out and recording more than 4,000 habitat patches for the Glanville fritillary butterfly on the Åland Islands. This work earned Hanski global fame, and enabled him to establish the Metapopulation Research Centre in Helsinki, which became one the world's leading focuses of ecological research. Hanski died of cancer in May 2016.

Key works

1991 *Metapopulation Dynamics*
1999 *Metapopulation Ecology*
2016 *Messages From Islands*

ORGANISMS CHANGE AND CONSTRUCT THE WORLD IN WHICH THEY LIVE
NICHE CONSTRUCTION

IN CONTEXT

KEY FIGURE
F. John Odling-Smee
(1935–)

BEFORE
1969 British biologist Conrad Waddington writes about ways in which animals change their environments, calling this "the exploitive system".

1983 Richard Lewontin, an American biologist, argues that organisms are active constructors of their own environments, in *Gene, Organism, and Environment*.

AFTER
2014 Canadian ecologist Blake Matthews outlines criteria for deciding whether an organism is a niche constructor.

All organisms alter the environment to cater to their own needs. Animals dig burrows, build nests, create shade from the sun, and create shelter from the wind to provide a more secure environment, while plants alter soil chemistry and cycle nutrients. When organisms modify their own and each other's place in the environment, this is "niche construction" – a term coined by British evolutionary biologist F. John Odling-Smee in 1988.

American evolutionary biologist Richard Lewontin had previously suggested that animals are not passive victims of natural selection.

Hares do not sit around constructing lynxes! But in the most important sense, they do.
Richard Lewontin

He argued that they actively construct and modify their environment, and affect their own evolution in the process: the lynx and the hare, for example, shape each other's evolution and shared environment by striving to outrun each other. Odling-Smee similarly argued that niche construction and "ecological inheritance" – when inherited resources and conditions such as altered soil chemistry are passed on to descendants – should be seen as evolutionary processes.

Levels of construction

Some common examples of niche construction are obvious, while others operate at a microscopic scale. Beavers build impressive dams across rivers, creating lakes and altering river courses. This alters the composition of the water and materials carried downstream, creates new habitats for other organisms to take advantage of, and also changes the composition of the river's plant and animal communities. British biologist Kevin Laland has suggested that, while a beaver's dam is clearly of great evolutionary and ecological importance, the impact of its dung may also be significant.

See also: Ecological niches 50–51 ▪ The ecosystem 134–137 ▪ Organisms and their environment 166 ▪ The ecological guild 176–177

Ecosystem engineers

Niche constructors have been described as "ecosystem engineers", a term coined in 1994 by scientists Clive Jones, John Lawton, and Moshe Shachak. They outlined two kinds of ecosystem engineers. The first, allogenic ecosystem engineers, change physical materials. Take, for example, beavers building dams, woodpeckers excavating nest holes, and people mining for gravel; these activities modify the availability of resources for other species. When woodpeckers abandon their holes, small birds and other animals move in. If water floods a gravel pit, ducks and dragonflies can colonize it.

Other ecosystem engineers are autogenic, which means that simply by growing, they provide new habitats for other plants and animals. A mature oak tree, for example, is a suitable environment for a broader range of insects, birds, and small mammals than an oak sapling. Likewise, a coral reef provides homes for more fish and crustaceans as it grows larger.

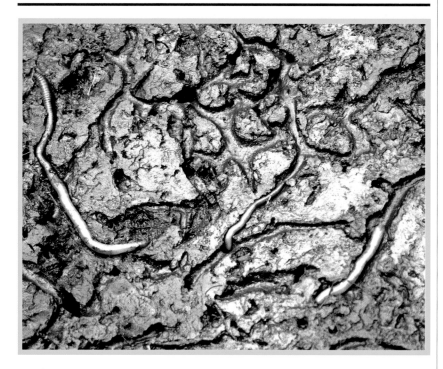

Earthworms leave castings that make them valuable natural fertilizers. They not only transform the soil for themselves but also help plants to grow.

Earthworms are highly effective niche constructors, constantly transforming the soil in which they live. They break down vegetable and mineral matter into particles small enough for plants to ingest. The worm casts they secrete are five times richer in usable nitrogen, have seven times the concentration of phosphates, and are about 11 times richer in potassium than the surrounding soil.

Similarly, microscopic diatoms living in seafloor sediments secrete chemicals that bind and stabilize the sand. In Canada's Bay of Fundy, for example, the changes diatoms make to the physical state of the seabed allow other organisms, such as mud shrimp, to colonize it.

British biologists Nancy Harrison and Michael Whitehouse have also suggested that when birds form mixed-species flocks – as many do outside of the breeding season – they are altering their relationship with competitors to find more food resources and gain more protection from predators. The complex social environment they create modifies their own ecology and behaviour.

In his explanation of niche construction, Odling-Smee pointed to ancient cyanobacteria, which produced oxygen as a by-product of photosynthesis more than 2 billion years ago. This was a key factor in the Great Oxygenation Event, which changed the composition of Earth's atmosphere and oceans, massively modifying our planet's environment. The oxygen boost helped to create the conditions for the evolution of much more complex life forms – including humans. ∎

A common starling in Arizona, US, takes advantage of a hole abandoned by a Gila woodpecker to make its own nest.

LOCAL COMMUNITIES THAT EXCHANGE COLONISTS

METACOMMUNITIES

IN CONTEXT

KEY FIGURE
Mathew Leibold (1956–)

BEFORE
1917 Arthur Tansley observes that two species of *Galium* plants grow differently in different soil patches.

1934 Georgy Gause develops the competitive exclusion principle stating that two species competing for the same key resource cannot coexist for long.

2001 Stephen Hubbell's "neutral theory" argues that biodiversity arises at random.

AFTER
2006 Mathew Leibold and fellow American ecologist Marcel Holyoak refine and develop the theory of metacommunities.

One of the limitations of traditional community ecology was that it tended to look at communities purely locally and take little account of what happens at different scales or across different places. Therefore, over the last few decades, ecologists have been developing theories of "meta" communities; the concept was summed up in 2004 in a key paper led by American ecologist Mathew Leibold.

The idea of metacommunities is linked to that of metapopulations. While studies of metapopulations examine the different patches where populations of the same species coexist, in metacommunity

See also: Competitive exclusion principle 52–53 ▪ The ecosystem 134–137 ▪ The neutral theory of biodiversity 152
▪ Metapopulations 186–187

Mountain goats in Colorado, US, live in a metacommunity of species in a mountain range – the Rocky Mountains – but within a population of goats on one single peak.

theory the different patches consist of entire communities that include a number of interacting species.

What is a metacommunity?

Metacommunities are essentially groups or sets of communities. The communities making up a metacommunity are separated in space, but they are not completely isolated and independent. They interact as various species move between them. For example, a metacommunity might consist of a set of separate forest communities, spread across a region. The various species within each patch of forest habitat interact as an independent community. However, certain species, including deer or rabbits, may migrate or disperse to another community in the metacommunity, moving to a different patch of forest in search of better opportunities to feed, shelter or breed. Differing types of habitat will influence this balance between interlinked and independent development. The theory of metacommunities provides a framework for studying how and why variations develop and their impact on biodiversity and population fluctuations.

Local v regional

A major advantage of looking at communities in this spatial way is that it may help resolve a number of seemingly contradictory observations. One ecologist's study, for instance, might look at the way species live and interact together in a small local community. This narrowly focused study finds that competition between species for resources is a crucial factor in the workings of the community. Another study might look at the picture across a larger community. This macro-study discovers that competition plays virtually no part. So which result is right? »

Wildlife crossings

Many different species cross naturally between separate habitat patches. This movement can be seasonal, as in annual migrations, or prompted by natural disasters, such as fire or flood, or may take place over long timespans. It creates connections that are often essential for the health and survival of species and communities, providing renewal or new resources at pivotal moments. Increasingly, however, manmade barriers, such as clearances for agriculture, road, railways, and urban sprawl, are breaking up this natural interflow from one habitat to another. The idea of providing wildlife with ways through is not new. For example, fishways for fish to bypass dams go back centuries. Wildlife crossings – from bridges for bears in Canada to tunnels for California's desert tortoises – are becoming an increasingly common feature of construction work. Thousands of crossings, among them bridges, viaducts, and underpasses – often planted with vegetation – have been built to conserve habitats and to avoid fatal collisions between animals and vehicles.

The answer may be that both are right, and the difference simply depends on scale. The benefit of metacommunity theory is that it allows ecologists to reconcile these differences. It enables them to look for explanations on both a local and regional scale.

A metacommunity might be a set of half a dozen deciduous trees within a park, with each tree an individual community. However, it could equally be all the deciduous forests in temperate zones all around the world. What metacommunity theory does is allow ecologists to work at any scale, at least in theory.

Umbrella framework

According to Mathew Leibold, the study of metacommunities brings together many seemingly disparate branches of ecology and apparently conflicting theories. It may make it easier, for example, to resolve the century-old debate between the "deterministic", niche-based theory of community ecology, in which species diversity is determined by each species' ecological niche, and "stochastic" (random) theory, which emphasizes the importance of chance colonization and ecological drift (random fluctuations in population sizes).

Metacommunity theory provides an umbrella framework for seeing how deterministic and stochastic processes can interact to form natural communities. It allows ecologists to state that patterns of biodiversity are determined both by local biological features, such as the balance of sun and shade in rock pools or variations in water quality in streams, and by regional stochastic processes, such as the spread of a species by freak storms or a die-off due to an epidemic. It also acknowledges that regional changes can be caused by the combined effect of local ones.

Finding metacommunities

One of the problems with Leibold's concept is that in practice it is not so straightforward to identify the separate components of a metacommunity. For the fish and other water creatures in different lakes within a lake district, for instance, each lake may clearly be a distinct community. However, for those birds able to fly between the lakes in minutes, the different lakes are all part of the same single community. This may explain why much of the continuing work and research on metacommunities

Metacommunity

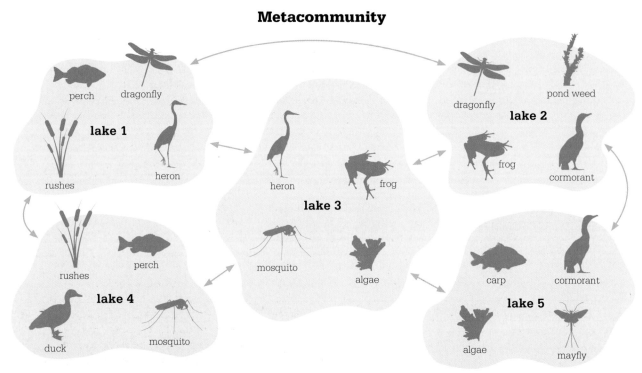

In this example of a metacommunity, arrows show how species move between lakes to feed or breed. Seeds and the spores of algae are dispersed by the wind.

Rockpools in a wave-cut platform form a metacommunity on Eysturoy in the Faroe Islands. The rockpools are separate between tides, but become joined as one when the tide comes in.

for a period of time following a storm, patches of fungal fruiting bodies that live just a few days or weeks, and even pitcher plants that, after dew or rain, provide a short-lived aquatic home for both bacteria and insects.

Blurred communities

Leibold's 2004 paper acknowledged that the metacommunities with blurred boundaries are perhaps the hardest to define. Coral reefs, for example, may look neatly separate, but many of the species that live among them swim freely and respond to a host of changing outside influences, such as shifts in ocean currents.

Since most of the world's life exists within such vaguely defined patches, theorists have attempted further clarification. Leibold and his colleagues have suggested two different ways of identifying metacommunities for study: distinct communities embedded within a "matrix" habitat, such as clearings in a forest rich in resources; and arbitrary sampling patches in a continuous habitat, such as a random circle of trees within a forest.

The work is still at an early stage. The world is entering a biodiversity crisis, and countless species and communities appear to be under threat from the effects of human activity. Metacommunity theory may, in time, help to provide a better understanding of how natural communities will respond, and how local changes to habitats may ripple through a region, either adversely or positively. ∎

has been theoretical and abstract rather than rooted in fieldwork. Some metacommunities are easy to identify, such as islands in an island group, or rockpools that are separate between tides but joined when the tide comes in. In their 2004 paper, Leibold and his colleagues acknowledged that local communities, or patches, do not always have clear boundaries that make them recognizably separate, and that different species may respond to things happening at a different scale. They identified three kinds of metacommunity: markedly separate patches; short-lived but distinct patches that appear in a habitat from time to time at varying size; and permanent patches with vague or "blurred" boundaries.

Distinct patches

The most obvious markedly separate patches are islands in the ocean. These are a convenient subject to

study and there is a vast literature on island biogeography, reaching back to Charles Darwin's famous study of variations between finches in the Galapagos Islands in the Pacific Ocean. Neatly separate patches make good subjects for study, which is why they have been popular with community ecologists. But, of course, birds and many other organisms blown across by the wind or washed in by the sea ensure that even island communities are never completely isolated. This is why some metacommunity studies focus on the space between the communities even where the patches are distinct, as they are with ponds and lakes, and analyse how species move between them.

Short-lived but distinct patches may be much harder to identify, simply because of their ephemeral nature. Nonetheless, ecologists have made metacommunity studies of holes in trees that fill with water

THE LIVI

EARTH

NG

Louis Agassiz shows that an **ice sheet** once covered Switzerland, and suggests that an **Ice Age occurred** in recent geological history.

↑

1840

Svante Arrhenius is the first to argue that **carbon dioxide emissions** can lead to **global warming**.

↑

1896

Vladimir Vernadsky's book *The Biosphere* explains how **atmospheric gases** are created by **biological processes**.

↑

1926

1869

↓

The father of biogeography Alfred Russel Wallace reports a clear **evolutionary division in fauna species** on neighbouring islands.

1912

↓

Alfred Wegener presents his theory that Earth was once a **single landmass** from which **continents drifted apart**.

1935

↓

Arthur Tansley coins the term "**ecosystem**" to describe an **interdependent community** of biological and non-biological components.

For centuries, scientists in the Western world tried to reconcile the findings of geologists and fossil hunters with literal interpretations of biblical stories about Creation and the Great Flood. In 1654, for example, Archbishop Ussher dated Earth's creation to 22 October 4004 BCE. A series of discoveries challenged this narrative and led to new ideas about the dynamic history of life on Earth.

Evidence in the rocks

Two Scottish geologists – James Hutton and Charles Lyell – advanced our understanding of Earth's age. In *Theory of the Earth* (1795), Hutton argued that the repeated cycles of sedimentation and erosion necessary to create thousands of metres of rock strata must indicate a truly ancient origin for the planet – an idea which Lyell developed further in the 1830s. Soon after, Swiss-American geologist Louis Agassiz proposed that the topography of some regions had been shaped by glaciations. Hutton and Lyell also noted that fossils of animals and plants vanished from the geological record. Lyell believed this to be evidence of extinction, challenging the prevailing belief that species were immutable.

Fossils also offered clues to movements of Earth's continents. German meteorologist Alfred Wegener noted that similar fossils could be found on both sides of the South Atlantic, even though they were thousands of miles apart. In his 1912 theory of continental drift, Wegener cited this as evidence that continents were once joined and had broken away. It was not until the 1960s that a viable mechanism was found for such movement. Geophysicists discovered patterns of magnetic anomalies running in parallel stripes either side of ocean ridges and identified the process of seafloor spreading – hot magma bubbling up through cracks in the oceanic crust and forming new crust as it cools and moves away. This gradual process shifts and shapes continents.

The birth of biogeography

In the Age of Exploration from the 16th century on, scientists began to study the geographical distribution of plants and animals. By the 1860s, Alfred Russel Wallace viewed these patterns, clearly defined by physical barriers such as mountains and

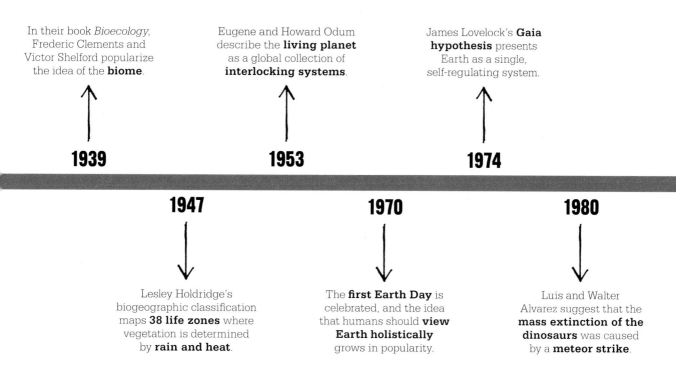

In their book *Bioecology*, Frederic Clements and Victor Shelford popularize the idea of the **biome**.

Eugene and Howard Odum describe the **living planet** as a global collection of **interlocking systems**.

James Lovelock's **Gaia hypothesis** presents Earth as a single, self-regulating system.

1939

1953

1974

1947

1970

1980

Lesley Holdridge's biogeographic classification maps **38 life zones** where vegetation is determined by **rain and heat**.

The **first Earth Day** is celebrated, and the idea that humans should **view Earth holistically** grows in popularity.

Luis and Walter Alvarez suggest that the **mass extinction of the dinosaurs** was caused by a **meteor strike**.

seas as a key supporting argument for evolution. Wallace noted, for example, the ocean straits that produced a sharp division between the flora and fauna of Australasia and Southeast Asia.

With a better understanding of Earth's biogeography, 20th-century ecologists divided the planet into biomes – broad communities of flora and fauna that interact in different habitats, such as tropical rainforests, deserts, or tundra. Botanist Leslie Holdridge refined the concept in 1947 with his life zone classification, in which he mapped zones based on the two crucial influences on vegetation: temperature and rainfall.

A "whole Earth" approach

The word "biosphere" was coined by Austrian geologist Edward Suess in 1875 to signify all the areas at or near the surface of the Earth where organic life can exist. In 1926, the Russian geochemist Vladimir Vernadsky explained the biosphere's close interaction with the planet's rock (lithosphere), water (hydrosphere) and air (atmosphere). This in turn led American biologist Eugene Odum to advocate a holistic approach to ecology. Odum argued that it was not possible to understand a single organism, or a group of organisms, without studying the ecosystem in which they live. He described this view as "the new ecology".

In 1974, British scientist James Lovelock advanced the Gaia hypothesis that the interaction of living and nonliving elements in the biosphere reveal Earth to be a complex, self-regulating system that perpetuates the conditions for life.

Almost two centuries earlier Hutton had articulated a similar idea – that biological and geological processes are interlinked and that Earth could be viewed as a superorganism. In Hutton's words, "The globe of this earth is not just a machine but also an organized body as it has a regenerative power."

Heading for extinction?

Life has survived on Earth for billions of years, despite the ravages of five mass extinctions. However, environmentalists now question whether it will survive another. Indeed, some contend that a sixth mass extinction has already started, as a result of human activity. Yet, if Lovelock's Gaia theory is correct, it seems likely that the planet will endure – even if humans and many other current life forms do not. ■

THE GLACIER WAS GOD'S GREAT PLOUGH

ANCIENT ICE AGES

IN CONTEXT

KEY FIGURE
Louis Agassiz (1807–73)

BEFORE
1795 Scottish geologist James Hutton argues that erratic boulders (rock fragments that are different from the underlying rock) in the Alps were transported by moving glaciers.

1818 In Sweden, naturalist Göran Wahlenburg publishes his theory that ice once covered Scandinavia.

1824 Danish–Norwegian mineralogist Jens Esmark theorizes that glaciers were once larger and thicker and had covered much of Norway and the adjacent sea floor.

AFTER
1938 Serbian mathematician Milutin Milanković publishes a theory to explain the recurrence of ice ages based on changes in Earth's orbit around the Sun.

I n the early 19th century, there were contradictory explanations for the development of Earth's landforms, plants, and animals. Supporters of catastrophism argued that a series of destructive shocks, such as the Great Flood described in the Bible, had re-formed the surface of the planet many times, reshaping existing mountains, lakes, and rivers and wiping out many plant and animal species. In contrast, followers of uniformitarianism contended that Earth's features were the result of continuous and uniform natural processes of erosion, sedimentation (the depositing of particles carried by fluid flows), and volcanism.

Detailed geological studies demonstrated that neither camp was right. They established that

Earth's history has been a process of slow change, punctuated by catastrophic events. The study of glaciers, and the landforms they create, informed these ideas. After observing parallel striations in rocks of the Swiss Alps, German–Swiss geologist Jean de Charpentier (or Johann von Charpentier) postulated that glaciers in the Alps had once been more extensive and had caused the scratches as they moved and their sediment cut into the rock. Geologist Jens Esmark drew similar conclusions in Norway.

Glacier movements

Swiss zoologist Louis Agassiz developed Charpentier's and Esmark's ideas further. In 1837, he proposed that vast sheets of ice had once covered much of the northern hemisphere, from the North Pole to the Mediterranean and Caspian coastlines. Agassiz also undertook some detailed studies of glacier movement in Switzerland and published his *Études sur les*

Animals enter Noah's ark in a depiction of the Great Flood described in the Bible. Catastrophists believed that the Great Flood was one of the formative shocks that shaped the geology of Earth.

See also: Evolution by natural selection 24–31 ■ Global warming 202–203 ■ The Keeling Curve 240–241 ■ Ozone depletion 260–261 ■ Spring creep 274–279

glaciers in 1840. The same year, he visited geologist William Buckland in Scotland to investigate glacial features there, prompting Scottish glaciologist James Forbes to begin similar research in the French Alps.

Some quarters, such as the Catholic Church, still argued that glacial striations had been caused by a great flood or that large silt and rock deposits had been transported by icebergs swept along by the flood. From the 1860s, however, there was wide support for Agassiz's glaciation theory and the idea that glaciers in the Swiss Alps and Norway had once extended much further. It was also accepted that a sheet of ice had once spread across Europe, and south from the Arctic through much of North America, with catastrophic implications for plants and animals.

By the late 1800s and early 1900s, as more expeditions to both Greenland and Antarctica were undertaken, it became known that both areas were still covered in ice.

Glaciers converge on Piz Argient, a mountain in the Swiss Alps. Like others in the Alps, these glaciers were once much more extensive than they are now, and they continue to shrink.

Aerial surveys in the 1920s and 1930s confirmed the extent of their vast ice sheets – now defined as areas of glacier ice exceeding 50,000 sq km (19,300 sq miles); ice caps, such as Iceland's Vatnajökull, are smaller.

Further evidence revealed that there had not been one single ice age, but at least five major ice ages in Earth's long history. The most recent, the Quaternary Ice Age, began 2.58 million years ago and is ongoing. In the last 750,000 years, there have been eight ice advances (glacial periods) and retreats (interglacial periods). During the last glacial period, which ended 10,000–15,000 years ago, ice sheets were up to 4km (2½ miles) thick, and the sea level was 120m (390 ft) lower. ■

Receding glaciers and bird migration

When the last glacial period began to end, around 26,500 years ago, Earth was much colder than it is today. Much of North America and northern Eurasia was covered with ice sheets. The environment was so harsh that most birds tended to live in subtropical and tropical regions where there was more food.

As temperatures began rising, the ice sheets started to shrink, uncovering a new landscape. Ice-free ground and short, wet summers were ideal for insects, and birds began to move in, too, to take advantage of this food supply. When days got shorter in autumn, some birds stayed on for the winter, but others returned south.

The distances flown by birds returning to their homes grew longer as the ice sheets retreated further, eventually developing into long-distance spring and autumn migrations between the tropics and northern latitudes. Common birds that undertake the journey include swallows, warblers, and cuckoos.

A male Baltimore oriole perches on a tree fern in Costa Rica. The species flies north to breed in March and returns to the tropics in August or September.

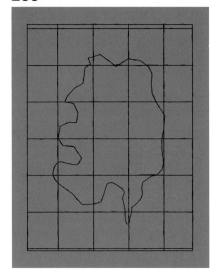

THERE IS NOTHING ON THE MAP TO MARK THE BOUNDARY LINE
BIOGEOGRAPHY

IN CONTEXT

KEY FIGURE
Alfred Russel Wallace
(1823–1913)

BEFORE
1831–36 Darwin's studies on the voyage of HMS *Beagle* confirm that many animals living in one area are not found in similar habitats elsewhere.

AFTER
1874 British zoologist Philip Sclater categorizes birds by zoogeographic regions.

1876 Alfred Russel Wallace publishes *The Geographical Distribution of Animals* – the first extensive publication on biogeography.

1975 Hungarian biogeographer Miklos Udvardy proposes dividing biogeographic realms into biogeographic provinces.

2015 Mexican evolutionary biologist J.J. Morrone proposes an International Code of Area Nomenclature for biogeography.

The places where animals and plants live often vary in a regular manner along geographic gradients of latitude, elevation, and habitat type. The study of this variation is known as biogeography. One branch (phytogeography) examines the distribution of plants, whereas the other (zoogeography) analyses the distribution of animals. British naturalist and biologist Alfred Russel Wallace is widely regarded as the "father of biogeography".

In the 18th century, as explorers recorded the plants and animals they saw, a picture of geographic change had begun to emerge. On the great 1831–36 expedition of HMS *Beagle*, Charles Darwin saw species of birds on the Falkland Islands that did not live on mainland South America, giant tortoises that were unique to the Galapagos Islands, and marsupials such as Australia's kangaroos. New pieces of the biogeographic jigsaw were falling into place.

Zoogeographic regions of the world

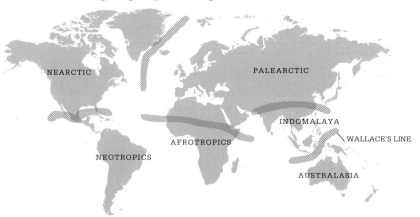

Wallace's six zoogeographic regions began with the line he proposed in 1859 to mark the division of fauna between southeast Asia and Australasia.

See also: Evolution by natural selection 24–31 ▪ Island biogeography 144–149
▪ The distribution of species over space and time 162–163 ▪ Biomes 206–209

Alfred Russel Wallace

Explorer, naturalist, biologist, geographer, and social reformer Alfred Russel Wallace left school at 14, and trained as a surveyor in London before becoming a teacher. He became fascinated with insects after meeting British entomologist Henry Bates. The pair ventured to the Amazon Basin in 1848 on a four-year collecting expedition. Trips to the Orinoco River and the Malay Archipelago followed. Wallace arrived at the same conclusion as Charles Darwin on the origin of species by natural selection, and they presented their papers jointly in 1858. A world authority on fauna distribution, Wallace also raised awareness about problems caused by human impact on the environment.

Key works

1869 *The Malay Archipelago*
1870 *Contributions to the Theory of Natural Selection*
1876 *The Geographical Distribution of Animals*
1878 *Tropical Nature, and Other Essays*
1880 *Island Life*

From 1848, Wallace conducted years of fieldwork in South America and southeast Asia. He researched the feeding and breeding behaviour and migratory habits of thousands of species, paying specific attention to animal distribution compared with the presence or absence of geographical barriers, such as seas between islands. He concluded that the number of organisms living in a community depends on the food available in that specific habitat.

Wallace's Line

During his 1854–62 expedition to the Malay Archipelago, Wallace collected an astonishing 126,000 specimens, many of them from species previously unknown to Western science, including 2 per cent of the world's bird species. He regarded biogeography as support for the theory of evolution by means of natural selection. One of Wallace's important findings was the marked difference in bird species either side of what was to become known as the Wallace Line, which runs along the Makassar Strait (between the islands of Borneo and Sulawesi) and the Lombok Strait (between

The whole of Siberia is in the Palearctic region, and the Siberian white birch trees depicted here are part of a subdivision called the East Siberian taiga.

Bali and Lombok); this separates Asian fauna from the Australasian. He found that larger mammals and most birds did not cross the line. For example, tigers and rhinos live only on the Asian side; babirusas, marsupials, and sulphur-breasted cockatoos only on the other side. He also highlighted the sharp differences between animals in North and South America.

In 1876, Wallace proposed six separate zoogeographic regions: Nearctic (North America); Neotropics (South America); Palaearctic (Europe, Africa north of the Sahara Desert, and Central, North, and East Asia); Afrotropics (Africa south of the Sahara Desert); Indomalaya (South and Southeast Asia); and Australasia (Australia, New Guinea, and New Zealand). Today Wallace's regions, with the addition of Oceania (the islands of the Pacific Ocean) and Antarctica, are known as biogeographic realms. ▪

GLOBAL WARMING ISN'T A PREDICTION. IT IS HAPPENING.
GLOBAL WARMING

IN CONTEXT

KEY FIGURE
Svante Arrhenius
(1859–1927)

BEFORE
1824 French physicist Joseph Fourier suggests that Earth's atmosphere traps the Sun's heat like a greenhouse.

1859 Irish physicist John Tyndall provides experimental evidence to support earlier hypotheses that atmospheric gases absorb radiant heat.

AFTER
1976 American scientist Charles Keeling proves that between 1959 and 1971 carbon dioxide levels in the atmosphere increased by about 3.4 per cent each year.

2006 In *Field Notes from a Catastrophe*, journalist Elizabeth Kolbert tells the stories of people and places impacted by climate change.

In 1896, Swedish chemist Svante Arrhenius became the first person to argue that carbon dioxide (CO_2) emissions caused by human beings could lead to global warming. Arrhenius thought that the average ground temperature could be influenced by carbon dioxide and other "greenhouse gases", as they are now known, and believed that increasing levels of CO_2 would lift Earth's temperature. More specifically, he estimated that if levels of carbon dioxide increased by 2.5 to 3 times, Arctic regions of the world would see temperature increases of 8–9°C (14–16°F).

The greenhouse effect

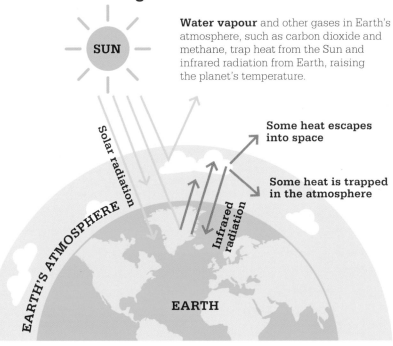

Water vapour and other gases in Earth's atmosphere, such as carbon dioxide and methane, trap heat from the Sun and infrared radiation from Earth, raising the planet's temperature.

Some heat escapes into space

Some heat is trapped in the atmosphere

SUN

Solar radiation

EARTH'S ATMOSPHERE

Infrared radiation

EARTH

See also: Environmental feedback loops 224–225 ▪ Renewable energy 300–305 ▪ The Green Movement 308–309 ▪ Halting climate change 316–321

Arrhenius was building on the work of scientists Joseph Fourier and John Tyndall earlier in the 19th century. Fourier had wondered why Earth was not a freezing wasteland, when the Sun was too far away to heat it to its current temperature. He knew that heated surfaces – such as the surface of Earth – emit thermal energy, and that the thermal energy radiating back into space should result in colder temperatures on Earth. Something was regulating the temperature, and Fourier theorized that Earth's atmosphere, made up of various gases, acted like a glass box, containing the air and keeping it warm. Fourier's hypothesis, although oversimplistic, led to the "greenhouse effect" theory of Earth's thermal regulation.

John Tyndall was the first to prove Fourier's greenhouse effect hypothesis. His experiments demonstrated how, when Earth cools down at night – by releasing the heat absorbed from the Sun during the day – atmospheric gases, especially water vapour, absorb the heat (radiation) and cause a greenhouse effect. This keeps Earth's temperature at an

> If the planet were a patient, we would have treated her long ago.
> **Prince Charles**

> The atmosphere may act like the glass of a greenhouse … [raising] the mean temperature of Earth's surface.
> **Nils Ekholm**
> *Swedish meteorologist*

average 15°C (59°F), although in recent decades human activities that release greenhouse gases have pushed this figure higher. For example, the 10 warmest years on record have occurred since 1998.

Fuelling a warming world

By 1904, Arrhenius had become concerned about the dramatic increase of CO_2 due to human actions – primarily through burning fossil fuels, such as coal and oil. He correctly predicted the influence that CO_2 emissions would have on global temperatures, but eventually came to the conclusion that an increase in global temperatures could have a beneficial effect on plant growth and food production.

The burning of fossil fuels has, in fact, increased CO_2 levels more quickly than Arrhenius expected, although the planet has warmed less than he predicted. Scientists understand now that global warming is having damaging effects on people and on the environment, and will continue to do so as long as long as emissions continue to increase. ▪

The effects of global warming

Since the end of the 19th century, carbon dioxide (CO_2) in the atmosphere has increased by about 25 per cent, and the average global temperature by around 0.5°C (0.9°F). Scientific evidence proves that these changes have contributed to melting glaciers and sea ice followed by rising sea levels – around 20 cm (8 in) since 1880 – as well as damage to coral reefs. Other phenomena include longer wildfire seasons, more extreme weather, and shifts in the ranges of animals and plants, leading to disease, extinction, and food shortages.

The extent to which global temperatures will increase depends on whether (and how rapidly) global CO_2 emissions diminish. Scientists predict that, at the current rate, this increase could range from 0.3°C to 4.6°C (0.5°–8°F) by 2100, with the greatest warming likely to occur in the Arctic regions.

The Perito Moreno glacier in Patagonia is one of the few glaciers that is still growing. The majority are slowly melting, causing sea levels to rise worldwide.

LIVING MATTER IS THE MOST POWERFUL GEOLOGICAL FORCE
THE BIOSPHERE

IN CONTEXT

KEY FIGURE
Vladimir Vernadsky
(1863–1945)

BEFORE
1785 Scottish geologist James Hutton proposes that in order to understand Earth, all its interactions should be studied.

1875 Austrian geologist Eduard Suess first uses the term "biosphere" to describe "the place on Earth's surface where life dwells".

AFTER
1928 In *Methodology of Systematics*, Russian zoologist Vladimir Beklemishe warns that humanity's future is irrevocably linked to the preservation of the biosphere.

1974 British scientist James Lovelock and American biologist Lynn Margulis first publish their Gaia hypothesis – the idea of Earth as a living entity.

Earth has four interacting subsystems: the lithosphere, Earth's rigid, rocky outer shell; the hydrosphere, which comprises all water on the planet's surface; the atmosphere, formed by layers of surrounding gases; and the biosphere – anywhere that supports life, from the ocean depths to the highest mountain tops.

The biosphere's origins are ancient: fossils of tiny single-celled microorganisms that date back 4.28 billion years suggest that it is almost as old as Earth itself. The biosphere extends into every land-and water-based environment, and reaches into extreme habitats, such as the intensely hot mineral-rich waters around hydrothermal vents. It is often divided into "biomes" – common major habitats, such as deserts, grasslands, oceans, tundra, and tropical rainforests.

Earth the superorganism
Ideas about the biosphere began to emerge in the 18th century, when the Scottish geologist James Hutton described Earth as a superorganism – a single living entity. A century later, Eduard Suess introduced the concept of the biosphere in *Das Antlitz der Erde* (*The Face of the Earth*). Suess explained that life is limited to a zone at Earth's surface and that plants are a good example of the interactions between the biosphere and other zones – they grow in the soil of the lithosphere, but their leaves breathe in the atmosphere.

In *The Biosphere* (1926), Russian geochemist Vladimir Vernadsky, who had met Suess in 1911, defined the concept in much more detail, outlining his view of life as a major geological force. Vernadsky was one of the first to recognize that atmospheric oxygen, nitrogen,

Man is becoming a more and more powerful geological force, and the change of his position on the planet coincided with this process.
Vladimir Vernadsky

See also: The ecosystem 134–137 ▪ Biodiversity and ecosystem function 156–157 ▪ A holistic view of Earth 210–211 ▪ The Gaia hypothesis 214–217

Over billions of years layers of cyanobacteria have fossilized to form stromatolites – mounds of sedimentary rock, as seen here at Hamelin Pool, Shark Bay, Western Australia.

and carbon dioxide result from biological processes, such as the respiration of plants and animals. He argued that living organisms reshape the planet as surely as physical forces, such as waves, wind, and rain. He also introduced the idea of three stages of Earth's development: first, the birth of the planet with the geosphere, in which only inanimate matter existed; secondly, the emergence of life in the biosphere; and finally the epoch in which human activity changed the planet forever – the noosphere.

Sphere interactions

Scientists believe the biosphere has constantly changed. Oxygen levels in the atmosphere began to rise at least 2.7 billion years ago, as microorganisms called cyanobacteria multiplied. As oxygen increased, more complex life forms evolved that would shape Earth in different ways, eroding and re-moulding its surface, and changing its chemical composition.

Gradually, elements of the biosphere became part of the lithosphere. Over millennia, dead corals created reefs in shallow tropical oceans. Similarly, the calcite skeletons of trillions of marine organisms fell to the ocean floor, fossilized, and formed limestone. ■

I look forward with great optimism. We live in a transition to the noosphere.
Vladimir Vernadsky

Vladimir Vernadsky

Born in 1863, Vladimir Vernadsky graduated from St Petersburg State University aged 22, and did postgraduate work in Italy and Germany, where he studied the optical, elastic, magnetic, thermal, and electrical properties of crystals. After the revolution in Russia in February 1917, Vernadsky became assistant Minister of Education in the provisional government. The following year, he founded the Ukrainian Academy of Science in Kiev. Although his book *The Biosphere* was not taken seriously by scientists outside Russia for many years, it later became one of the founding documents of Gaia theory.

In the 1930s, Vernadsky advocated the use of nuclear power, and played an advisory role in the development of the Soviet atomic bomb project. He died in 1945.

Key works

1924 *Geochemistry*
1926 *The Biosphere*
1943 "The Biosphere and the Noosphere"
1944 "Problems of Biochemistry"

THE SYSTEM OF NATURE

BIOMES

IN CONTEXT

KEY FIGURES
Frederic Clements
(1874–1945),
Victor Shelford (1877–1968)

BEFORE
1793 Alexander von Humboldt
coins the word "association"
to sum up the mix of plant
types that occurs in a
particular habitat.

1866 Ernst Haeckel poses
the idea of the biotope, the
living space for a range
of plants and animals.

AFTER
1966 Leslie Holdridge
champions the idea of life
zones based on the biological
effects of temperature and
rainfall variations.

1973 German–Russian
botanist Heinrich Walter
creates a biome system that
considers seasonal variations.

Different parts of the world have varying patterns of plant and animal life, but there are usually similarities over vast areas. These are called biomes, and each one is a large geographical region with its own distinctive plant and animal community and ecosystem. The idea of the biome was first popularized by plant ecologist Frederic Clements and zoologist Victor Shelford in the US, in their key book *Bioecology* (1939), although its origins date back earlier.

The biome concept took shape as ideas on plant succession and community ecology developed. Clements identified "formations", large plant communities, which led

See also: The distribution of species over space and time 162–163 ▪ Climax community 172–173 ▪ Open community theory 174–175 ▪ Biogeography 200–201

to his idea of climax communities in 1916. The same year, Clements used the term "biome" to describe biotic communities – all the interacting organisms within a specific habitat.

Like-minded thinkers

Clements was not the only one thinking along these lines. Zoologist Victor Shelford was working towards the same idea. The pair began to meet over the next 20 years, while pursuing their own research, to see how they could combine the worlds of plants and animals. Clements studied plant biomes in Colorado with his wife, the eminent botanist Edith Clements. Meanwhile, Shelford compiled the *Naturalist's Guide to the Americas* (1926) – the first major geographical summary of wildlife in the Americas, in which he talked about "biota". This book laid much of the foundation for later findings.

Ways of looking at interactions in ecological communities took a major step forward when British »

The Mongolian steppe belongs to the same grassland biome as the prairies in North America. Despite being on separate continents, they are linked by their climate, animals, and plants.

The **geographic spread** of plants is **determined** mainly **by climate**.

Different plants flourish in each **climatic region**.

The **major types** of plants in each region **match precipitation and temperature patterns** closely.

The major plant types can be used to divide the world into broad natural zones called biomes, which reflect variations in climate.

Threatened coral reef biomes

Coral reefs are such bountiful habitats that they are often seen as the tropical rainforests of the sea. They support a quarter of all marine species and provide livelihoods for half a billion people. Yet they are now facing catastrophe. Half of all reefs have been lost in the last 30 years, and some experts estimate that 90 per cent will be gone over the next 30 years. The main global threats are ocean acidification and global warming. As seas warm, stressed corals expel the algae they rely on for food. They stop growing, lose their colour, and often die in what is called a coral bleaching event. Such events are becoming ever more frequent. There are local threats, too, including overfishing, both for the table and for aquariums. Even more seriously, to catch fish for aquariums, sodium cyanide is often squirted into the water to temporarily immobilize the fish, and this kills corals. More brutally, fish for the table are often caught by throwing dynamite into the water. This kills fish, making them easy to scoop up in vast numbers, but it also blasts coral reefs apart.

botanist Arthur Tansley introduced the term "ecosystem" in 1935. When Clements and Shelford published the results of their collaboration in 1939, they were not making a sudden breakthrough – rather it was a consolidation of ideas that had been taking shape over a long time.

The collaboration between botany and zoology was crucial. Only by looking at the totality of the natural world with its dynamic interactions could scientists hope to get a full picture, and Clements defined a biome as "an organic unit comprising all the species of plants and animals at home in a particular habitat". Even so, biomes have come to be defined principally by vegetation type.

The most important feature of biomes is that they link vegetation and plant communities across the world. There are tropical forests, for example, in every continent, but most tree species appear only in one continent. So, the range of trees within the Amazonian forests is completely different from the range of trees in the forests of Indonesia. Yet both areas are identifiable as tropical forest, because the trees have features in common.

Since *Bioecology* first appeared, there have been countless attempts to define what a biome is, and many different ways of classifying them. Biomes provide a simple way of understanding global vegetation patterns, but when looked at closely they present a crude way of grouping ecosystems. There is no single accepted classification system, and the only division everyone seems to agree on is that between terrestrial (land-based) and aquatic (water-based) biomes. Many of the same biomes crop up in most systems, such as the polar biome, tundra, rainforest, grasslands, and deserts, but there is no agreed definition and there are marked variations.

The climate factor

The one common factor in all biome classifications has been climate, although other "abiotic" factors can also play a part. Climate determines the form of plant growth best suited to a region, and plants that grow in a certain way are restricted to particular climates. The leaves of deciduous trees are broad, with a large surface for light absorption, but little resistance to drying out or frost. Conifer tree needles, on the other hand, are narrow and can survive the harshest frosts. Desert shrubs often have very thin leaves, or no leaves at all, to resist drying out. Biogeographers acknowledge climate's key role when they talk about "tropical" rainforests and "temperate" grasslands.

Terrestrial biomes of the world

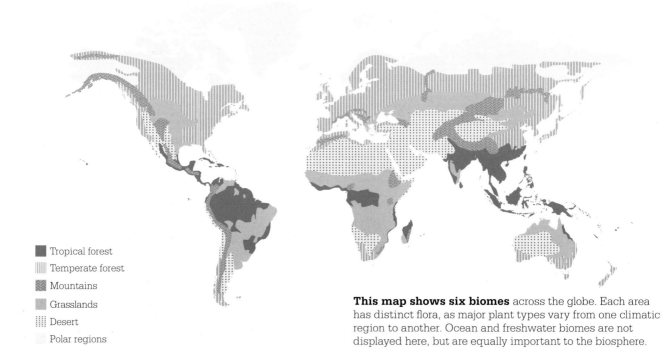

■ Tropical forest
▥ Temperate forest
▨ Mountains
▤ Grasslands
▥ Desert
☐ Polar regions

This map shows six biomes across the globe. Each area has distinct flora, as major plant types vary from one climatic region to another. Ocean and freshwater biomes are not displayed here, but are equally important to the biosphere.

Tropical rainforest is the hottest and wettest biome and covers 7 per cent of Earth's surface. One of the oldest biomes, it also contains far more animal and plant species than any other biome.

Very few species have identical climate requirements. Even among varieties of the same plants, there are variations. The sugar maple of eastern North America, for example, is slightly more tolerant of winter cold than its cousin the silver maple. Although the areas where both trees grow overlap, the sugar maple can be seen far over the Canadian border, whereas the silver maple flourishes as far south as Texas. Since biomes give only an approximate picture of plant and animal distribution, ecologists are constantly devising new systems of classification.

Rain, heat, and evolution

One of the most widely recognized systems of classification is the life zones system devised by American botanist Leslie Holdridge in 1947, and updated in 1967. His system is based on the assumption that two key factors, rain and heat, determine vegetation type in each region. He created a graphic depiction of 38 life zones in a pyramid. The three sides of the pyramid represent three axes: rain, temperature, and evapotranspiration (which depends on both rain and temperature). Using these axes, he could plot hexagons showing regions that also reflect humidity, latitude, and altitude.

American plant ecologist Robert Whittaker devised a much simpler graph, with average temperature on one axis and annual rainfall on the other. With these two variables plotted against each other, he was able to divide the graph into nine biomes – from tropical rainforest (the hottest and wettest) through to tundra (the coldest and driest).

Underpinning all these systems is the idea of convergent evolution, which argues that species develop similar traits as they adapt to similar environments. Insects, birds, bats, and pterosaurs all developed wings independently to occupy air space. Different biomes are therefore assumed to develop corresponding life forms in response to similar environmental conditions. However, in recent decades, it has been noted that species can evolve differently in the same biome and also that different stable biomes can develop in an identical climate. While central to understanding life, biomes remain a complex and elusive concept. ∎

Ecozones

Biomes are about identifying the similar forms that life takes in response to particular regional conditions such as climate, soil, and topography. However, there are other methods of dividing the world in ecological terms. In 1973, Hungarian biologist Miklos Udvardy came up with the concept of biogeographic realms; this system was then further developed in a scheme by the World Wildlife Fund. The BBC later replaced the term "biogeographic realm" with "ecozone". Biogeographic realms divide the whole planet according to the evolutionary history of plants and animals. The ways in which continents have split apart and drifted means that species have evolved variously in different parts of the world. Ecozones are therefore based on identifying this diversification. Australasia, for example, is a single ecozone, because marsupials evolved there in isolation from other mammals in the rest of the world.

The short-beaked echidna is one of the most widespread native mammals in the Australasian ecozone. They live in a range of habitats from desert to rainforest.

WE TAKE NATURE'S SERVICES FOR GRANTED BECAUSE WE DON'T PAY FOR THEM
A HOLISTIC VIEW OF EARTH

IN CONTEXT

KEY FIGURE
Eugene Odum (1913–2002)

BEFORE
1905 In *Research Methods in Ecology*, American botanist Frederic Clements writes about plant communities and how they change over time.

1935 Arthur Tansley, a British botanist, proposes the term "ecosystem" to describe a community of plants, animals, soil minerals, water, and air.

AFTER
1954 Eugene and Howard Odum's study of the coral Eniwetok Atoll in the Pacific Ocean applies the principles of holistic ecology.

1974 British environmentalist James Lovelock and American biologist Lynn Margulis first publish their Gaia hypothesis. It states that Earth is a self-regulating system that maintains the conditions necessary for life on our planet.

he American ecologist Eugene Odum was not the first scientist to write about ecology, but in the 1950s he proposed that it deserved to be a discipline in its own right. Until then, ecology was viewed as a relatively insignificant subdivision of the biological sciences – the poor relation of biology, zoology, and botany. However, Odum believed passionately that studying plant and animal species in isolation

Salt marshes, such as these on the coast near Porthmadog, North Wales, form their own ecosystem, with the seawater and its nutrients providing a unique habitat for wildlife.

could never lead to a full knowledge of the living world. He argued that it was more important to study the places and roles that the species held in their community, rather than simply finding out more about what they were. Odum's

See also: The ecosystem 134–137 ▪ Macroecology 185 ▪ The peaceful coexistence of humankind and nature 297 ▪ The Green Movement 308–309

new approach to the subject – first set out in his 1953 book *The Fundamentals of Ecology* – revolutionized the purpose and influence of ecological research.

The "new ecology"

The holistic view of Earth involves studying the systems of organisms as a whole. As Odum explained, one organism, or any one group of organisms, cannot be understood without studying the ecosystem in which it lives. The holistic approach examines all the roles played by each member of an ecosystem, and how that system interacts with others. Climate, geology, water and mineral input, and human activity all affect – and are affected by – a multitude of living communities.

Odum was writing in the 1950s and '60s, when there was a growing awareness of the environmental destruction wrought by humanity. The role of people was a crucial part of "systems ecology", as he called his idea. Odum wanted humans to be sympathetic allies with the natural world – collaborators rather than manipulators – and his views of an all-embracing ecology did much to inspire the first Earth Day, which was celebrated in 1970.

The holistic concept of Odum's "new ecology" deals with Earth as a whole, bringing together physics, chemistry, botany, zoology, geology, and meteorology. The fundamental assumptions of ecology are that the ecosystem is the basic unit of nature, that biological diversity increases the ability of ecosystems to survive, and that the whole is greater than the sum of its parts. Systems in the natural world – whether they are groups of cells in an animal's body,

> …ecology has been badly presented and has been broken into too many antagonistic subdivisions.
> **Eugene Odum**

the whole animal, or the ecosystem in which the animal lives – are able to self-regulate to provide stability.

Integrated investigation

A holistic study of a lake ecosystem would involve looking at all the inputs into the lake and its margins as well as all the outputs, including energy, water, minerals, and nutrients. It would also consider any human inputs. The study would examine the roles played by both producer organisms, such as plants and algae, and consumers such as herbivores and carnivores. The holistic approach also examines changes over time, in which developments that benefit some organisms in the short-term might lead to a lack of diversity in the future. For example, although trout thrive in warmish, alkaline waters, if those waters become too warm or acidic due to ecological change, the fish can no longer breed.

Odum's holistic approach leaves a legacy of a far more detailed appreciation of what is happening in an ecosystem than a series of individual species studies. ▪

Earth Day

After witnessing a horrific oil spill in Santa Barbara, California, in 1969, US Senator Gaylord Nelson decided to focus on growing worries about pollution during a national "teach-in" on the environment. He could not have envisaged the size of the movement he would inspire. On 22 April 1970, 20 million Americans took part in the first Earth Day, with rallies, marches, and lectures taking place nationwide. Such was the effect of the protests that later that year the Clean Air, Clean Water, and Endangered Species acts became law, and the Environmental Protection Agency was established in the US that December. Earth Day became a global phenomenon, with 200 million people participating in 141 countries in 1990 – and built momentum for the 1992 UN Earth Summit in Rio de Janeiro. Earth Day celebrations are held every April, with a different theme each time. In 2018, the focus was on ending global pollution by plastics.

The first Earth Day on 22 April 1970 saw crowds such as this one in Philadelphia, Pennsylvania, gather across the US to protest against pollution and the use of pesticides.

PLATE TECTONICS IS NOT ALL HAVOC AND DESTRUCTION

MOVING CONTINENTS AND EVOLUTION

IN CONTEXT

KEY FIGURE
Alfred Wegener (1880–1930)

BEFORE
1596 Abraham Ortelius, a Dutch scholar, is one of several geographers who observe that the two sides of the Atlantic Ocean seem to "fit" each other.

AFTER
1929 British geologist Arthur Holmes proposes that convection in Earth's mantle drives continental drift.

1943 George Gaylord Simpson dismisses fossil evidence for continental drift and argues for "stable continents".

1962 American geologist Harry Hess explains how the seafloor spreads, by molten magma rising from below.

2015 A group of Australian scientists propose that periods of rapid evolution in the oceans were triggered by collisions between tectonic plates.

The surface of Earth is constantly moving, very slowly, and has been doing so for more than three billion years. The lithosphere (Earth's crust and upper mantle) is divided into seven large sections and many smaller ones, called tectonic plates. Where plates meet, the type of movement determines the nature of the boundary. Where plates push against each other, new mountains are created. If plates pull apart, new crust forms on the ocean floor.

The first inkling that the continents may not have always been in their current positions came in the late 16th century. European explorers sailing to the Americas saw from their newly-created maps that the coastlines on each side of the Atlantic Ocean mirrored each other. Later, geologists found strong structural and geological similarities between the Caledonian-era mountains of Northern Europe and the Appalachian Mountains of North America.

Lookalike fossils

There are various examples of fossil finds straddling different continents that can only be explained by continental movement – since the

This fossilized head of the extinct reptile *Cynognathus crateronotus* was found in southern Africa. The same fossils occur in South America: evidence that the two continents were once one.

animals or plants concerned would have been unable to cross the ocean divide. These include *Cynognathus crateronotus*, a mammal-like reptile that lived over 200 million years ago in southern Africa and eastern South America. *Glossopteris*, a genus of woody trees, grew in South America, South Africa, Australia, India, and Antarctica, but nowhere else, around 300 million years ago.

To German geophysicist Alfred Wegener, such fossil patterns indicated that these continents had once been joined together. In 1915, he published his theory that all the continents were once a single land mass, "Pangaea", which has since broken up and drifted apart.

See also: Island biogeography 144–149 ▪ The distribution of species over space and time 162–163 ▪ Macroecology 185 ▪ Metapopulations 186–187 ▪ Biogeography 200–201

Three types of plate boundary

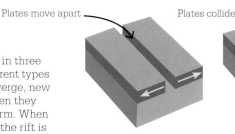
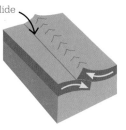
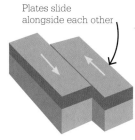

Plates move apart

Plates collide

Plates slide alongside each other

Tectonic plates can move in three different ways, forming different types of boundary. When plates diverge, new oceanic crust is formed. When they converge, new mountains form. When plates slide past each other, the rift is known as a transform fault.

Divergent

Convergent

Transform

Wegener's theory was not well received at first. In 1943, George Gaylord Simpson, one of the most influential palaeontologists in the US, criticized the theory. He argued that the fossil record could be explained by static continents linked and unlinked by periodic flooding.

Evidence and evolution

Despite early doubts, evidence for the plate tectonics theory grew. A series of discoveries established that the seafloor was spreading and that new oceanic crust was constantly being created. We now understand that the movement of the tectonic plates is driven by convection currents carrying heat from deep inside the planet to the surface.

Once Wegener's theory was accepted, the fossil evidence made much better sense. Continental drift has had a profound influence on how species have evolved. For example, if a continent splits apart, the two separated populations of a species can start to evolve in completely different directions. On the other hand, if two continents collide, or a bridge of land forms between them, different species begin to mix and compete, and some may become extinct as a result. ▪

The forces which displace continents are the same as those which produce great fold-mountain ranges.
Alfred Wegener

Marsupials are strongly identified with Australia, yet they evolved in America and are still also found there.

Marsupials in America and Australia

Marsupials are non-placental mammals whose young complete their gestation feeding from their mother's teats, typically in a pouch on the belly. Now found only in the Americas (mainly South and Central) and Australia, they are thought to have evolved in North America 100 million years ago. They spread to South America and diversified into many different species.

Several groups later moved into what is now Antarctica and on into southern Australia. It is thought that they travelled via a belt of vegetation straddling the three areas, which were once all part of the southern landmass called Gondwana.

By 55 million years ago, the continents had separated, and marsupial species began to evolve differently. The only known Antarctic marsupial fossils, found on Seymour Island in 40-million-year-old rocks, resemble South American marsupials of the same period, but not those of Australia.

LIFE CHANGES EARTH TO ITS OWN PURPOSES

THE GAIA HYPOTHESIS

IN CONTEXT

KEY FIGURE
James Lovelock (1919–)

BEFORE
1935 British botanist Arthur Tansley uses "ecosystem" to describe an interdependent community of biological and nonbiological components.

1953 In *Fundamentals of Ecology*, American ecologist Eugene Odum describes Earth as a collection of interlocking ecosystems.

AFTER
1985 In the US, the first conference on the Gaia hypothesis is held, entitled, "Is the Earth a Living Organism?"

2004 James Lovelock voices his support for nuclear power over renewable energy.

I n 1979, British scientist James Lovelock's book *Gaia: A New Look at Life on Earth* presented his Gaia hypothesis to a general readership. In essence, Lovelock claimed that Earth is a single, self-regulating system, in which living and non-living elements combine to promote life. The book quickly became a bestseller, and caught the imagination of the growing Green movement, offering a fresh approach to environmentalism.

What Lovelock proposed was not without precedent. In the 1920s, Vladimir Vernadsky, a Russian scientist, had developed the idea of the biosphere, the zone of Earth that holds all living organisms, and suggested that it could be seen as a single entity in which organic and

See also: The ecosystem 134–137 ▪ Evolutionary stable state 154–155
▪ The biosphere 204–205 ▪ A holistic view of Earth 210–211

Evolution is a tightly coupled dance, with life and the material environment as partners. From the dance emerges the entity Gaia.
James Lovelock

inorganic elements interact. The British botanist Arthur Tansley then took this idea further in the 1930s, with his concept of an "ecosystem" that regulates itself into a state of equilibrium.

Tansley's theory was at the heart of Lovelock's hypothesis: that all living organisms and their environment form one complex super-ecosystem that regulates and balances conditions to sustain life on Earth. The idea first occurred to Lovelock in the late 1960s, but it was after discussing it with US microbiologist Lynn Margulis that it began to take shape. Together, they presented the hypothesis in a paper in 1974, giving it a name suggested by the writer William Golding – Gaia, after the ancient Greek Earth goddess. Lovelock and Margulis portrayed Earth as a living entity, composed of the biosphere, living organisms; the

A stone relief shows Gaia, the Greek goddess of the Earth. The nonscientific name chosen by Lovelock for his hypothesis initially hindered its acceptance by many scientists.

pedosphere, the surface layer of the Earth; the hydrosphere, the bodies of water on the Earth's surface; and the atmosphere, the gases surrounding the Earth. These spheres and their complex interactions maintain the Earth in "homeostasis". This concept is borrowed from physiology, which describes the stable internal conditions, such as temperature and chemical composition, that allow organisms to function optimally. They are controlled by self-regulating mechanisms that react to change in those conditions. Lovelock's use of the word homeostasis reinforced the implication that the Earth, or Gaia, is a living entity.

Keeping the balance

The hint of mysticism in the Gaia principle chimed with the "New Age" thinking of the time. This helped to popularize the idea, but it also led to a negative reception from the scientific establishment. However, behind the Earth "goddess" metaphor was a serious science-based hypothesis that »

James Lovelock

Inspired by writers such as Jules Verne and H.G. Wells, James Lovelock, born in 1919, was fascinated by science and invention from an early age. He graduated in chemistry from Manchester University in 1941. Lovelock was a conscientious objector during World War II and worked for the National Institute for Medical Research in London. In 1948, he received his PhD in medicine, and then spent time in the US on a Rockefeller fellowship. After returning to Britain in 1955, he turned his attention to inventions, notably the electron-capture detector (ECD), which detects trace atoms in a gas sample. In the 1960s and 1970s, he held visiting professorships in Houston, Texas, and Reading, England, during which time he developed the Gaia hypothesis. In 2003, Lovelock was made a Companion of Honour by Queen Elizabeth II.

Key works

1988 *The Ages of Gaia*
1991 *Gaia: The Practical Science of Planetary Medicine*
2009 *The Vanishing Face of Gaia: A Final Warning*

Daisyworld

At first, scientists criticized the Gaia hypothesis for its supposed implication that the ecosystems in the biosphere could collectively influence Earth's environment. So to enhance the plausibility of the Gaia theory, in 1983 James Lovelock and fellow British scientist Andrew Watson produced "Daisyworld", a simple explanatory model.

Daisyworld is a barren planet, orbiting a sun. As the intensity of the sun's rays increases, black daisies start to grow. They absorb heat and warm the planet's surface to the point where white daisies can thrive. They, in turn, reflect the sun's energy, so cooling the ground. The two kinds of daisy reach a point of equilibrium, whereby they regulate the temperature of the planet. When the sun's heat increases further, the white daisies, able to reflect the sunlight and stay cool, replace the black daisies. Finally, the sun heats up so much that even the white daisies can no longer survive.

In the Gaia hypothesis, Earth, the only known planet to support life, is itself a "superorganism", where the sea, land, and atmosphere work together to maintain the right living conditions.

the interactions of living organisms and their physical surroundings – including the cycles of oxygen, carbon, nitrogen, and sulphur – form a dynamic system that stabilizes the environment.

According to Lovelock, Gaia is controlled by the action of "feedback loops", which are the checks and balances that compensate for disturbances in the system, bringing it back into equilibrium. To function well, life on Earth depends on a particular balance of variables such as water, temperature, oxygen, acidity, and

salinity in its environment. When these are constant, Earth is in a stable state of homeostasis, but if the balance is disturbed, the planet encourages the organisms that will restore the equilibrium, while being hostile to those that reinforce the disturbance. The organic components of the Earth system do not simply react to changes in their environment, but control and regulate them.

These feedback mechanisms operate in a complex global network of interconnected natural cycles, to maintain the optimum conditions for the organisms within them. They can resist change, but only to a certain extent. A big enough disturbance can push the system to a "tipping point", where, with the balance of its components altered, it is likely to settle into a very

different state of equilibrium. Such a tipping point, argued Lovelock, occurred about 2.5 billion years ago, at the end of the Archean Eon, when oxygen first appeared on Earth. At this time, Earth was a hot, acidic place in which methane-producing bacteria were the only life that thrived. Bacteria capable of photosynthesis then evolved, which created an atmosphere that was

If there were a nuclear war, and humanity were wiped out, Earth would breathe a sigh of relief.
James Lovelock

conducive to more complex forms of life. Eventually, the equilibrium conditions that exist on Earth today were established.

Saving the planet

As Lovelock elaborated on the theme, the scientific establishment gradually began to accept the Gaia hypothesis. In the 1980s, a series of "Gaia conferences" attracted scientists from many different disciplines, willing to explore the mechanisms involved in regulating Earth's environment to achieve homeostasis. Later, more attention was devoted to looking at the implications of the hypothesis in the face of climate change. Human activity had been shown to disturb Gaia's system, but the issue was now whether its regulatory mechanisms could withstand further pressure – or whether Earth was facing another irreversible tipping point.

Environmentalists, who had been among the first to embrace Gaia, reacted with dismay to the theory that the human species may precipitate a catastrophic change in Earth's equilibrium. The rallying cry of Green activists became "Save the planet!" but this was at odds with the fundamental idea of Gaia. Although the destruction of natural habitats, the excessive burning of fossil fuels, the depletion of biodiversity, and other human-made threats were likely to have severe consequences for many species – including humans – the planet, according to the Gaia hypothesis, will survive and find a new equilibrium. ■

Nuclear power stations produce plentiful "clean" energy, but also toxic waste. James Lovelock believes Earth is able to absorb and overcome the waste's radioactive effects.

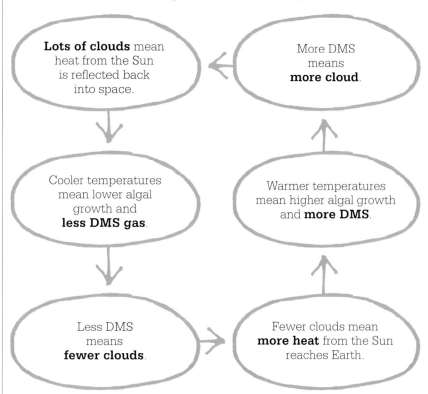

An algal feedback loop

Lots of clouds mean heat from the Sun is reflected back into space.

More DMS means **more cloud**.

Cooler temperatures mean lower algal growth and **less DMS gas**.

Warmer temperatures mean higher algal growth and **more DMS**.

Less DMS means **fewer clouds**.

Fewer clouds mean **more heat** from the Sun reaches Earth.

In Gaia theory, feedback loops keep Earth in balance. One example is the effect that sea algae called coccolithophores have on keeping the planet's climate in check. When the algae die, they release a gas, dimethyl sulphide (DMS), that helps to create clouds.

65 MILLION YEARS AGO SOMETHING KILLED HALF OF ALL THE LIFE ON THE EARTH

MASS EXTINCTIONS

Brainse Fhionnglaise
Finglas Library
Tel: (01) 834 4906

IN CONTEXT

KEY FIGURE
Luis Alvarez (1911–88)

BEFORE
1953 American geologists
Allan O. Kelly and Frank
Dachille suggest in their book
Target: Earth that a meteor
impact may have been
responsible for the extinction
of the dinosaurs.

AFTER
1991 The Chicxulub Crater
in the north of the Yucatan
Peninsula in southeastern
Mexico is proposed as the site
of a massive comet or meteor
impact at the end of the
Cretaceous period.

2010 An international panel
of scientists agrees that the
Chicxulub impact led to
the Cretaceous–Paleogene
mass extinction, around
65 million years ago.

There have been five periods
in Earth's history when
abnormally large numbers
of multicellular organisms have
died off in a relatively short time.
These mass extinctions are defined
by the loss of multicellular animals
and plants because their fossils are
far easier to detect than those of
single-celled organisms.

The general ("background") rate
of extinction is between one and five
species a year. Fossil records show,
for example, the extinction of two to
five families of marine animals every
million years. This is far exceeded
during mass extinctions, which
always mark the boundary between
two geological periods. Scientists
do not understand all the factors
responsible for these events, though
they are agreed on some. Increased

The meteor that hit Earth at the end
of the Cretaceous period was travelling
at 64,000 kph (40,000 mph). Its power
was a billion times greater than the
Hiroshima atomic bomb.

Mass extinction events from 499 million years ago to the present

Late Devonian
A rapid drop in sea level
is one of a number of
possible causes for the
loss of 70–80 per cent
of animal species.

Triassic
Climate change or
an asteroid hit are
potential causes for
the extinction of round
75 per cent of species.

HOLOCENE
(OR ANTHROPOCENE) PERIOD
100,000 YEARS AGO–PRESENT

ORDOVICIAN 485–444	SILURIAŃ 444–419	DEVONIAN 419–359	CARBONIFEROUS 359–299	PERMIAN 299–252	TRIASSIC 252–201	JURASSIC 201–145	CRETACEOUS (K) 145–66	PALAEOGENE 66–23	NEOGENE 23–03

Ordovician
Global cooling
leads to the
extinction of
85 per cent of
marine life.

Permian
Huge volcanic
activity helps to
wipe out 96 per cent
of all marine species.

Cretaceous
A meteor strike and
volcanic activity drive
up to 80 per cent of
animals, including
most dinosaurs,
to extinction.

See also: Ancient ice ages 198–199 ▪ Moving continents and evolution 212–213
▪ The Gaia hypothesis 214–217 ▪ Ocean acidification 281

> All geologic history is full
> of the beginnings and the
> ends of species—of their
> first and last days.
> **Hugh Miller**
> *Scottish geologist*

volcanic activity, changes in the composition of the atmosphere and oceans, climate change, sea level rises and falls, tectonic movement of the continents, and meteor impacts are all likely causes. Some scientists suggest we have now entered a sixth mass extinction, this time the result of human activity.

End of the dinosaurs

The mass extinction that scientists understand best is also the most recent, around 66 million years ago. Geologists refer to it as the K-Pg extinction event, as it occurred at the end of the Cretaceous and start of the Palaeogene periods. Although an extraterrestrial origin was first suggested for the event in the 1950s, this was not taken seriously until two discoveries, in Europe and North America.

In 1980 a team of scientists working in Italy, including physicist Luis Alvarez and his geologist son, Walter Alvarez, discovered a clay layer between Cretaceous and Paleogene deposits. Examination of the clay revealed that it contained the mineral iridium, which is rare

on Earth but common in asteroids. The discovery led to the Alvarez Hypothesis, which proposed that the extinction at the end of the Cretaceous period was caused by a catastrophic meteor strike. The location of the impact was still a mystery, until 11 years later, when a massive crater 170 km (106 miles) across on Mexico's Yucatan Peninsula was found to date from the time of the extinction.

The scientific consensus is that a massive comet or asteroid struck Earth, producing a blast of radiation and a destructive megatsunami more than 100 m (328 ft) high. The radiation would have killed animals nearby, and the megatsunami would have obliterated coastal regions around the Gulf of Mexico. The main damage, however, would be more gradual. A vast cloud of soot and dust would have spread through the atmosphere, blocking out sunlight for several years. Plants died because they could no longer photosynthesize, and algae in coral reefs also succumbed, disrupting food chains worldwide. The »

> We have very strong physical
> and chemical evidence for
> a large impact … the
> extinction coincides with
> the impact to a precision
> of a centimetre or better.
> **Walter Alvarez**

Luis Alvarez

Considered one of the greatest physicists of the 20th century, Luis Alvarez was born in San Francisco in 1911. He graduated from the University of Chicago in 1936 and went on to work at the Radiation Laboratory in the University of California, Berkeley. There he helped to develop nuclear reactors and, during World War II, nuclear weapons. He witnessed the atomic bombing of Hiroshima and helped to build a plutonium bomb.

After the war, Alvarez developed the liquid hydrogen bubble chamber, used to discover new subatomic particles. For this, in 1968 he was awarded the Nobel Prize for Physics. Later he provided the calculations to back up the Alvarez Hypothesis of mass extinction caused by a meteor strike. He died in 1988.

Key works

1980 "Extraterrestrial Cause for the Cretaceous–Tertiary Extinction", *Science*
1985 "The Hydrogen Bubble Chamber and the Strange Resonances"
1987 *Alvarez: Adventures of a Physicst*

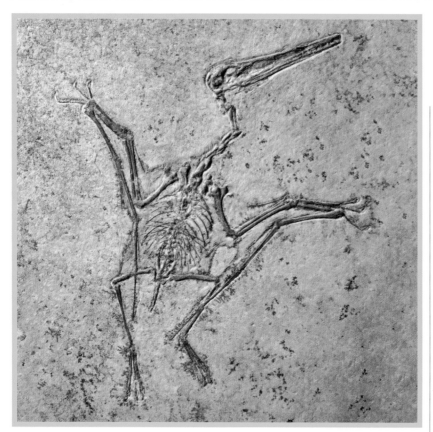

Although many flying dinosaurs survived the K-Pg mass extinction at the end of the Cretaceous period, all pterosaurs perished, ending their 162-million-year stay on Earth.

impact would have also released sulphuric acid into the atmosphere, which produced acid rain, acidifying the oceans and killing off marine life. Around the same time, a huge amount of volcanic activity flooded 500,000 sq km (193,000 sq miles) of southern India with lava, forming the Deccan Plateau and further changing the climate and atmosphere.

The K-Pg event is best known for the extinction of all non-flying dinosaurs. It was also responsible for the death of nearly all four-legged animals (tetrapods) that weighed more than 25 kg (55 lb). An exception were crocodiles, which may have survived because they are ectotherms (cold-blooded animals), able to survive for a long time without food. Dinosaurs were endotherms (warm-blooded animals),

with a fast metabolism that demanded regular meals. Many plant species died because they could not photosynthesize, leaving herbivorous dinosaurs with little vegetation to eat, while predatory species starved for lack of prey. In contrast, fungi, which do not depend on photosynthesis, proliferated.

In the oceans, phytoplankton, a vital food source that also relied on photosynthesis, died out. Creatures that fed on phytoplankton then faced extinction. These included cephalopods, such as belemnites and ammonites, and the marine reptiles known as the mosasaurs and the sauropterygians.

Marine annihilation

The earliest mass extinction, and the second-most catastrophic, occurred when our planet cooled

dramatically, towards the end of the Ordovician period, around 444 million years ago. At this time, most organisms on Earth lived in the oceans. As the supercontinent Gondwana drifted slowly over the South Pole, a giant ice cap formed, lowering global temperatures. Much of the planet's water became "locked up" as ice, depressing sea levels and reducing the area of Earth's surface covered by ocean.

As a result, marine organisms living in shallow continental-shelf water suffered particularly high rates of extinction. In at least two peak die-off periods, separated by hundreds of thousands of years, nearly 85 per cent of marine species died out, including brachiopods, bryozoans, trilobites, graptolites, and echinoderms.

Slow extinction

By the Late Devonian period, around 359 million years ago, the continents had been colonized by

The current extinction has its own novel cause: not an asteroid or a massive volcanic eruption but "one weedy species".
Elizabeth Kolbert
American journalist

plants and insects, and massive organic reefs thrived in the oceans. The continents of Euramerica and Gondwana were converging into what would become Pangaea – the last of the supercontinents. In this period, a succession of extinctions – possibly as many as seven – took place over a longer timescale than any other mass extinction event, possibly up to 25 million years.

The extinctions may have had many causes, including reduced oxygen in the oceans, falling sea levels, atmospheric changes, the draining of water produced by the spread of plants, and asteroid impacts. Most organisms lived in the oceans, and shallow seas were worst affected, with many reef-building organisms, brachiopods, trilobites, and the last of the graptolite species dying off. Around 75 per cent of marine species died, and it would be another 100 million years before corals re-established themselves on a large scale.

"The Great Dying"

The most dramatic mass extinction took place at the end of the Permian period, 252 million years ago. Also

> Modelled future extinction rates are projected to be 10,000 times Earth's historical geological background rate.
> **Ron Wagler**
> *American academic*

known as "The Great Dying", it resulted in the loss of 96 per cent of marine species and 70 per cent of land-living vertebrates. Insects suffered the only mass extinction in their history, and the last of the trilobites, which had been in decline for millions of years, disappeared from the fossil record.

Potential causes for the mass extinction include asteroid impact and oxygen depletion in the oceans. The extinction also coincided with one of the biggest periods of volcanic activity in

Earth's history. The eruptions, which lasted nearly 1 million years, flooded more than 2 million sq km (0.8 million sq miles) of ancient Siberia with basalt lava. The resulting build up of greenhouse gases would have transformed the atmosphere of Earth, likely resulting in severe global warming and contributing to species extinction.

Phased losses

All life today is descended from the small minority of species that remained at the start of the Triassic period. During the period's final 18 million years, ending about 201 million years ago, at least half of all animal species known to be living at that time were wiped out in two or three extinction phases. Climate change caused by more basalt eruptions and an asteroid impact have been cited as causes. In the seas, many reptiles, cephalopods, molluscs, and reef-building organisms died out. On land, most of the reptile-like archosaurs and many large amphibians became extinct. The loss of the archosaurs, in particular, opened up ecological niches that the dinosaurs would fill. ∎

The sixth extinction

Some ecologists have estimated the current rate of extinction of animals and plants at 100–1,000 times the natural background rate, with most of the increase due directly or indirectly to human activities. They argue that this is evidence the world is already in the middle of the Holocene extinction, named for the present geological epoch. Many species of animals and plants have been lost since the start of the Industrial Revolution in the 18th century

These losses have been driven by habitat change, climate change, overfishing, overhunting, ocean acidification, air pollution, and the introduction of animals that disrupt food chains. American ecologist E.O. Wilson, known as "the father of biodiversity", believes that if the species die-off continues at the present rate, half of all higher life forms will be extinct by 2100. Stuart Pimm, a British–American biologist and modern extinctions expert, is more cautious, claiming that we are on the cusp of such an event and can still act to stop it.

Sudan, the last male northern white rhinocerous, died in 2018 (two females remain). Poaching has taken the species to the edge of extinction.

BURNING ALL FUEL RESERVES WILL INITIATE THE RUNAWAY GREENHOUSE
ENVIRONMENTAL FEEDBACK LOOPS

IN CONTEXT

KEY FIGURE
James Hansen (1941–)

BEFORE
1875 In the book *Climate and Change,* Scottish scientist James Croll describes the climate-warming feedback effect of melting ice.

1965 Canadian biologist Charles Krebs discovers the "fence effect", showing vole populations protected from foxes rocketing, then crashing.

1969 American planetary scientist Andrew Ingersoll highlights the "runaway greenhouse effect" that caused the planet Venus to heat up.

AFTER
2018 Ecologists in Alaska predict that the accelerating release of methane from formerly frozen lakes will increase global warming.

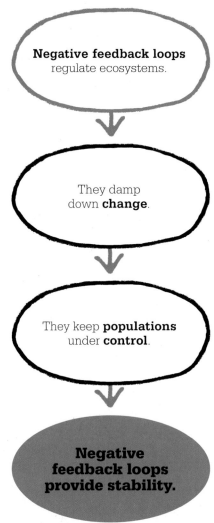

Negative feedback loops regulate ecosystems.

They damp down **change**.

They keep **populations** under **control**.

Negative feedback loops provide stability.

All parts of an ecosystem are interdependent. Any species or habitat change will feed back into the system, and affect the whole of that system, including the part where it all started. In other words, the feedback travels around in a loop.

In some situations, change is kept in check by the loop. For example, if aphids suddenly multiply, they provide more food for ladybirds, leading to an increase in the number of ladybirds. But with more ladybirds feeding on the aphids, aphid numbers drop again. This is negative feedback and it helps to maintain the status quo.

In other cases, feedback can accelerate change. Shrubs, for example, may begin to take over from grass on newly colonized land, casting their shade over the grass, depriving it of sunlight, and slowing its growth. The shrubs now have more water and nutrients, so they prosper at the expense of the grass. This is positive feedback and it is inherently destabilizing.

Ideas about feedback loops first developed early in the 20th century. They were based on the work of two mathematicians – Alfred Lotka (1880–1949) in the US and Italian

See also: Predator–prey equations 44–49 ▪ Competitive exclusion principle 52–53 ▪ Global warming 202–203 ▪ Halting climate change 316–321

Vito Volterra (1860–1940) – who independently devised equations based on the interaction between predators and prey. Their equations showed that a prey population will grow rapidly when the number of predators drops, while the predator population will drop when prey numbers drop, because the predators go hungry. The result is a constant cycle of falling and rising predator and prey populations.

Balancing the system

The predator–prey cycles identified by Lotka and Volterra were focused on the interaction between single predator and prey species. Since their studies, the theory of feedback loops has developed to embrace entire ecosystems. Ecologists now think that negative feedback loops are of central importance for the functioning of all ecosystems, keeping every part of them naturally within the bounds of sustainability. Populations can never swell for long beyond the carrying capacity of the rest of the system to support them. Thus, negative feedback regulates an ecosystem and keeps it stable.

In a healthy ecosystem, a repeating fluctuation in numbers between prey, such as rabbits, and predators, such as foxes, is an example of a negative feedback loop balancing the system.

Positive feedback interferes with a balanced ecosystem. If there is a surplus of resources, or a lack of predators, a population can grow freely. A bigger population leads to more births, and so an acceleration of the growth in population.

Equally, positive feedback can result in an accelerated contraction of a population. If fish stocks decline in a lagoon, for instance, local people may resort to importing canned food. Pollution from the dumps where the cans are thrown away can seep into the lagoon, killing the fish – and encouraging the locals to import even more of the damaging cans. And yet, positive feedback loops can sometimes set off a chain of events that becomes a "virtuous" circle. For example, if shrubs are planted in unstable soil, their roots may stabilize the soil, allowing both the shrubs and soil to thrive. ▪

Feedback loops and climate change

In recent years, accelerating and decelerating warming trends have brought the idea of feedback loops to the fore in climate change science. In 1988, the climate scientist James Hansen spoke at a US congressional hearing of the rises in global temperature caused by human activity. He has since voiced the belief that the continued burning of fossil fuels could set in train a series of calamitous positive feedbacks on Earth's climate, leading to the "runaway greenhouse" he describes in his 2009 book *Storms of My Grandchildren*.

One warming feedback loop is created by the melting of polar ice caps, as newly exposed land and water absorb the heat that the ice once reflected back into the atmosphere. The melting of Siberian permafrost is another warming loop. As temperature rises melt the permafrost, huge amounts of methane, a greenhouse gas, could be released into the atmosphere, accelerating global warming.

Arctic areas such as Greenland have seen a reduction in summer ice of 72 per cent since 1980. The warming of the atmosphere and rising sea levels are part of the resulting positive feedback loop.

THE HUM
FACTOR

AN

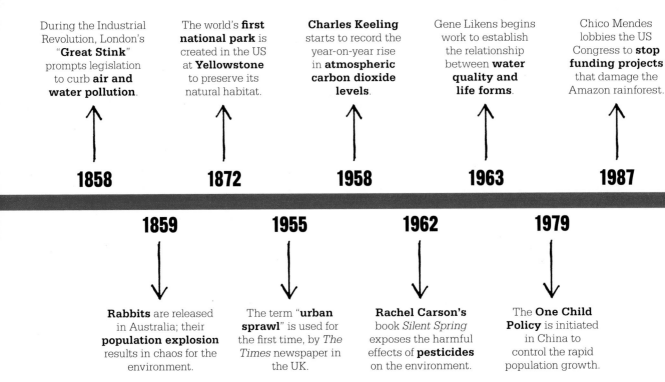

During the Industrial Revolution, London's "**Great Stink**" prompts legislation to curb **air and water pollution**.

1858

The world's **first national park** is created in the US at **Yellowstone** to preserve its natural habitat.

1872

Charles Keeling starts to record the year-on-year rise in **atmospheric carbon dioxide levels**.

1958

Gene Likens begins work to establish the relationship between **water quality and life forms**.

1963

Chico Mendes lobbies the US Congress to **stop funding projects** that damage the Amazon rainforest.

1987

1859

Rabbits are released in Australia; their **population explosion** results in chaos for the environment.

1955

The term "**urban sprawl**" is used for the first time, by *The Times* newspaper in the UK.

1962

Rachel Carson's book *Silent Spring* exposes the harmful effects of **pesticides** on the environment.

1979

The **One Child Policy** is initiated in China to control the rapid population growth.

R aw sewage produced by millions of Londoners once poured into the River Thames for decades, until the stench of the effluent became so bad that in 1858 action was demanded. When a new system of sewers, pumping stations, and treatment works revolutionized the city's sanitation, deaths and illness from cholera and other bacterial infections fell dramatically, and the river became much cleaner.

Human activity has always altered the environment, but its impact increased dramatically in the mid-18th century with the Industrial Revolution that began in Britain, and spread to Europe, North America, and beyond. The negative effects can be broadly divided into pollution, and destruction of resources and habitats.

Scottish-American environmentalist John Muir was one of the first to identify habitat degradation and destruction as problems, and in 1890 he won protection for the Yosemite Valley in California. However, despite a steady increase in protected natural environments, in the 20th century, the destructive pressures of human development have grown ever more powerful.

Trees and climate change

Forest has been especially hard-hit, mainly due to the dual demands of timber required for construction and fuel, and land cleared for agriculture and development. An estimated 140,000 sq km (54,000 sq miles) of tropical rainforest – which contains the greatest biodiversity – is cleared each year. Scientists will never

know how many forest-dwelling species died out before they could be "discovered".

Deforestation also contributes to global climate change. As trees photosynthesize, they absorb carbon dioxide and release oxygen. However, less forest means that more CO_2 stays in the atmosphere, fuelling the greenhouse effect and global warming.

Carbon and other greenhouse gases are emitted from cars and factories burning fossil fuels. Since 1958, American scientist Charles Keeling's measurements of atmospheric CO_2 have shown that CO_2 emissions are increasing at an ever-faster rate. While a minority of scientists maintain that human activity is not responsible, climate change has warmed the continents. The consequences, including trees

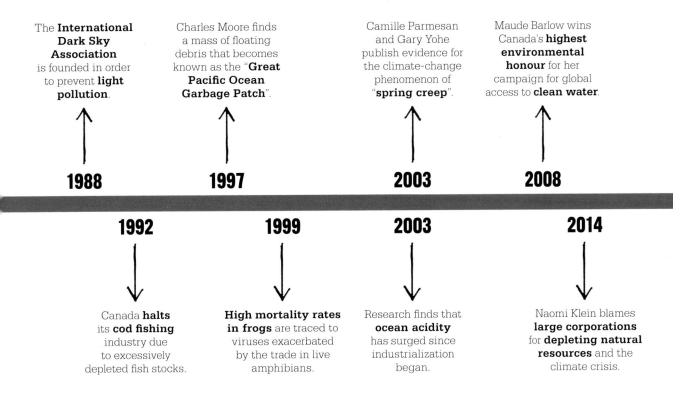

The **International Dark Sky Association** is founded in order to prevent **light pollution**.

Charles Moore finds a mass of floating debris that becomes known as the "**Great Pacific Ocean Garbage Patch**".

Camille Parmesan and Gary Yohe publish evidence for the climate-change phenomenon of "**spring creep**".

Maude Barlow wins Canada's **highest environmental honour** for her campaign for global access to **clean water**.

1988

1997

2003

2008

1992

1999

2003

2014

Canada **halts** its **cod fishing** industry due to excessively depleted fish stocks.

High mortality rates in frogs are traced to viruses exacerbated by the trade in live amphibians.

Research finds that **ocean acidity** has surged since industrialization began.

Naomi Klein blames **large corporations** for **depleting natural resources** and the climate crisis.

coming into leaf and flowers blooming earlier in spring, may benefit some organisms but could prove disastrous for others.

Toxic controls

The introduction of pesticides, such as DDT, to increase crop harvests proved to be an environmental disaster: they eradicated useful invertebrates as well as harmful ones; they caused cancers in humans; and rendered birds of prey infertile. Rachel Carson's 1962 book *Silent Spring* highlighted many of these issues, and caused a partial rethink of pesticide use. The work of several other ecologists has resulted in legislative controls to mitigate the environmental impact.

When Gene Likens and his team investigated why previously fish-rich lakes had died, they found that the culprit was acid rain, caused by emissions of sulphur dioxide and nitrogen oxide from industrial chimneys. As a result, legislation to control the emissions was passed in the US and Europe. After US chemists Frank Rowland and Mario Molina showed that chlorofluorocarbons (CFCs) destroy atmospheric ozone, the use of CFCs was banned worldwide in 1989.

Light pollution, which affects beach-nesting turtles, bats, and migrating birds, has proved harder to control. The International Dark-Sky Association is at the forefront of campaigns for environmentally responsible lighting.

Diminished resources

Garrett Hardin, an American ecologist, warned of the dangers of overpopulation in 1968, when the global population was 3.6 billion. By 2018, it had swollen to 7.6 billion, and although the growth rate has slowed considerably, the ever-increasing consumption of natural resources has led to depleted stocks of wood, fossil fuels, minerals, and even fish. The collapse of the once bountiful cod fishery off Newfoundland in 1992 highlighted the vulnerability of our food chain to overfishing and led the Canadian government to impose an indefinite moratorium on fishing on the Grand Banks.

Clean water is one of the most fundamental requirements for society but almost 1 billion people do not have access to it. A lethal combination of climate change and population growth in some developing regions threatens to increase this number. ∎

ENVIRONMENTAL POLLUTION IS AN INCURABLE DISEASE

POLLUTION

IN CONTEXT

KEY FIGURE
Emma Johnston (1973–)

BEFORE
1272 King Edward I of
England bans the burning
of sea coal in London because
of the smoke it produces.

19th century Coal-burning
during Britain's Industrial
Revolution stunts children's
growth and raises death rates
from respiratory diseases.

AFTER
1956 The Clean Air Act is
introduced in the UK, bringing
the thick smogs that plagued
its major cities to an end.

1963 The Clean Air Act
is passed in the US.

1972 The Clean Water Act
is fully ratified in the US.

1984 Toxic gas leaks from the
Union Carbide India factory
in Bhopal kill thousands and
injure many more.

Effects of pollution on health

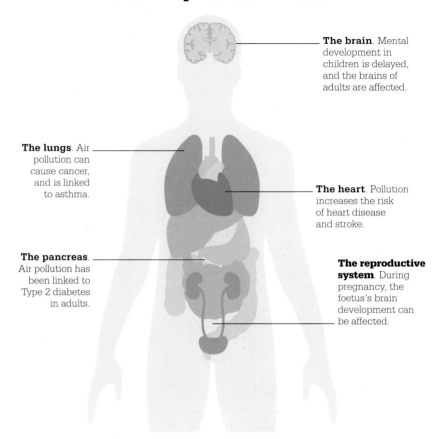

The brain. Mental
development in
children is delayed,
and the brains of
adults are affected.

The lungs. Air
pollution can
cause cancer,
and is linked
to asthma.

The heart. Pollution
increases the risk
of heart disease
and stroke.

The pancreas.
Air pollution has
been linked to
Type 2 diabetes
in adults.

**The reproductive
system**. During
pregnancy, the
foetus's brain
development can
be affected.

Polluted air and water cause the deaths of millions of
people every year. This illustration describes the specific
damage caused to different organs of the human body.

Pollution comes in many
forms, ranging from toxins
in the air to rubbish at the
bottom of the sea. Any substances
or forms of energy that spoil the
quality of the atmosphere, oceans,
water, or soil are pollutants. They
may be chemicals or biological
contaminants (including human
waste), products (such as plastics),
or noise, light, or heat. The effects
of pollution on life of all kinds can
be far-reaching, spreading
thousands of miles beyond its
original source. Pollution can
spread through the food chain and

be carried through air and water,
affecting all life. Contaminants such
as plastics can facilitate the invasion
of non-indigenous species, as
discovered by Australian marine
biologist Emma Johnston. There is
also a direct effect on human health:
it is estimated that exposure to
polluted air, water, and soil caused
9 million premature deaths – one
in six of all deaths – in 2015.

Pollution through the ages
Human-made pollution has a long
history. The presence of soot on
the walls of caves, dating back

thousands of years, indicates
that early humans generated air
pollution from their fires. Analysis
of 2,500-year-old ice cores in
Greenland has shown evidence of
air pollution from copper smelting
thousands of miles away, in the
centre of the Roman Empire.
However, such impacts were
on a small scale. With the start
of the Industrial Revolution in
Europe, air and water pollution
became serious. Factory chimneys
pumped smoke out into the air;
toxic chemicals poured into rivers.
Cities expanded quickly and had

See also: Pesticides 242–247 ▪ Acid rain 248–249 ▪ Light pollution 252–253 ▪ A plastic wasteland 284–285 ▪ The water crisis 286–291 ▪ Waste disposal 330–331

> Air pollution control systems still lag behind economic development.
> **Bob O'Keefe**

no sanitation. The River Thames, in London, was both the source of water for domestic use and the outlet for untreated human sewage. Disease spread, river fish were wiped out, and the smell was sometimes unbearable. Other urban

Of the world's 20 worst cities for air pollution, 14 are in India. In Delhi, thick smog in November 2017 reduced the air quality to the equivalent of smoking 50 cigarettes a day.

centres fared little better: similarly unsanitary conditions were recorded in Berlin in 1870, for example.

In the United States, the first two cities to enact laws to ensure clean air were Chicago and Cincinnati, in 1881. By that time, the manure from 3 million horses pulling wagons in North American cities was seeping into water supplies and producing plagues of disease-causing flies. As horses were gradually replaced by the internal combustion engine, smog from cars and trucks became a major issue. London's Great Smog of 1952, described as a "pea-souper" for the colour of the filthy air, killed more than 4,000 people.

Air pollution
The result of harmful substances being released into the atmosphere, such as gases or small particles called aerosols, air pollution can have natural sources, such as volcanoes or wildfires, but is mainly caused by human activity. The main air pollutants are emissions »

The "Great Stink"

By the early 19th century, London's Thames was the most polluted river in the world. Industrial pollution and human effluent emptied into it from thousands of drains. People complained, but the government did nothing. In 1855, the scientist Michael Faraday lambasted politicians for their inaction, to no avail. However, they got the message three years later, when a hot summer contributed to the "Great Stink" of 1858. The Houses of Parliament, being adjacent to the Thames, were badly affected, and legislation was suddenly enacted in a mere 18 days.

Civil engineer Joseph Bazalgette was commissioned to design a new sewage system. It was based on six interceptor sewers, 160 km (100 miles) long, which flowed to new treatment works. Most of London was connected to it within a decade. Much of the sewage system is still in operation today, more than 150 years later.

This cartoon, published in *Punch* magazine in July 1858, was entitled "The Silent Highwayman". People at the time attributed the spread of cholera to the bad river smells.

> Pollution is one of the biggest problems we are facing globally, with horrible future costs to society.
> **Maria Neira**
> *World Health Organization*

from fossil-fuel-burning power stations, factories, motor vehicles, the burning of wood and dung for heat and cooking fuel, and methane from cattle, landfill sites, and fertilized fields. Poor air quality damages human health and crops, and some fossil-fuel emissions cause acid rain, which has killed forests and fish in thousands of lakes.

Orcas may become extinct as a result of PCB (polychlorinated biphenyl) pollutants. The compound becomes more concentrated higher in the food chain, and orcas are apex predators.

The World Health Organization (WHO) estimates that nine out of ten people worldwide are breathing polluted air, causing widespread illness and allergies. Furthermore, some aerosols, depending on the composition and colour of the particles, block the amount of solar radiation reaching Earth's surface, thus having a cooling effect on the planet. Efforts to reduce air pollution can therefore make the effects of global warming worse.

Rivers, lakes, and seas
Surface water, groundwater, and the oceans become contaminated by toxic chemicals from industry, from chemical run-off from farmland, from general rubbish such as plastics, and from human waste.

Some rivers and lakes are so polluted that they can support no life at all, deprive communities of fresh water and food, and carry a risk of waterborne diseases, such as polio, cholera, dysentery, and typhoid. The WHO estimates that 2 billion people worldwide are drinking water contaminated with human waste, resulting in the deaths of 500,000 people a year.

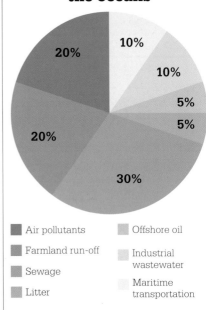

Pollutants entering the oceans

- Air pollutants
- Farmland run-off
- Sewage
- Litter
- Offshore oil
- Industrial wastewater
- Maritime transportation

In the oceans, the most acutely destructive pollution has resulted from disasters involving oil tankers and oil terminals. When the *Exxon Valdez* supertanker broke up on rocks off the coast of Alaska in 1989, 50 million litres (11 million gallons) of crude oil were released into the North Pacific. The oil smothered or

Emma Johnston

Born in 1973, Australian marine biologist Emma Johnston was interested in the oceans from an early age. She gained her PhD in marine biology in 2002 and, in 2017, became the Dean of Science at the University of New South Wales (UNSW), and Head of the UNSW's Applied Marine and Estuarine Ecology Lab, which investigates human impacts on marine ecosystems.

Johnston discovered how non-native species can invade waterways in coastal areas by adhering to rafts of plastic pollution floating on the oceans.

She has also studied marine communities in the Antarctic, developed new biomonitoring techniques, and advised agencies on the management of estuarine biodiversity.

Key works

2009 "Contaminants reduce the richness and evenness of marine communities", *Environmental Pollution*
2017 "Building 'blue': an eco-engineering framework for foreshore developments", *Journal of Environmental Management*

---

poisoned an estimated 250,000 seabirds, 2,800 sea otters, 300 harbour seals, 250 bald eagles, and 22 killer whales. Billions of salmon and herring eggs also died. Further catastrophic damage followed in 1991, during the Iraq War, when Iraqi forces opened the valves of an offshore oil terminal and released at least 1,700 million litres (380 million gallons) into the Persian Gulf. The long-term effects of such disasters are still unfolding and have yet to be fully understood.

Much of our non-degradable products ends up in the oceans. Since the 1950s, around 8.3 billion tonnes of plastic has been produced, of which only a fifth has been recycled or incinerated. Each year, a staggering 8 million tonnes of plastic reaches the oceans, and is responsible for the deaths of huge numbers of marine animals.

Intangible pollutants

Pollution in the form of energy, be it light, noise, or heat, can be just as intrusive as physical waste or chemical emissions. Light pollution from buildings, streetlights, vehicles, and advertising hoardings was first described as a problem in New York in the 1920s. It can cause problems for nocturnal wildlife, for example, because predator–prey relations are interrupted. Excessive noise can be highly disturbing in cities, on flight paths, and near factories and roads. But it also affects wildlife in subtler ways. There is evidence that some birds now sing at night because their song can be heard more clearly than during the day.

Waste heat, too, can be damaging. When water from rivers or the sea is used as a coolant in factories or power stations, the hot water that is returned to the source

In 2015, pollution caused three times as many deaths as AIDS, tuberculosis, and malaria combined.
Philip Landrigan

is a form of thermal pollution. It can kill fish and alter the composition of the food chain, reducing biodiversity.

Nuclear energy is sometimes viewed as "cleaner" than fossil-fuel energy, because it does not produce greenhouse gases, but it does result in waste that remains radioactive for thousands or millions of years. The industry also bears the inherent risk of accidental damage. An explosion at the Chernobyl nuclear power plant in Ukraine in 1986 killed dozens of people and spread radiation across Western Europe. The slowly dwindling effects of contamination on the ecosystem and human health are predicted to last a century.

Mitigation measures

Tackling the problem of pollution is a huge challenge, and involves both cleaning up existing pollution and making changes to reduce the rate at which we add to it. Key aspects of this include replacing fossil fuels with sustainable energy, more recycling and reuse, and the replacement of non-degradables with degradable materials. This will take time and, ultimately, demands a fundamental shift in our culture of consumption. ∎

GOD CANNOT SAVE THESE TREES FROM FOOLS

ENDANGERED HABITATS

IN CONTEXT

KEY FIGURE
John Muir (1838–1914)

BEFORE
1872 Yellowstone, in the states of Wyoming, Montana, and Idaho, is declared a national park – the first in the world.

AFTER
1948 The International Union for Conservation of Nature (IUCN), a partnership of governments and civil society organizations, is founded.

1961 The World Wide Fund for Nature (WWF), initially known as the World Wildlife Fund, is formed, to protect endangered species and habitats.

1971 The Man and the Biosphere Programme (MAB) is founded by the United Nations, to promote sustainable development. It has a global network of Biosphere Reserves.

T he origin of the movement to conserve natural habitats is usually credited to the Scottish–American naturalist John Muir, described as the "father of the national parks". He was one of the first to realize that in order to survive, wild places needed legal protection. Of the many types of natural habitat on Earth, some are more fragile than others, but each faces different threats, whether anthropogenic (human-made) or from natural causes, or both, and many are critically endangered.

Habitats have, of course, always been affected by destructive natural events. Every year, lightning strikes trigger large grassland and forest

See also: Human activity and biodiversity 92–95 ▪ Biodiversity hotspots 96–97
▪ Biomes 206–209 ▪ Deforestation 254–259 ▪ Environmental ethics 306–307

Yosemite National Park was created in 1890, thanks to the efforts of John Muir. The park is famed for its glaciers, waterfalls, and granite rock formations, such as the El Capitan monolith.

fires. Hurricanes and rivers in flood can wreak havoc. Storm surges may produce inundations of the sea, turning freshwater wetlands saline. About 66 million years ago, the Chicxulub meteor impact in Mexico produced a dust cloud so great that it stopped sunlight from reaching Earth's surface. Plants struggled to photosynthesize, and many animals, including the dinosaurs, became extinct.

Nor is human influence an exclusively recent issue. Throughout history, people have modified their environment. Deforestation, for example, is not a new problem. In Europe, the clearance of forests for agriculture and construction began thousands of years ago, and a similar pattern followed in North America.

However, the impact of modern-day humans on the environment is unprecedented. In the past 200 years, the human population has exploded. This has fuelled the rapid growth of cities, the development of large-scale industry based on the extraction of fossil fuels and raw materials, a growing agricultural demand to feed more people, and conflict and war. All these have taken their toll on the natural world.

Fragile ecoregions

A concept that is now often used to identify the major habitat types on Earth is that of the ecoregion – smaller than a biome, with a more detailed gauge of biodiversity. Ecoregions are defined as large units of land or water containing a geographically distinct mix of species, natural communities, and environmental conditions. Some examples include deserts, tropical »

John Muir

Born in Scotland in 1838, John Muir developed a passion for nature as a boy. He moved with his family to Wisconsin, US at the age of 11. In 1867, he had an accident in which he lost his sight temporarily, after which he "saw the world in a new light". An accomplished botanist, geologist, and glaciologist, Muir visited the Yosemite Valley in California in 1868, and later determined to preserve it from the scourge of domestic sheep (which he called "hoofed locusts"). In 1903, Muir took President Theodore Roosevelt on a guided tour through the Yosemite Valley, and their three-day trip inspired Roosevelt to create the US Forestry Service and, in 1916, to form the National Conservation Commission. Until his death in 1914, Muir continued to advocate for the conservation of land such as Mount Rainier, which became a National Park in 1899.

Key works

1874 *Studies in the Sierra*
1901 *Our National Parks*
1911 *My First Summer in the Sierra*

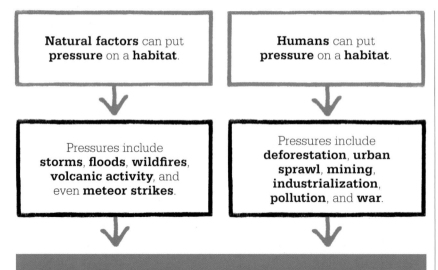

| Natural factors can put pressure on a habitat. | Humans can put pressure on a habitat. |

↓ ↓

| Pressures include **storms**, **floods**, **wildfires**, **volcanic activity**, and even **meteor strikes**. | Pressures include **deforestation**, **urban sprawl**, **mining**, **industrialization**, **pollution**, and **war**. |

↓ ↓

The habitat becomes endangered

rainforests, temperate coniferous forests, lakes, mangrove swamps, and coral reefs. Of these, coral reefs and tropical rainforests are under particular threat from humankind.

Rainforest clearance
Despite covering only 6 per cent of Earth's land surface, tropical rainforests represent the greatest biomass of any terrestrial ecoregion, and are home to about 80 per cent of land species. Every year, some 140,000 sq km (54,000 sq miles) of tropical rainforest is cleared – the equivalent of a football pitch every second. Logging is carried out for firewood and construction materials, and is also driven by the demand for roads, settlements, and agriculture.

Globally, the rainforests that are under greatest threat are in West Africa, Central America, and southeast Asia. Indeed, only about 30 per cent of the lowland rainforest in Borneo now survives . In the Amazon Basin, home to nearly one-third of the world's rainforest, much of the clearance is for agriculture, especially ranching.

Once deforestation starts, the problem quickly gets worse. When rain falls on a forested slope, it is mostly absorbed by vegetation. But when the slope is cleared, the rainfall erodes the soil, making it useless for agriculture and impossible to replant. It produces silt run-off into rivers and lakes, killing fish, and increases the risk of flooding. The destruction of any forest reduces its capacity to

absorb the greenhouse gas carbon dioxide, thereby contributing to the acceleration of climate change.

Loss of coral reefs
Coral reefs are important ecoregions and yet are especially endangered. They support about 25 per cent of the planet's marine species, and are also nurseries for billions of fish. Two-thirds of the world's reefs are under threat, and about a quarter of them are likely damaged beyond repair. Possibly the biggest threat to coral reefs is increased acidity caused by a greater uptake of CO_2 from the atmosphere. This impedes the ability of many sea creatures to build their shells, and induces coral "bleaching", which is a step on the way to the reef dying. In addition, coral reefs are being destroyed by overfishing, and by harmful practices such as cyanide and blast fishing, and bottom trawling. Sediment resulting from coastal development blocks the sunlight that reefs need. Chemical pollution, coral mining, and careless tourism all add to the burden on this highly sensitive habitat.

Wide-ranging impacts
All over the world, diverse natural habitats are critically threatened by human activity. Tropical deciduous dry forest is easier to clear than rainforest, and on Madagascar, where dry forest was widespread, less than 8 per cent now remains. At one time, tallgrass prairie stretched across the US Midwest, but only 3 per cent of it is left: the rest has been converted to farmland. Many wetlands have been drained for

Palm oil trees are being planted on a large scale in Indonesia and Malaysia, where this is one of the main drivers of deforestation. Orangutans are among the species endangered as a result.

Protected areas

National parks, wilderness areas, nature reserves, and sites of special scientific interest (SSSIs) are all types of protected habitats. Within these areas, interference with the natural environment is prohibited or limited by some kind of legal framework. They must cover a specified expanse of land or sea, but they vary greatly in size and in the level of protection given. Just over 10 per cent of Earth's land is protected, but only 1.7 per cent of the oceans; although marine reserves are essential, they require local and national governments to agree on key issues such as fishing rights.

Marae Moana, the largest protected area on Earth, covers 2 million sq km (772,000 sq miles) around the Cook Islands in the Pacific Ocean. It is home to sea turtles, at least 136 species of corals, and 21 kinds of whales and dolphins. The biggest land reserve is the Northeast Greenland National Park, which covers almost 1 million sq km (386,000 sq miles) of ice sheet and tundra.

Wetlands and intertidal zones are important for marine invertebrates and migratory shorebirds, but in many parts of the world they have been drained for industry and ports.

agriculture or urban development; others are irreversibly damaged by pollution. Nutrient run-off from agricultural fertilizers has spoiled many lakes and rivers. In many countries, intertidal zones have been destroyed by the building of ports. Coastal development has been largely responsible for the loss of 35 per cent of mangrove swamps. In the tropics and subtropics, overgrazing by domestic animals such as goats has converted an estimated 9 million sq km (3.5 million sq miles) of seasonally dry grassland and scrub into desert.

Halting the decline

The destruction of these habitats is not only a loss in terms of natural beauty and biodiversity, but also creates serious problems for people: for example, poorer water quality, declining fish stocks, crashes in populations of pollinators, flooding from increased rainwater run-off, and faster build-up of greenhouse gases. Conservation is now paramount, and ecologists work to refine their understanding of the best ways to go about it.

Appropriate measures depend on the situation, and range from the creation of protected reserves or "corridors", to link areas that have become fragmented, to schemes to recreate lost habitat. Sustainable sources of fuel and timber for those who are otherwise dependant on forest wood are also important, as is banning the trade in rainforest hardwood. Since the impact of habitat destruction is global, international agreements and cooperation are crucial. ■

In every walk with nature, one receives far more than he needs.
John Muir

Muskoxen are Arctic herd animals whose numbers were severely depleted in the 19th century by hunting. They now live on reserves in Alaska, Norway, and Siberia.

WE ARE SEEING THE BEGINNINGS OF A RAPIDLY CHANGING PLANET
THE KEELING CURVE

IN CONTEXT

KEY FIGURE
Charles Keeling (1928–2005)

BEFORE
1896 The Swedish chemist Svante Arrhenius is the first to estimate the extent to which atmospheric CO_2 could increase Earth's temperature.

1938 Comparing historic temperature data and CO_2 measurements, the British engineer and scientist Guy Stewart Callendar concludes that the increase in CO_2 is responsible for the warming of the atmosphere.

AFTER
2002 The European Space Agency's ENVISAT satellite begins to produce up to 5,000 readings of greenhouse gases every day.

2014 NASA's Orbiting Carbon Observatory generates up to 100,000 high-precision measurements daily.

The Keeling Curve, named after Charles Keeling, an American scientist, charts the daily record of atmospheric carbon dioxide (CO_2), measured in parts per million by volume (ppmv), in a series dating back to 1958. It shows two things: the natural seasonal respiration of Earth and the year-on-year rise in atmospheric CO_2. Atmospheric CO_2 is significant because carbon dioxide is the most important of the greenhouse gases, which trap warmth in Earth's atmosphere. More molecules of CO_2 and other greenhouse gases cause

We were witnessing for the first time nature's withdrawing CO_2 from the air for plant growth during summer and returning it each succeeding winter.
Charles Keeling

more heat to be trapped, leading to an overall increase in temperature and global climate change.

Measuring CO_2 levels
Since the start of the Industrial Revolution in the late 1700s, human activity has produced increasing emissions of CO_2. This is largely due to burning fossil fuels, while forest clearance for agriculture and development has resulted in less vegetation absorbing CO_2 through photosynthesis. Many scientists once believed that excess CO_2 would be absorbed by the oceans. Others disagreed, but there was little hard evidence either way.

Charles Keeling was not the first to propose a link between atmospheric warming and CO_2 emissions. Others had measured CO_2 levels but had produced only "snapshots" in time rather than a long-term dataset. Keeling knew that a long study was needed to prove the link. In 1956, he took up a post at the Scripps Institution of Oceanography in San Diego, California, and obtained funds to establish CO_2 monitoring stations at remote locations 3,000m (9,843ft) up on Mauna Loa, Hawaii, and at the South Pole. By 1960, Keeling

See also: Global warming 202–203 ▪ The biosphere 204–205 ▪ Environmental feedback loops 224–225 ▪ Halting climate change 316–321

CO₂ analysis in icecaps

Scientists can measure past concentrations of carbon dioxide by analysing bubbles of air trapped in Antarctic and Greenland ice sheets. This evidence indicates that there have been several cycles of variation over the past 400,000 years. These range from lower readings in the most severe glaciations – when glaciers actually formed – to higher readings during warmer, interglacial periods. The increase since the start of the Industrial Revolution has been matched by the average global temperature. This has risen by 0.07°C (0.13°F) per decade since 1880 and 0.17°C (0.31°F) per decade since 1970.

The Intergovernmental Panel on Climate Change (IPCC) warns that unless the world's governments reduce greenhouse gas emissions dramatically, by the year 2100 average temperatures could be around 4.3°C (7.7°F) higher than they were before the Industrial Revolution. Such an increase would cause both a marked rise in sea levels and more extreme weather, which would result in people having to abandon some regions of the world entirely.

Mauna Loa in Hawaii is an ideal site for an atmospheric research station. The high altitude and remote location of the volcano ensure that the air is largely unaffected by humans or vegetation.

was sure that he had a long enough series of records to detect a year-on-year increase.

Seasonal changes

Although South Pole funding ended in 1964, Mauna Loa has produced data from 1958 onwards. Plotted on a graph, the measurements became known as the Keeling Curve. It is, in fact, a series of annual curves, reflecting seasonal changes. During spring and summer in the northern hemisphere, as new foliage takes more CO₂ out of the atmosphere, the global concentration of the gas declines, reaching a low point in September. It increases again in the northern autumn as leaves fall and photosynthesis declines. Plant growth in the southern hemisphere later in the year does not make up for the loss, because most of Earth's vegetative cover is in the north.

Ancient air bubbles trapped in polar ice cores reveal that, over the past 11,000 years, average CO₂ concentrations were 275–285 ppmv, but increased sharply from the mid-19th century. In 1958, the level was 316 ppmv. It rose steadily at a rate of 1.3–1.4 ppmv each year until the mid-1970s, then increased by about 2 ppmv each year. By spring 2018 it had hit 411 ppmv, almost 1.5 times higher than pre-industrial levels. ▪

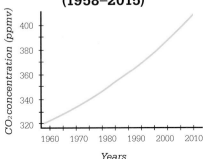

Mauna Loa CO₂ record (1958–2015)

CO₂ concentration (ppmv)

400
380
360
340
320

1960 1970 1980 1990 2000 2010

Years

The Keeling Curve of steadily rising CO₂ levels is clearly shown on a graph plotting results from the continuous monitoring of atmospheric carbon dioxide (CO₂) at Mauna Loa, Hawaii.

Bubbles in an ice core provide a sample of the atmosphere going back centuries. Scientists measure the CO₂ in the trapped air bubbles.

THE CHEMICAL BARRAGE HAS BEEN HURLED AGAINST THE FABRIC OF LIFE

THE LEGACY OF PESTICIDES

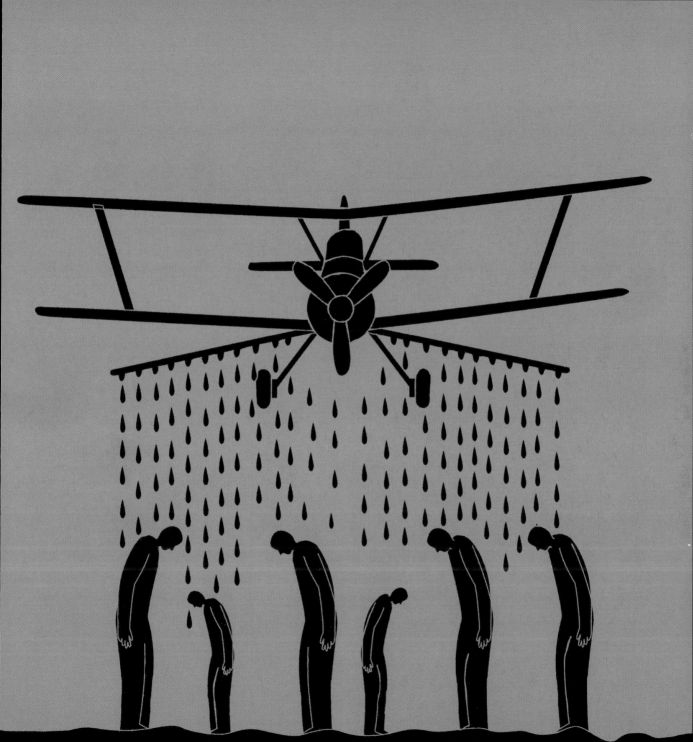

IN CONTEXT

KEY FIGURE
Rachel Carson (1907–64)

BEFORE
1854 Henry David Thoreau's book *Walden* describes a social experiment to live the simple life in tune with nature. It is seen as an inspiration for the environmentalist movement.

1949 *A Sand County Almanac* by Aldo Leopold proposes a deep ecology of people living in harmony with the land.

AFTER
1970 The USA establishes the Environmental Protection Agency (EPA).

1989 Bill McKibben's book *The End of Nature* highlights the dangers of global warming.

2006 The documentary *An Inconvenient Truth* records ex-US vice president Al Gore's efforts to educate the public about climate change.

rguably the most revered and influential book ever published on the subject of environmentalism, *Silent Spring* garnered a huge amount of publicity when it was released in 1962. It galvanised the fledgling conservation movement, forced legislative change, and, perhaps most significantly, championed the right of the public to question those in power and hold them to account.

However, the author of this ground-breaking work was far from the typical "eco-warrior" – a term that was unheard of when the book was first published. On the contrary, Rachel Carson was a quiet, scholarly woman, with a masters degree in zoology and 20 years' service as an aquatic biologist in the United States. Most of all, she was an exceptional writer, able to fuse scientific fact with compelling narrative.

Dying wildlife

Like many great and influential works, *Silent Spring* began in a very personal way. In January 1958, Carson's friend Olga Huckins sent her a letter that she had originally tried to have published in the

Spraying insecticide such as DDT whether indoors or outside, has been – and in some places still is – a common method of controlling the mosquitoes that transmit malaria.

Boston Herald. It spoke about aerial spraying of a mixture of fuel oil and a chemical compound named DDT (Dichloro-diphenyl-trichloroethane), in the vicinity of her small bird sanctuary in Michigan. The morning after the spraying, Huckins found several birds dead on her property and hoped that Carson might know someone in Washington who could stop further spraying. Carson was outraged and resolved to help. For more than a decade she had been aware of troubling incidents in which

Rachel Carson

Born in 1907, Rachel Carson grew up on a farm in Pennsylvania, where she developed a love of nature. She won a scholarship to Pennsylvania College for Women and later gained a masters in zoology. Growing up in a land-locked state, Carson dreamed of the ocean; it became an enduring passion, and she went to work as an aquatic biologist with the US Fish and Wildlife Service (FWS).

Carson wrote and published many educational brochures and eventually became the US Fish and Wildlife Service's editor-in-chief. From 1941 onwards, she

wrote books about marine biology, most notably *The Sea Around Us*, which won the National Book Award, and was a national best seller. This success enabled Carson to write full time and she began work on *Silent Spring* in 1958. In 1960, Carson was diagnosed with breast cancer; she died in 1964.

Key works

1941 *Under the Sea Wind*
1951 *The Sea Around Us*
1955 *The Edge of the Sea*
1962 *Silent Spring*

See also: Human activity and biodiversity 92–95 ▪ Animal ecology 106–113 ▪ The ecosystem 134–137 ▪ A holistic view of Earth 210–211 ▪ Human devastation of Earth 299 ▪ Environmental ethics 306–307

indiscriminate spraying of DDT had been killing wildlife. Carson swiftly approached the editor of the *New Yorker*, E.B. White, suggesting that the magazine run a piece about the growing concern around synthetic pesticides and their effect on non-target organisms. The editor suggested she write the article herself. Reluctantly, Carson began research on what she at first called "the poison book". It went on to shake the world.

The chemical future

Silent Spring's impact needs to be seen against the backdrop of the time in which it was published. Although academics and scientists had already voiced concerns about synthetic pesticides, the public was oblivious to this issue.

Synthesized pesticides had been in use since the 1920s but had advanced significantly during World War II, powered by military-funded research. During the 1950s, the popular notion was that they could solve the world's problems of famine and sickness by killing pests that destroyed crops and transmitted disease. Advertising campaigns of chemical giants such as Union Carbide, DuPont, Mobil, and Shell spread this message to a huge audience. *Silent Spring* aimed to challenge the received wisdom, arguing that the so-called scientific progress enjoyed in post-war America would come at a huge price for the environment.

The most notorious of the pesticides, and the one most associated with *Silent Spring*, was DDT. It was first synthesized in the late 19th century, but in 1939, Swiss chemist Paul Hermann Müller realized that it could be

used to kill a wide range of insects, due to its pervasive action as a nerve poison. It was used during World War II to control insects that destroyed vital food crops as well as those which transmitted malaria, typhus, and dengue fever to combat troops.

DDT proved cheap to produce, highly effective, and at first appeared to pose no threat to human beings. After the war, with the chemical in plentiful supply, its use in agriculture was an obvious next step. With its wide range of apparently safe applications, it must have seemed like a panacea to farmers, who happily sprayed it on their crops, often without the use of masks or protective clothing, because they did not fully appreciate the powerful toxicity of this dangerous chemical compound.

After DDT came a whole host of synthetic agrochemicals, including aldrin, dieldrin, endrin, parathion, malathion, captan, and 2,4-D. Used in conjunction with fertilizers made out of surplus nitrogen that was no longer needed to make explosives, these chemicals enabled the »

No one since [Silent Spring] would be able to sell pollution as the necessary underside of progress so easily.
H. Patricia Hynes

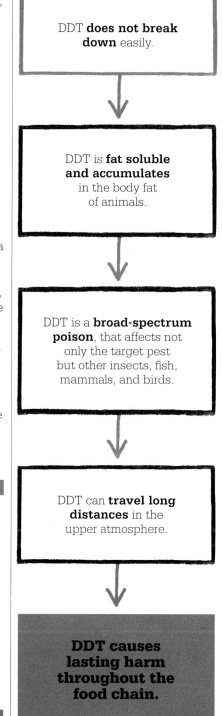
DDT **does not break down** easily.

DDT is **fat soluble and accumulates** in the body fat of animals.

DDT is a **broad-spectrum poison**, that affects not only the target pest but other insects, fish, mammals, and birds.

DDT can **travel long distances** in the upper atmosphere.

DDT causes lasting harm throughout the food chain.

A persistent poison

DDT (Dichloro-diphenyl-trichloroethane) belongs to a group of pesticides called organochlorides. It kills insects on contact by interfering with their nerve impulses. The compound is fat soluble and is deposited in the tissues of animals exposed to it, either directly or by eating contaminated food. Repeated exposure to DDT results in it building up in the body fat and becoming toxic.

DDT also biomagnifies up the food chain. Humans are susceptible to poisoning from regular exposure to DDT and while the effects of small amounts in the environment are unknown, it has been associated with cancer, infertility, miscarriage, and diabetes. It is now banned in western countries, but studies carried out by the US Center for Disease Control in 2003–4 found DDT or its breakdown product (DDE) in the blood of 99 per cent of people tested.

DDT biomagnification in the food chain

TERTIARY CONSUMERS	13.8 PPM
SECONDARY CONSUMERS	2.07 PPM
PRIMARY CONSUMERS	0.23 PPM
PRODUCERS	0.04 PPM

Organisms higher in the food chain suffer from the impacts of DDT the most. In producers, the poison only represents 0.04 PPM (parts per million), but the concentration increases with each step up the food chain. By the time tertiary consumers are involved, levels are high enough to have toxic effects.

A spray as indiscriminate as DDT can upset the economy of nature… Ninety per cent of all insects are good, and if they are killed, things go out of kilter right away.
Edwin Way Teale

intensification of farming. The chemical age had dawned and by 1952, there were almost 10,000 separate new pesticide products registered with the US Department of Agriculture (USDA).

Raising awareness

Carson was not the first person to notice the harmful effects of DDT. There were a few early dissenters, including nature writer Edwin Way Teale, who warned that a spray with the indiscriminate impact of DDT could upset the balance of nature. In 1945, the director of the US Fish and Wildlife Service (FWS), Dr Clarence Cottam,

stated that caution was essential in the use of DDT because the true impact of the product was not yet fully understood. The following year, Fred Bishop, writing in the *American Journal of Public Health,* stressed that DDT must not be allowed to get into foods or be ingested by accident.

Various scientific studies and reports also raised concerns. For example, in 1945 the US government published a study that found traces of DDT in the milk of cows sprayed with the chemical. It recommended that farmers use "safe alternative substitute insecticides" to control flies and lice on cattle. Carson's

> They should
> not be called
> insecticides
> but biocides.
> **Rachel Carson**

longstanding position as editor-in-chief at the FWS until 1952 meant that she had access to a great many of these reports; she found them to be very disturbing reading.

Since the research was rather scattered and by no means accessible for the general reader, Carson resolved to gather what material she could find and present it in a way that the ordinary non-scientist could understand. As she made progress with the writing of *Silent Spring*, it became clear to her that she had a moral duty to make the information public. As well as documenting the hazards of indiscriminate pesticide use, Carson dared to suggest that the chemical companies were putting profits before people and that the government might even be colluding with them, knowingly or otherwise, by failing to regulate the industry effectively.

The response from the US chemical industry was predictable. At first, they tried to sue Carson, her publishers, and *The New*

After DDT was banned in many countries, osprey populations – which had declined significantly from the 1940s – began to recover. Ospreys had eaten small animals affected by DDT.

Yorker – which had published a serialization of the book. However, Carson was prepared for this kind of response. She knew the book would be controversial and seen as threatening by the chemical industry. Therefore, as well as meticulously tracking and recording her research – which had been gained from government bodies, her contacts in research institutions, and other reputable sources – she also had the manuscript reviewed by scientists and experts.

When suing Carson did not work, the chemical companies launched a campaign to bring her into disrepute, stooping to personal attacks such as depicting Carson as a "hysterical" cat-loving woman, who was ill-equipped to write such a book. The smear campaign backfired, merely increasing the sales of *Silent Spring*.

New policies

Notable scientists supported Carson's findings and US President John F. Kennedy invited her to testify before a Congressional Committee in 1963. She called for new policies that would serve to

> Man is part of nature
> and his war against
> nature is inevitably a
> war against himself.
> **Rachel Carson**

protect the environment. The Committee released a report entitled "The Uses of Pesticides", which broadly supported Carson's book. Inspired by Carson, activists continued to lobby government until in 1972, a decade after *Silent Spring* was first published, DDT was banned in the US. Other countries followed, although some retain it to control mosquitoes.

The legacy of *Silent Spring* is greater than the banning of DDT. It demonstrated to industry giants and government the power of an educated public. ∎

A LONG JOURNEY FROM DISCOVERY TO POLITICAL ACTION
ACID RAIN

IN CONTEXT

KEY FIGURE
Gene Likens (1935–)

BEFORE
1667 The corrosive effect of polluted city air on limestone and marble is noted by the English diarist John Evelyn.

1852 British chemist Robert Angus Smith argues that industrial pollution causes the acidic rainfall that damages buildings. He is the first person to call it "acid rain".

AFTER
1980 The US Congress passes the Acid Deposition Act, undertaking an extensive 18-year research programme into acid rain.

1990 An amendment to the US Clean Air Act (passed originally in 1963) establishes a system that is designed to effectively control emissions of sulphur dioxide and nitrogen oxide.

The effects of acid rain on stone were noticed as long ago as the 17th century in England, and in Norway in the 19th century. However, it was not until American ecologist Gene Likens carried out in-depth studies in an area of rural New Hampshire that the phenomenon came to be properly understood.

From 1963 onwards, freshwater ecologist Likens and his team studied the relationship between water quality and life forms in the Hubbard Brook drainage basin in New Hampshire. They discovered that the rainfall there was unusually acidic. Acidity, as expressed by pH (potential of hydrogen), ranges from 0 (most acidic) through 7 (neutral), to 14 (least acidic). Most fish and other aquatic animals fare best in water with pH values of 6–8, but Likens found values of 4 – too acid for fish, frogs, and the insects they eat to survive. He set up monitoring stations around New England, which showed that acid rain and snowfall were widespread in the densely populated and heavily industrialized north-eastern states. Likens's systematic work persuaded the US government to introduce laws to control emissions of the chemicals responsible for acid rain.

Effects of acid rain
When fossil fuels are burned in power stations and factories, sulphur dioxide (SO_2) and nitrogen oxide spew out of their chimneys.

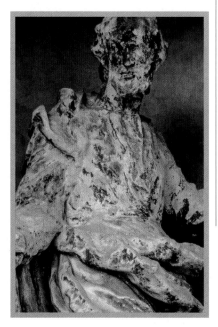

Acid rain had been wearing away stonework – such as this statue in the churchyard of St Peter's and St Paul's, Krakow, Poland – for hundreds of years before the phenomenon was understood.

See also: Endangered habitats 236–239 ▪ Pesticides 242–247 ▪ Deforestation 254–259 ▪ Depletion of natural resources 262–265 ▪ Ocean acidification 281

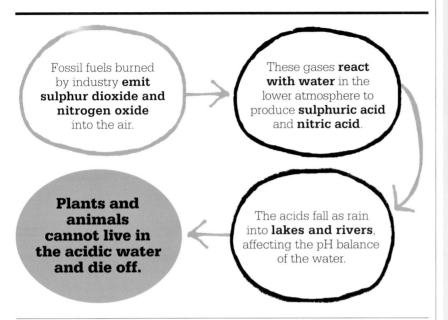

Fossil fuels burned by industry **emit sulphur dioxide and nitrogen oxide** into the air.

These gases **react with water** in the lower atmosphere to produce **sulphuric acid** and **nitric acid**.

The acids fall as rain into **lakes and rivers**, affecting the pH balance of the water.

Plants and animals cannot live in the acidic water and die off.

Gene Likens

Likens was born in Indiana in 1935. After studying zoology at university, he was appointed assistant professor at Dartmouth College in New Hampshire. In 1963, with fellow scientists F. Herbert Bormann, Noye Johnson, and Robert Pierce, he began systematic research into the water, minerals, and life forms in the Hubbard Brook basin. In 1968, his studies recorded the widespread prevalence of acid rain, the product of emissions from factories in the Midwest. The team's work in the area over many years was described as one of the world's most thorough studies of how air pollution and land use has shaped a drainage basin. Likens's work on deforestation, land use, and sustainability led to a change in policy by the US Forestry Service. It also helped to shape the amended Clean Air Act in 1990. Likens was awarded the National Medal of Science in 2001.

Key works

1985 *An Ecosystem Approach to Aquatic Ecology: Mirror Lake and its Environment*
1991 *Limnological Analyses*

Spreading through the lower atmosphere, these gases react with water to produce dilute sulphuric acid (H_2SO_4) and nitric acid (HNO_3). These weak acids fall as rain and enter rivers and lakes, making them more acidic. Increased acidity stresses animals and plants. Water snails disappear, fish eggs fail to hatch, and insects and the frogs that eat them die. Eventually, lakes will not support any life.

We experienced eight years of denial, but that's not unusual in environmental issues.
Gene Likens

By the early 1970s, thousands of lakes in Scandinavia had lost their fish and were virtually dead. By 1984, Brooktrout Lake and others in the Adirondack Mountains, New York, were devoid of fish. Acid rain also leaches harmful aluminium from the soil, and acidic clouds and fog harm plants, reducing their ability to photosynthesize, leading to death.

Emission control

In the 1970s and 1980s, other areas badly affected by acid rain included the "Black Triangle" of Czechoslovakia, Germany, and Poland, where large areas of forest died. Thanks to Likens's work, stricter controls were brought in after 1990. Scrubber systems that extract SO_2 were fitted to power station chimneys with great success. Emissions of the gas were cut by almost half in the US, and by two thirds in Europe. Fish began to return to lakes and rivers. However, the problem of acid rain still blights parts of Russia, China, and India. ▪

A FINITE WORLD CAN SUPPORT ONLY A FINITE POPULATION
OVERPOPULATION

IN CONTEXT

KEY FIGURE
Garrett Hardin (1915–2003)

BEFORE
1798 Thomas Malthus forecasts that continued population growth will exhaust global food supplies by the mid-19th century.

1833 In *Two Lectures on the Checks to Population*, British economist William Forster Lloyd discusses overpopulation, using the example of common land, which is less productive if too many cattle graze it.

AFTER
1974 A United Nations conference in Bucharest creates the UN's first World Population Plan of Action.

2013 British social geographer Danny Dorling outlines in *Population 10 Billion* why it is unlikely the world's population will ever reach that number, contrary to UN estimates.

I
n 1968, two scientists in the US issued dire warnings about overpopulation. Ecologist Garrett Hardin predicted that Earth's resources would soon be used up and environmental damage would increase. In *The Tragedy of the Commons*, he cited examples of several major global crises that had been caused by overpopulation: the destruction of fish stocks by overfishing; the draining of lakes by over-extraction of groundwater for irrigation; deforestation; pollution of air, land, and sea; and species

extinction. Hardin himself proposed a controversial solution to the problem of overpopulation, arguing that the government should deny welfare assistance to people who bred "excessively", as this would prevent further births. Biologist Paul Ehrlich similarly advocated population control in *The Population Bomb*, with warnings that human numbers would soon reach a point where mass starvation would ensue.

Growth and decline

For most of human history, the world's population had grown only slowly. It began to increase more rapidly in Western Europe and the United States in the early years of the Industrial Revolution, when British economist Thomas Malthus warned of a future famine. His fears, however, proved premature as food production increased more quickly than many expected. Life expectancy also fell in the new industrial cities, due to infectious diseases. It rose again with better

Ragpickers Court (1879) by William Allen Rogers shows a poor Italian neighbourhood in New York City. Such overcrowding allowed diseases to spread through poverty-stricken areas.

See also: Human activity and biodiversity 92–95 ▪ The Verhulst equation 164–165 ▪ Depletion of natural resources 262–265 ▪ Urban sprawl 282–283

World population growth, 1750–2100

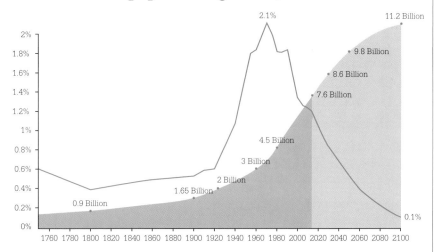

This graph plots a comparison between the annual growth rate of the world population and the total population in absolute numbers. The data for the years after 2017 is a projection.

——— Annual growth rate as a percentage of the world population

▨ World population in numbers

healthcare and nutrition, cleaner water, and more rights for workers. By 1924, there were 2 billion people in the world, and by 1960 there were 3 billion, with most growth occurring in the developing countries of Latin America, Africa, and south and east Asia.

A slowing birthrate
In Europe and North America in the 20th century, wider access to birth control, better education, and more women entering the labour market resulted in lower birth rates. This phenomenon is now being replicated for women everywhere. Although the world's population passed 4 billion in 1974, 5 billion in 1987, 6 billion in 2000, and 7 billion in 2011, the annual rate of increase peaked near the end of the 1960s at 2.5 per cent a year. Populations are

still growing quickly in some parts of the developing world, but the trend is not as rapid as it once was. It took just 11 years for the world's population to rise from 6 billion to 7 billion, but the increase to 8 billion is forecast to be 13 years, then another 25 years to reach 9 billion. The UN forecasts a peak of 11.2 billion in 2100.

Despite the slowing growth, challenges remain. In 2009, a UN report warned that the world would need to produce 70 per cent more food by 2050 to feed its extra population, thereby putting more pressure on land, water, and energy resources. Future population growth is also likely to aggravate many environmental problems, such as pollution, and rising levels of atmospheric greenhouse gases, fuelling global climate change. ▪

China's one-child family planning policy

Until the 1960s, China encouraged families to have as many children as possible, and the population rose from 540 million in 1949 to 940 million in 1976. However, the government soon became concerned about the demand on resources. In 1978, scientist and politician Song Jian calculated that China's ideal population was between 650 and 700 million people, and in 1979, his projections led the government to create a new policy limiting couples to one child per family.

This one-child policy was enforced more strictly in urban areas than in the countryside; in some regions a second child was permitted if the first was a girl. In the cities, however, women were forced to abort second children, and in 1983 alone, 21 million women were forced to undergo sterilization. The policy was relaxed in 2015, but the government still only allows two children per family.

A 1994 poster of a smiling mother and daughter promotes China's one-child policy. Many baby girls were abandoned or killed so that their parents could try for a son.

DARK SKIES ARE NOW BLOTTED OUT
LIGHT POLLUTION

IN CONTEXT

KEY FIGURE
Franz Hölker

BEFORE
1000 CE The first organized system of street lighting (by oil lamps) is introduced in Muslim Spain.

1792 Scottish-born engineer William Murdock invents the gas light. Over the next half century, many cities introduce gas street lighting.

1879 American inventor Thomas Edison demonstrates the first commercially viable electric light bulb.

1976 High-brightness, high-efficiency, LED lights are introduced.

AFTER
2050 The date by which Hölker and others predict that, with the global population set to exceed 9 billion, Earth's total illuminated area will have doubled since 2016.

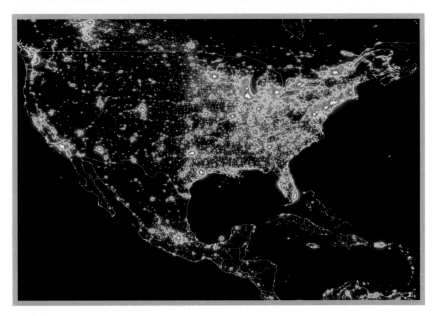

According to some ecologists, light pollution – the amount of artificially generated light in the world – could be the most damaging pollutant of all. Around 80 per cent of humanity lives under skies saturated with light. In 2017, a major German study of light pollution, carried out by ecologist Franz Hölker and others using satellite data, showed that the area of Earth illuminated artificially grew 9 per cent between 2012 and 2016. The brightening is most

A map of light pollution across North America (white and red indicate where it is highest, black where it is lowest) explains why 99 per cent of Americans cannot see the Milky Way.

intense in industrializing countries in South America, Africa, and Asia, but it also continues to increase in the already well-lit countries of Europe and in the US.

Astronomers were among the first to notice light pollution because it interfered with their ability to see

See also: Environmental feedback loops 224–225 ▪ Spring creep 274–279
▪ Man and Biosphere programme 310–311

> Dark areas are being lost in places where nocturnal animals, insects, and plants have adapted to darkness over billions of years.
> **Franz Hölker**

celestial objects in the night sky. In 1988, American astronomers Tim Hunter and David Crawford founded the International Dark-Sky Association to protect the night skies from light pollution. It was the first organization of its kind.

Since then, studies have examined the effects of light pollution on plants and animals, which rely on the cycles of light and dark to govern life-sustaining behaviours such as nourishment, sleep, protection from predators, and even reproduction. Such research reveals a raft of ill-effects. One study showed that trees in Europe are budding more than a week earlier than they were in the 1990s; this alters their period of growth, and may mean that they fail to drop their leaves and fruit and enter the dormant phase in time to avoid damage over the winter.

Vicious circle

Light pollution also has a detrimental effect on animals. Lights on tall towers, for example, draw migrating birds, causing them to crash into the towers and into power lines.

Artificial light can also damage birds' immune systems. Studies have found that house sparrows infected with the West Nile virus carried the virus twice as long when kept under dim light than when kept in the dark – doubling the time in which mosquitoes could bite them and pass on the virus.

Ill-effects on animals can have a knock-on effect on plants. When moths, which are attracted to light, are repeatedly drawn to artificial sources, not only can they be killed by exhaustion (because the light is never extinguished), or by the heat generated, but they also become more vulnerable to predators, which can spot them more easily.

The decline in moth numbers has a knock-on effect on the plants that they help to pollinate, which then affects seed yield. In some places, seed yield has declined by as much as 30 per cent. Researchers who studied a Swiss flower meadow under street lights found that nocturnal visits from pollinators declined by two-thirds. ▪

> The solution is simple – turn off unnecessary lights, use only the amount of light needed for the task at hand, and shield all lighting so it shines down where it is needed.
> **Tim Hunter**

The effect on turtles

Light pollution is a major problem for nesting sea turtles, which need to lay their eggs on land, as the embryos breathe through the permeable shells. Females need dark, sandy beaches on which to lay their eggs. If there are bright lights from beach resorts, street lights, or housing, they will look to nest elsewhere. If a whole stretch of coastline is illuminated, they may lay their eggs in inferior habitats or even deposit their eggs at sea, where their offspring will die.

Such problems may be the reason for the reduction in sea turtle populations. Scientists believe that hatchlings move towards the brightest light. In natural conditions, this will be moonlight shining on the ocean, but if there is artificial lighting inland, the hatchlings wander towards that and get run over by traffic, eaten by predators, or caught in fencing. Solutions include getting people and businesses to turn off lights at night or use "turtle-safe" lighting, which is virtually invisible to turtles.

Olive ridley sea turtle hatchlings make their way towards the sea at Boca del Cielo Turtle Research Station, Mexico.

I AM
FIGHTING FOR
HUMANITY

DEFORESTATION

IN CONTEXT

KEY FIGURE
Chico Mendes (1944–88)

BEFORE
1100–1500 Temperate forest is cleared across large parts of western and central Europe.

1600–1900 Forests are cut down in North America to make room for agriculture.

Late 1970s Tropical rainforest clearance, mostly for ranching, accelerates dramatically.

AFTER
2008 The UN launches its Reducing Emissions from Deforestation and Degradation (REDD) incentive programme.

2010 The US converts $21m (£16m) of Brazil's debt into a fund that will protect Brazil's coastal rainforest.

2015 The UN Paris Agreement sets targets for planting trees to offset the threat of climate change and global warming.

Chico Mendes fought to **save** the **tropical rainforest** in Brazil.

⬇

His **local actions** helped to reduce **global CO_2 emissions.**

⬇

Mendes realized that he had had a global effect: "I am fighting for humanity."

Deforestation is the removal of forest or woodland for conversion to non-forest use. This can be conversion to agricultural land, including cattle ranches, or development for housing, industry, or transport. Forest may be degraded without being destroyed completely, when valuable mature trees, such as teak, are selectively logged or some trees are cut down to create a road. This can have a disproportionate negative effect on the biodiversity of the forest, even though most trees are left standing. Another form of deforestation is the clearance of primary forest and its replacement with monoculture plantations, such as palm oil, as has happened extensively in Indonesia.

Deforestation can impact all kinds of forest habitat, but tropical rainforest – tropical moist broadleaf forest that grows between the Tropic of Cancer and the Tropic of Capricorn – is the most severely affected. Concern for the rainforest

Polluting smoke swirls up as rainforest burns to make way for agriculture in Brazil. It is estimated that Brazil clears 1.1 million hectares. (2.7 million acres) of rainforest a year.

By felling trees ... men bring upon future generations two calamities at once: want of fuel and scarcity of water.
Alexander von Humboldt
19th-century German explorer

See also: Biodiversity and ecosystem function 156–157 ▪ Climate and vegetation 168–169 ▪ Global warming 202–203 ▪ A holistic view of Earth 210–211

was first raised in the 1970s when activist Chico Mendes – who went on to become a founding member of Brazil's National Council of Rubber Tappers – called on the Brazilian government to establish forest reserves, from which local people could extract natural products, such as nuts, fruits, and fibres, sustainably. Mendes's campaign, which eventually cost him his life, highlighted the ecological damage wreaked by forest clearance.

Human need

The human race has used trees from its earliest days. In Neolithic times, they were cut down for fuel and to construct shelters and fencing. Five-thousand-year-old stone axes for chopping wood have been found, as well as axe factories from the same era in Europe and North America. During the Middle Ages, however, as human populations expanded rapidly in western Europe between 1100 and 1500, extensive deforestation took place. Forests were cleared to make way for agriculture, and wood was used to build homes and boats, and to make bows, tools, and other implements.

Trees were cut down on an industrial scale in central Europe and England to produce charcoal, which became an important fuel (until replaced by coal) because it burns at higher temperatures than wood. An early example of sustainable production was practised in England, where many woods were managed as coppices whose trees were partially cut back and then allowed to regrow to create a cyclical supply of charcoal. Even so, by the 17th century England had to import

wood for shipbuilding from the Baltic nations and New England in the US.

Primeval forest clearance accelerated globally between 1850 and 1920, with the biggest losses in North America, the Russian empire, and South Asia. In the 20th century, the focus shifted to the tropics, especially to tropical rainforest, half of which has been destroyed since 1947, with the proportion of the land that it covers having fallen from 14 per cent to 6 per cent.

It is estimated that an area equivalent to 27 football pitches is lost from forests globally each minute. Some regions have been hit harder than others. In the Philippines, for instance, 93 per cent of tropical broadleaf forest has been removed; 92 per cent of Atlantic forest in Brazil has gone; 92 per cent of temperate coniferous forest in south-west China has disappeared; and 90 per cent of dry broadleaf forest in California has been cleared.

Effects on biodiversity

Recent estimates suggest that almost half of all forest clearance is carried out by subsistence »

We are unable to remain silent in the face of so much injustice.
Chico Mendes

Chico Mendes

Born in 1944, the son of one of the 50,000-strong "Rubber Army" who tapped rubber for use in the Allied war effort in World War II, Mendes started work as a rubber tapper at the age of nine. Influenced by priests from the progressive Liberation Theology movement, he helped to found a branch of the Workers' Party and became leader of the Rubber Tappers' Union.

As large areas of Brazil's rainforest were cleared to make way for cattle ranches, Mendes publicized the tappers' fight to save the forest. He went to Washington, D.C. to persuade the World Bank and US Congress that cattle-ranching projects should not be funded. Instead, Mendes proposed that forest areas be protected as "extractive reserves" – public land managed by local communities with the right to harvest forest products sustainably. Cattle ranchers saw his movement as a threat, and one, Darcy Alves, shot him dead in 1988. After Mendes's death, the first of many such reserves was established, covering 1 million hectares (2.5 million acres) of forest around Xapuri.

farmers, and a third by commercial interests. Urban development, logging for the best-quality timber, mining and quarrying, and trees cut for firewood account for any remaining deforestation. In every case, the environment suffers. Biodiversity is particularly impacted, as only a small number of species of mammals, birds, and invertebrates can live on grassland or a palm oil plantation, and even fewer in industrial or urban settings. Human conflicts also blight forest, the worst example being the Agent Orange chemical used to defoliate trees during the Vietnam War.

The rainforest

Destruction of the rainforest poses a severe threat to global biodiversity, as it has been estimated that between half and

Replacing trees with human settlements destabilizes the soil on slopes, and mudslides, such as this catastrophic event in Sierra Leone in 2017, are more likely to occur.

> I became an ecologist long before I had ever heard the word.
> **Chico Mendes**

two-thirds of the world's plants and animals live in this environment. Between 1.5 million and 1.8 million species – mostly insects, followed by plants and vertebrates – have already been identified in rainforests, and many others have yet to be discovered and described. In Borneo, Indonesia, for example, an area of just 0.5 sq km (0.2 sq mile) may contain more species of tree than the combined landmass of Europe and North America. Such biodiversity is vitally important to

humans – not least because most new medicines are derived from plants, and so the eradication of the rainforest's rich store destroys potential cures for disease.

Rainforests, together with all other trees and woodland, also act like a sponge for rainfall. Tree roots drink up moisture and limit surface run-off. When forest is cut or burned, the soil is leached of many of its nutrients. If it covers a slope, the soil will wash away, leaving the land unfit for growing any kind of plants. Deep gullies may undermine trees that have not been cut, bringing them down. After heavy rains, catastrophic mudslides, which happen with increasing frequency, sweep down the slope, destroying everything in their path – including human settlements. In May 2014, for example, heavy rainfall on the deforested slopes of the Caribbean island of Hispaniola caused mudslides and floods that killed more than 2,000 people. Conversely, in extended periods of dry weather,

Deforestation in the Brazilian Amazon

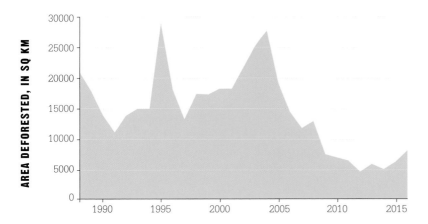

AREA DEFORESTED, IN SQ KM

The depleting rainforest cover in the Amazon Basin is a global concern. The land is now being stripped of trees at a rate of 8,000 sq km (3,000 sq miles) annually.

exposed soil dries out faster than tree-covered areas, making it more prone to wind erosion.

Fuelling global warming

Burning wood or forests adds carbon dioxide (CO_2) to the atmosphere. By contrast, living plants of all kinds reduce CO_2, as they absorb carbon, taking up the greenhouse gas to perform photosynthesis, thus countering the damaging impact of human activities. Globally, forests suck up 2.4 billion tonnes (2.65 billion tons) of CO_2 every year. Environmentalists and climatologists worry that removing large tracts of tropical forest could be disastrous.

Reforesting Earth

Currently, about 31 per cent of Earth's land surface is covered by forests, but that figure is rapidly decreasing in some parts of the world. However, there are regions, including Europe, where forest areas are gradually expanding. Measures to restrict deforestation include payments to communities for conserving forest, and the creation of extractive reserves, where local people can harvest products sustainably.

Globally, alternative sources of fuel need to be found, along with new ways to develop less land-hungry forms of agriculture. A few nations are taking the lead in reforestation programmes. For example, a project in which people from 500 villages have planted 150 million mangrove trees on the coast of Senegal will restore mangrove forests to boost fishing and shield rice paddies from the influx of salt water. The Chinese aimed to plant 6.6 million hectares (16.3 million acres) in 2018, equal to the area of Ireland; in 2000, the proportion of China covered by forest had fallen to 19 per cent, but the target is to increase this to 23 per cent by 2020 and 26 per cent in 2035. ■

The first African woman to win a Nobel Peace Prize (2004), Wangari Maathai initiated a community-based tree planting programme to reverse erosion and desertification in Kenya.

Reforesting the Amazon

About 17 per cent of rainforest in the Amazon Basin has been lost since the mid-1970s. At the United Nations Paris Climate Summit in 2015, Brazil pledged to restore 12 million hectares (nearly 30 million acres) by the year 2030. In 2017, Conservation International, in partnership with the Brazilian government, launched the area's biggest reforestation programme to date, under which Amazonas state will gain 73 million trees through seeding and planting.

Local communities are being enrolled to implement the scheme, using a technique called *muvuca*. This involves spreading the seeds of more than 200 native forest species over every square metre of land. Less labour-intensive than traditional tree-planting, the method can reforest land quickly, delivering up to 2,500 plants per hectare. In addition to the seeding programme, some planting will enrich secondary forest and return pasture land to forest.

THE HOLE IN THE OZONE LAYER IS A KIND OF SKYWRITING
OZONE DEPLETION

IN CONTEXT

KEY FIGURE
Joseph Farman (1930–2013)

BEFORE
1974 American chemists Frank "Sherry" Rowland and Mario Molina suggest that chlorofluorocarbons (CFCs) destroy atmospheric ozone.

1976 The US National Academy of Sciences declares that ozone depletion is a reality.

AFTER
1987 The Montreal Protocol on Substances that Deplete the Ozone Layer, a global treaty to phase out CFCs and similar chemicals, is agreed.

1989 Montreal's worldwide ban on the production of CFCs comes into effect (ratified by the EU and 196 states to date).

2050 The year in which ozone over the Antarctic is predicted to return to pre-1980 levels; however, other harmful emissions may delay recovery.

In 1982, a team of scientists working for the British Antarctic Survey (BAS) discovered that ozone levels above the Antarctic had fallen dramatically. Ozone (O_3, a colourless gas in the stratosphere, 20–30 km (12–18 miles) above Earth's surface, forms the "ozone layer", a protective shield that absorbs most of the Sun's ultraviolet (UV) radiation. Without it, more of the Sun's harmful radiation would reach the surface.

Since the mid-1970s, there has been a 4 per cent decrease in the amount of ozone in the stratosphere. An even bigger decrease has been seen above the poles, particularly

Joe Farman [made] one of the most important geophysical discoveries of the 20th century.
John Pyle and Neil Harris
Atmospheric scientists,
University of Cambridge

in spring. Over Antarctica, ozone measurements have been down by 70 per cent compared with 1975. Over the Arctic, levels have fallen by nearly 30 per cent. This effect became known as "the ozone hole", although it is better described as "the ozone depression", since it is a thinning of the ozone layer rather than a complete hole.

Antarctic discovery
British geophysicist Joe Farman was one of the team who made the discovery in 1982. BAS teams had been collecting atmospheric data at the Halley Research Station in Antarctica since 1957. Their work was poorly funded, and they relied on dated instruments such as the Dobson meter – a rudimentary machine that worked properly only when wrapped in a duvet.

When Farman first noticed the drop in ozone levels, he found it hard to believe, and thought there must be a problem with his Dobson meter. He ordered a new instrument for the next year – and it recorded an even bigger dip. The following year, the dip was bigger again. The year after, his team took their measurements 1,000 km (620 miles) from Halley. Again, there was a

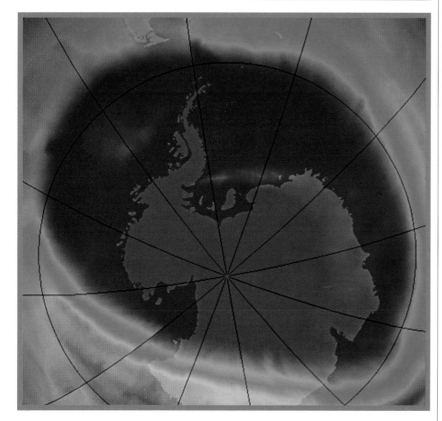

A **NASA image** of the "ozone hole" over Antarctica in 2014. The blue area shows where there is least ozone. The amount of ozone in Earth's stratosphere overall has stabilized since about 2000.

large dip. Farman decided it was time to publish, and a paper written by him and his colleagues Brian Gardiner and Jon Shanklin appeared in the journal *Nature* in 1985.

Reaction and response

Most scientists greeted Farman's discovery with alarm: the potential increase in UV radiation would make skin cancers, cataracts, and sunburn far more prevalent.

What could be done? One reason for ozone depletion had been identified in 1974 by American scientists Frank Rowland and Mario Molina. They had concluded that gases containing chlorine – including the chlorofluorocarbons (CFCs) used in aerosol sprays and halogen refrigerants – were, in the presence of UV light, reacting with ozone in the stratosphere and breaking down the gas. A few countries, including the US, banned the use of these products, but most were yet to be convinced.

When ozone levels continued to fall throughout the 1980s, opinion gradually changed. Consequently, in 1987, the Montreal Protocol for a global ban was agreed. The ozone layer is showing signs of recovery, and it is hoped that by 2075, stratospheric ozone will return to 1975 levels. ▪

CFCs

Chlorofluorocarbons (CFCs) are chemicals made up of carbon, chlorine, and fluorine atoms. They are non-toxic, non-flammable, and extremely stable. Their low reactivity makes them very useful, but is also the reason why they are so destructive. They can survive for over 100 years, which gives them time to diffuse into the stratosphere. There, they are broken down by the intense UV light to release chlorine, which reacts with ozone to form oxygen.

CFCs were first produced in 1928, and were used as coolants for refrigerators. They were later used in a wide range of aerosol products, for example insect sprays, hair conditioners, and spray paints.

The replacements for CFCs were hydrochlorofluorocarbons (HCFCs), which also deplete the ozone layer, although to a much lesser extent, and hydrofluorocarbons (HFCs). HCFCs will be phased out by 2020. HFCs do not harm the ozone layer at all – but they are very powerful greenhouse gases, and so in 2016 it was agreed that, from 2019, they too would be phased out.

Aerosol products such as insect repellents were widely available from the 1950s. The damaging effects of the CFCs they contained were not known until the 1970s.

WE NEEDED A MANDATE FOR CHANGE

DEPLETION OF NATURAL RESOURCES

IN CONTEXT

KEY FIGURE
Naomi Klein (1970–)

BEFORE
1972 The UN's Conference on the Human Environment calls for an international approach to environmental protection.

1980 The World Conservation Strategy, launched in 35 countries, introduces the concept of sustainability.

1992 At the UN Earth Summit in Rio de Janeiro, member states produce "Agenda 21", which outlines plans for managing resources in the 21st century.

AFTER
2015 The UN Sustainable Development Summit sets out 17 sustainable development goals and launches a bold global agenda, adopted by 193 member states.

In *This Changes Everything* (2014), Naomi Klein railed against the way that governments and corporations deplete natural resources. "Ethical oil", she maintains, is not just a contradiction in terms, "it's an outrage". A Canadian citizen, Klein has campaigned against the exploitation of the Athabasca tar sands, the largest of three major oil sand deposits in western Canada. The oil sand deposits lie under thousands of square miles of coniferous forests. The open-pit extraction of oil from tar sands is particularly harmful to the environment. Vast acres of forest are cleared, and ponds of pollutants

See also: Deforestation 254–259 ▪ Overfishing 266–269 ▪ The water crisis 286–291 ▪ Humankind's dominance over nature 296 ▪ Human devastation of Earth 299

Extracting crude oil from Canada's tar sands is notoriously harmful to the environment. It accounts for a tenth of Canada's annual greenhouse gas emissions.

are left behind. These can leak into the land, rivers, and groundwater, killing fish, migrating birds, and other animals.

Global action

By the 1980s, the environmental effects of industrialization and depletion of the Earth's resources were already becoming a matter of concern. The United Nations (UN) created a World Commission on Environment and Development, which published a report in 1987 called "Our Common Future". Contributing experts, including scientists, agriculturalists, foreign ministers, technologists, and economists, made it clear that the future of humans relied on balancing ecology and economics

in a way that was sustainable and fair for all nations around the world. Key areas in the struggle for a sustainably managed Earth are use of fossil fuels, deforestation, and water management.

Five years later, at the 1992 Earth Summit in Rio de Janeiro, 172 nations signed environmental resolutions. Among them was Agenda 21, a plan for governments to work together to protect natural resources and the environment. However, implementing changes has proved difficult, and subsequent Earth summits have called for better international cooperation in order to achieve goals set.

Peak oil

Fossil fuels are among the world's most highly prized resources. People have become increasingly reliant on oil, squandering it to create a lifestyle that is ultimately unsustainable. The oil crises of the 1970s highlighted how dependent »

Naomi Klein

Born in Montreal, Canada in 1970 to politically active parents, Klein developed a sophisticated understanding of the way the world works while still young. Her first job was on a Toronto newspaper, *The Globe and Mail*. Her debut book *No Logo,* criticizing globalization and corporate greed, was a bestseller. Her second, *The Shock Doctrine*, attacked neoliberalism. Klein then began campaigning against corporate interests taking priority over the environment and the interests of humanity. Her book *This Changes Everything* was later made into a film. Klein's many campaigns included a protest against the construction of the Keystone XL pipeline – a symbol in the battle against fossil fuel use and climate change. In November 2016, she was awarded Australia's Sydney Peace Prize.

Key works

2000 *No Logo*
2007 *The Shock Doctrine: The Rise of Disaster Capitalism*
2014 *This Changes Everything: Capitalism vs The Climate*

> ... the conservation of natural resources is the fundamental problem. Unless we solve that problem it will avail us little to solve all others.
> **Theodore Roosevelt**

industrialized nations were on an economically viable supply. With this came, too, the realization that oil is a finite resource. Scientists had already considered the problem and calculated the date when the supply of oil would peak, before it ran out or became uneconomical to extract. In 1974 the peak oil date was predicted to be 1995, with the caveat that there were several potential variables and unknowns such as consumption rates, available technology, and reserves yet to be discovered. In the early 21st century, new dates were given, some extending the timeline for oil to 2030 or beyond. In 2011, however, US environmentalist Bill McKibben declared that calculating a peak oil date was pointless; if all known oil reserves were burned, the carbon produced would be five times the amount required to heat the planet by 2°C (3.6°F) – the "safe" temperature limit that climatologists had worked out in 2009. The science has evolved, but the predicted risks of burning fossil fuels remain dire.

Saving trees

Forests are a valuable natural asset that Earth cannot afford to lose. Their diminished numbers pose a significant threat for the climate; trees are "carbon sinks", meaning they take in carbon dioxide and use it to fuel growth. This then prevents carbon from contributing to global warming. Trees are a renewable resource, and people, businesses, and nations often plant them to offset fossil fuel use, but not in sufficient numbers. According to Friends of the Earth, the annual loss of forests worldwide directly causes 15 per cent of global greenhouse gas emissions.

Rainforests, estimated to contain 50 per cent of the world's species, are particularly vulnerable to deforestation. Around 17 per cent of the Amazon rainforest alone has been lost in the past 50 years. As "Our Common Future" suggested, part of the problem is that developing countries can earn money from large corporations if they clear rainforests for mining, logging, and cash crops. In Indonesia, for example, intensive deforestation took place to make way for palm oil plantations. Greenpeace reports that the amount of Indonesian rainforest logged, burned, or degraded in the last 50 years is equivalent in area to twice the size of Germany. The UN and other bodies now offer developing countries technical advice and financial incentives to manage their forests in a more sustainable way.

Deteriorating soil

Topsoil is perhaps one of the world's most undervalued resources. This vast ecosystem,

Easter Island

The fate of the ancient people of Easter Island illustrates the importance of managing natural resources. Once a thriving community of 12,000 people who erected enormous stone monuments, they had dwindled to just a couple of thousand by the time Europeans discovered the island in 1722.

Mismanagement of a fragile ecosystem, especially mass deforestation, and warring between tribes, had been the cause of their demise. The giant heads, or *moai*, are made of stone, but logs were needed as rollers to transport them from the quarries to ceremonial sites. As the island's many palms were cut down, there was no wood left for fishing canoes, which led to many people starving to death.

The final tragedy came in 1862, with the arrival of slave traders, who captured 1,500 islanders and took them to Peru, where almost all of them died. The 15 islanders who eventually managed to make it home unwittingly introduced smallpox to the island. By 1877, only 111 inhabitants survived.

Some 887 *moai* cover the slopes of Rano Raraku, Easter Island's volcanic crater, the source of the stones from which the statues were carved.

Thick forests like the one in this 15th-century painting by Italian artist Paolo Uccello are returning to Europe, where they have grown by 17 million ha (42 million acres) since the 1990s.

composed of animals, microbes, plant roots, and minerals, is a complex and delicate structure that is slow to form and easily lost. The World Wildlife Fund estimates that half of the world's topsoil has been eroded by wind and rain in the last 150 years. Particles then collect in streams and rivers, clogging them with sediment. Soil loss occurs due to overgrazing, removal of hedges, and use of agrochemicals that affect the soil structure. Meaures such as resting the soil, terracing, dams, and strategic planting can help. In the village of Aamdanda in Nepal, for example, steep-sided slopes are stabilized with broom grass. The plant binds the soil; it is also a fodder crop and is used to make brooms, which the villagers sell.

Water pressures
Clean drinking water is a limited resource. Water covers around 75 per cent of Earth's surface, but 97.5 per cent of it is salt water. Of the remaining 2.5 per cent, most is locked away in glaciers or in deep underground aquifers. Only one-hundredth of 1 per cent of all the water in the world is readily available for human use. Drinking water is also not distributed equally, being naturally scarcer in hot, arid areas of the world than in temperate zones.

Population pressures and wealth also have an impact on water supplies. The UN believes everyone should have access to at least 50 litres (88 pints) of fresh water a day, but people in sub-Saharan Africa manage on 10 litres (21 pints) a day, while the average American enjoys 350 litres (almost 740 pints).

Around the world, water sources are also being bought up by large corporations. Some scientists warn that, if our current usage patterns continue and population rates grow at their current rate, by 2030 global demand for clean water will exceed supply by 40 per cent.

Future plans
New strategies are evidently required to save the world from human destruction. Transition engineering, an emerging multi-disciplinary field, may help. It aims to use existing businesses, organizations, and systems to find innovative ways to minimize environmental impacts and manage resources.

Some progress is being made, in part thanks to campaigning by people like Naomi Klein. A number of European and Asian countries, including the UK, have decided to phase out fossil-fuel vehicles. In other areas, however, socio-economic and political problems remain obstacles to reform. As "Our Common Future" stated, meeting humanity's goals and aspirations responsibly "will require the active support of us all". ∎

You have to think in terms of the survival of human society … it is not only the magnitude of change, it's the pace at which it changes.
Benjamin Horton
British geographer

BIGGER AND BIGGER BOATS CHASING SMALLER AND FEWER FISH

OVERFISHING

IN CONTEXT

KEY FIGURE
John Crosbie (1931–)

BEFORE
1946 The International Whaling Commission is set up to review and control whaling, reversing a dramatic decline after centuries of hunting.

1972 Overfishing and a strong El Niño cause Peru's coastal anchovy fisheries to crash – a blow to the national economy.

AFTER
2000 The World Wildlife Fund places cod on its endangered species list and launches a UK Oceans Recovery Campaign.

2001 Jeremy Jackson and other marine biologists trace the history of overfishing.

2010 The UNESCO Aichi Biodiversity Target 11 calls for a tenth of coastal and marine areas to be protected by 2020.

In 1992, one piece of legislation changed the ecological, socioeconomic, and cultural structure of Canada's Atlantic Maritime provinces. John Crosbie, the Federal Minister of Fisheries and Oceans, placed a moratorium on the Atlantic cod fishery; no further cod could be harvested from the ocean. His ruling was a necessity; the volume of northern cod was down to 1 per cent of previous levels. The region had been overfished to the point where recovery could not occur if fishing were allowed to continue. Crosbie called it the toughest political moment of his career. The decision put thousands of Canadians out of

See also: A holistic view of Earth 210–211 ▪ Pollution 230–235 ▪ Human devastation of Earth 299 ▪ Sustainable Biosphere Initiative 322–323

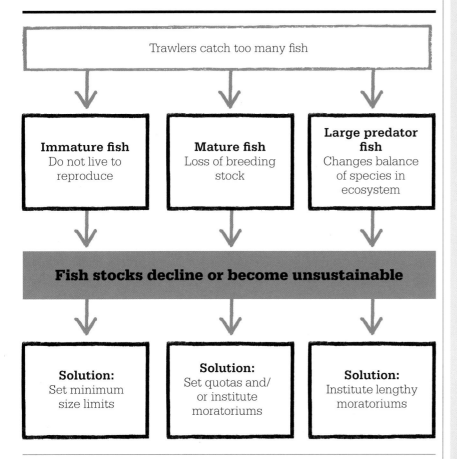

Trawlers catch too many fish

| **Immature fish** Do not live to reproduce | **Mature fish** Loss of breeding stock | **Large predator fish** Changes balance of species in ecosystem |

Fish stocks decline or become unsustainable

| **Solution:** Set minimum size limits | **Solution:** Set quotas and/ or institute moratoriums | **Solution:** Institute lengthy moratoriums |

Marine reserves

A promising tool for fish management is the creation of marine protected areas (MPAs), which legally protect fish stocks and ecosystems. MPAs cover around 3.5 per cent of the world's oceans, but only 1.6 per cent of MPAs are the strongest "no-take zones" where fishing, extraction of materials, dumping, drilling, and dredging are banned. One meta-analysis of scientific studies showed that the volume of diverse fish species is on average 670 per cent greater in fully protected "no-take" marine reserves than in areas that have no protection, and 343 per cent greater than in partially protected MPAs. No-take zones effectively preserve and restore damaged ecosystems, too; coral reefs in protected zones of the Pacific Line Islands recovered from an El Niño event within a decade, but those in unprotected areas did not. Some studies suggest that legally enforced reserves may even help to replenish fisheries outside their borders.

Big-eye trevallies are among the many species in the Malpelo Fauna and Flora Sanctuary, the biggest no-fishing zone in the Eastern Tropical Pacific, noted for its sharks.

jobs. For 500 years, the cod fishery had supported Maritime residents, particularly in Newfoundland.

The 1992 moratorium was initially supposed to last only two years, but, with the stocks not yet recovered, it is still largely in place. From around 2005 to 2015, the volume of northern cod rose by about 30 per cent each year along Newfoundland's northeast coast, although stocks further south did not recover as fast. In 2017 and 2018, however, cod numbers declined sharply, and the overall stocks are still too low to support large-scale fishing. Climate change has contributed to the problem: higher temperatures have created conditions in which both the cod and its food sources struggle to survive. A further blow to Newfoundland's fishermen – who largely turned to catching shrimp and crab – is that where cod numbers improved, cod began eating the shrimp. The ecosystem cannot support both a large-scale shrimp and crustacean industry, and large-scale cod fishing.

A sustainable harvest

The Newfoundland problem demonstrates the complexities of fishery management, which often relies on the concept of maximum sustainable yield: the volume of fish harvested from the sea should be »

Disrupting the ecosystem

Large-scale fishing operations disturb the balance of marine ecosystems in various ways, depleting the target fish species, upsetting the food chain, and damaging the marine environment.

Physical impact of fishing gear

Bycatch

Harvest mortality

Incidental mortality

Habitat modification or destruction

Discarded bycatch and offal

Decline in mean trophic level

Biological interactions

Altered ecosystem structure and function

I didn't take the fish from the goddamn water.
John Crosbie

The management strategies adopted depend on the nature of the problem. If fish are being taken before they are mature, this will limit the stock's future ability to reproduce at a maximum level and keep their numbers replenished. Placing minimum size limits on fish can help control this type of overfishing. If too many mature fish are being caught, this could leave too few to reproduce and replenish the present population. In this case, moratoriums and quotas are among measures that can help. Finally, ecosystem overfishing occurs when a fishery is so depleted that the ecosystem itself changes and is no longer able to support the fish stock at a sustainable level. It generally occurs when large predatory fish are overfished, allowing populations of smaller forage fish to increase and alter the entire ecosystem. This happened in the North Atlantic cod fishery: without the cod to keep them in check, the cod's three main food sources – shrimp, crab and capelin fish – all increased in numbers.

The overfishing problem is now compounded by climate change and pollution, which are also affecting ocean ecosystems. The consequences could be dire. If global warming continues, it will

equal to the volume replenished through reproduction. This is usually achieved through quotas, which limit the number of fish that can be brought in during a season. Quotas can curb unsustainable fishing: for example, 16 per cent of fish stocks in American waters were overfished in 2015, down from 25 per cent in 2000. However, the quota system can encourage fishermen to take the largest fish possible, and to throw back smaller fish, which frequently die from the stress of being caught. In many cases, quotas are also not set at a truly sustainable limit; commercial fishermen often have considerable lobbying power, and focus on the short-term economic gains of catching more fish rather than on long-term sustainability. Fishery management can be further complicated by factors such as the open access nature of the ocean, illegal fishing, and an absence of regulation and oversight.

A worldwide crisis

Overfishing is now a global issue, with more than 30 per cent of the world's fisheries harvested beyond their biological limits, and 90 per cent of fish stocks currently at their limits or overfished. Sustainable management is now essential if fisheries are to continue to provide jobs and meet consumer demand.

cause higher ocean temperatures, sea ice will melt further, and wind and ocean current patterns will change. As a result, nutrients from the upper ocean will be transferred to the deep ocean, starving marine ecosystems and reducing photosynthesis by phytoplankton, which serve as the base food in the ocean food chain. Within three centuries – by 2300 – the world's fisheries could be 20 per cent less productive, and between 50 to 60 per cent less productive in the North Atlantic and western Pacific. The predictions, calculated in 2018 by scientists at the University of California, Irvine, are based on extreme global warming – a 9.6°C (17°F) increase, but their models show that it is a possibility.

Finding new solutions

Seafood consumption has risen from 9.9kg (21.8lb) per capita annually in the 1960s to more than

A deep-sea salmon farm, built in China, begins its journey to Norway. The huge, semi-submersible cylinder aqua-farming platform is designed to produce 1.5 million salmon a year.

20kg in 2016. Global demand is predicted to reach around 236 million tonnes by 2030. Aquaculture, the farming of fish and seafood, has begun to meet much of the demand and has the potential to reduce the pressures on wild fish stocks. However, aquaculture has its own problems. Nutrients and solids added to the water can cause the environment to degrade. The build-up of organic matter from many fish in a farm can change the sediment chemistry, which then has an impact on the surrounding water. Fish may also escape, thereby introducing alien species or diseases into the outside freshwater or marine environment.

While fish farming helps to meet demand, overfishing still poses huge dangers for the health of the world's marine ecosystems, and the economic future of many nations. The Canadian moratorium severely disrupted the economy and culture of Newfoundland and neighbouring maritime provinces. To avoid such crises, more governments will have to develop sustainable fishing practices, and protect the health of ecosystems and fish stocks. ∎

Effects of pollution

Two main types of pollution damage marine ecosystems. Run-off from fertilizers is a common problem: the nitrogen and phosphorus that many contain produce algal blooms (overgrowths of algae, or phytoplankton), which later die. As they decompose, they take up oxygen, creating a "dead zone" in the water that cannot sustain life. Because fish must leave such water or perish, juvenile fish living close to the shore are at risk before they move into the open ocean. In 2017, the annual dead zone in the Gulf of Mexico was more than 8,500 sq miles (22,000 sq km). Plastic pollution is another threat, as fish both eat it, and get caught in nets and debris. Estimates suggest there are more than 5 trillion pieces of plastic in the ocean, with over 8 million tonnes added each year. If plastic pollution continues unchecked, the volume of plastic in the ocean will exceed that of fish by 2050.

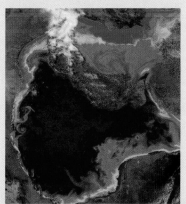

Thick blooms of phytoplankton appear in red on this satellite image of the Gulf of Mexico. Bacteria break down decaying algae, releasing CO_2 and absorbing essential oxygen.

THE INTRODUCTION OF A FEW RABBITS COULD DO LITTLE HARM

INVASIVE SPECIES

IN CONTEXT

KEY FIGURE
**Ryan M. Keane,
Michael J. Crawley** (1949–)

BEFORE
1951 The International Plant
Protection Convention is set up
to prevent the introduction and
spread of pests of plants and
plant products as a result of
international trade. It is
adopted in many countries.

1958 *The Ecology of Invasions
by Plants and Animals* by
British ecologist Charles
Elton is the first book to be
published on invasion biology.

AFTER
2014 Studies of some of
the "world's worst" invasive
species by ecologists at
Queen's University, Belfast,
and Stellenbosch University,
South Africa, reveal that the
ecological impacts of these
species could be predicted
from their behaviour.

Some of the greatest damage to ecosystems is caused by invasive species. These are plants, animals, or other organisms that are not native to an ecosystem but introduced largely through human action, either deliberately or by accident. They can become competitors, predators, parasites, and hybridizers of native plants and animals, ultimately threatening the survival of those species.

The rise of the rabbit
One of the most notable species invasions has been that of the European rabbit in Australia. It began in 1788, when 11 ships

See also: Predator–prey equations 44–49 ▪ Non-consumptive effects of predators on their prey 76–77 ▪ Human activity and biodiversity 92–95 ▪ The food chain 132–133 ▪ The ecosystem 134–137 ▪ Chaotic population change 184

The harlequin ladybird is the world's most invasive ladybird. In the UK, where it was first seen in 2004, it is reportedly responsible for the decline of seven native ladybird species.

Spread of rabbits in Australia

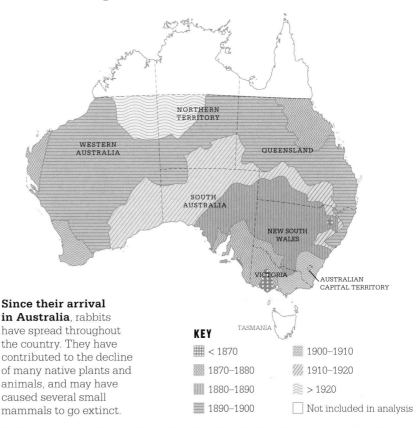

Since their arrival in Australia, rabbits have spread throughout the country. They have contributed to the decline of many native plants and animals, and may have caused several small mammals to go extinct.

KEY

▦ < 1870	▨ 1900–1910
▨ 1870–1880	▨ 1910–1920
▥ 1880–1890	≋ > 1920
▤ 1890–1900	☐ Not included in analysis

landed at Botany Bay from Britain, to establish the first Australian penal colony. On board the "First Fleet", along with more than 1,000 people, including convicts and emigrants, were six European rabbits, brought along for food.

By the 1840s, rabbits had become a staple food in Australia, and were contained within stone enclosures. All this changed in 1859 when a settler, Thomas Austin, imported 12 pairs of European rabbits and released them on his estate near Geelong in Victoria. Twenty years later, rabbits had migrated to South Australia and Queensland, and then in the next two decades to Western Australia. By 1920, the rabbit population was 10 billion.

Rabbits appear to be innocuous creatures, but they have wreaked havoc on Australia's native species, competing with them for resources such as grass, herbs, roots, and seeds, and degrading the land. They become particularly troublesome during a drought, when they eat anything they can find in order to stay alive.

There have been several attempts to control the feral population, from rabbit-proof fences stretching more than 3,200 km (2,000 miles) to the more successful introduction of the myxoma virus and the rabbit calicivirus, in 1950 and 1995 respectively. The resulting disease has proved the most effective way of controlling their numbers and protecting native species.

The secrets of success

As invasive species have spread throughout the world, scientists have tried to determine what makes some of these species so successful, and how to control them without accidentally introducing additional ecosystem problems. Despite being hampered by the lack of comparative data on those invasive species that fail to succeed, scientists have developed a number of theories to explain the success of certain species in non-native environments, including the resource availability hypothesis, the evolution of increased competitive ability hypothesis, and the enemy release hypothesis.

In general, species success depends on a variety of genetic, ecological, and demographic factors. The resource availability hypothesis, first proposed in 1985 »

The zebra mussel

The case of the zebra mussel demonstrates the diverse ways to approach invasive species control, and the challenges that result. Zebra mussels are small, fingernail-sized molluscs with a dark-striped shell. The mussel is native to Eurasia but was discovered in the Great Lakes area of North America in 1988, probably carried there in ballast water discharged from ships travelling from Europe. Since then, zebra mussels have spread throughout the midwestern United States, and have been found as far west as California.

The zebra mussels attach themselves to clams and other mussels, filtering out algae that the native species need for food to survive. They also clog water intake pipes used for power plants and drinking water supplies. Current control mechanisms include chemicals, hot water, and filtering systems. While each has had some success, none of these solutions has been capable of safely eradicating the mussels. As a result, they continue to spread throughout the waterways of the US.

We are seeing one of the great historical convulsions in the world's fauna and flora.
Charles Elton

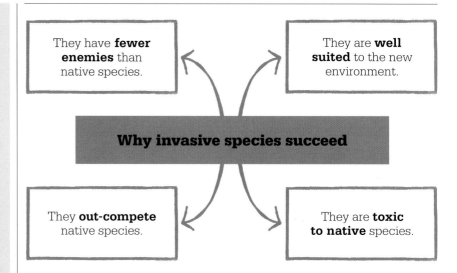

| They have **fewer enemies** than native species. | They are **well suited** to the new environment. |

Why invasive species succeed

| They **out-compete** native species. | They are **toxic to native** species. |

by the ecologists Phyllis Coley, John Bryant, and F. Stuart Chapin, argues that an invasive species thrives because it is already well-suited to its new environment and can take advantage of any surpluses in resources. The evolution of increased competitive ability hypothesis, published by ecologists Bernd Blossey and Rolf Nötzold in 1995, suggests that invasive plants facing fewer herbivores in their naturalized environment can allocate more resources to reproduction and survival and so out-compete the native species. The enemy release hypothesis, set out by ecologists Ryan M. Keane and Michael J. Crawley in their 2002 article "Exotic Plant Invasions and the Enemy Release Hypothesis", argues that the invasive species has fewer enemies in its naturalized environments, and so can spread further. The reality is that the success of invasive species is likely due to many mechanisms working together.

Plant invaders

One plant that appears to support multiple hypotheses about the success of invasive species is garlic mustard (*Alliara petiolata*). Native to Europe, western and central Asia, and northwestern Africa, it was brought to North America by early settlers to use in cooking and medicines, and rapidly spread. Continued infestation has affected the growth rate of tree seedlings and reduced the native plant diversity, leading to changes in the forest ecosystems invaded.

In its native range, garlic mustard is consumed by as many as 69 insect species, but none of these is present in North America. This lack of predation and the plant's invasive success provide

Garlic mustard is highly invasive in North America, inhibiting other plants. In its native habitat, it is considered an attractive wildflower, although it can have a strong smell.

Since their introduction to Australia in 1935, cane toads have out-competed native frogs because they reproduce far more quickly.

support for the enemy release hypothesis. Garlic mustard also successfully competes with native plants for resources, fulfilling the resource availability hypothesis. The plant even exudes secondary compounds that may "attack" native plants by inhibiting their germination and growth. This supports the "novel weapons" hypothesis, proposed by ecologists Wendy M. Ridenour and Ragan M. Callaway in 2004, which posits that invasive species have biochemical weapons that give them a key advantage over native species.

The art of control

Successful invasive species are extremely difficult to control and almost impossible to eradicate. If the species is a plant, the most obvious way to remove it is to pull it up or cut it down, but such methods are highly labour intensive, especially over a wide area. The use of chemicals to destroy invasive species is often successful, but it can also kill native species and undermine soil health, with the added threat of harm to humans.

One frequently used method of control, known as biological control, or "biocontrol", pits an invasive species' own enemies against it. In an early success, the cactus moth was introduced to Australia from South America in 1926 to feed on the prickly pear. This plant had itself been introduced in the 1770s and was choking farmland in New South Wales and Queensland. By the early 1930s, most prickly pears had been eradicated.

Not all biological controls are effective, and some measures have had disastrous consequences. For example, in 1935 cane toads were introduced to Australia to control the invasive greyback cane beetle, which was destroying sugar cane fields. The cane toad had been effective in controlling beetles in Hawaii, so the assumption was that it would be equally successful in Australia. However, greyback cane beetles feed primarily at the top of sugar cane stalks, which is out of reach for the cane toads. A lack of understanding of the different environments favoured by the two creatures meant that the cane toad was the wrong choice as a biological control. By the time the mistake was realized, the toad had spread throughout Australia, poisoning any predator species that tried to eat the toxic amphibian.

Even when biological controls curb an invasive species, they may create imbalances in ecosystems or the economy of local communities. Regulators are, therefore, often hesitant to support biological controls without extensive prior research. No magic bullet exists that can control every invasive species. They are dependent on complex ecosystem interactions, and scientists continue to design field experiments to test their hypotheses of how invasive species function in the wild. ∎

Now is the time to take action. The costs to habitats and the economy are … out of control.
Bruce Babbitt
US Secretary of the Interior, 1993–2001

AS TEMPERATURES INCREASE, THE DELICATELY BALANCED SYSTEM FALLS INTO DISARRAY

SPRING CREEP

IN CONTEXT

KEY FIGURE
Camille Parmesan (1961–)

BEFORE
1997 A group of American scientists publishes evidence of a longer plant growing season at northern high latitudes in 1981–91.

2002 Naturalist Richard Fitter reveals that the first flowering date of 385 plant species in the UK has advanced by 4.5 days in the previous decade.

AFTER
2006 Jonathan Banks, from the American Clean Air Task Force, is the first person to use the term "season creep" to describe the increasingly early onset of the seasons as a result of climate change.

2014 In the US, the National Climate Assessment confirms long-term trends towards shorter, milder winters and earlier spring thaws.

We are seeing change happen much faster than I thought it would 10 years ago.
Camille Parmesan

Most scientists now agree that climate change, driven by an increase in greenhouse gases, is raising the global mean temperature. The Intergovernmental Panel on Climate Change (IPCC) cites an increase of 1°C (1.8°F) since 1880, though in some regions the warming has been even more marked. This warming has affected both plant and animal behaviour, and the IPCC forecasts a further increase of 1.4–5.5°C (2.5–9.9°F) during the next 100 years.

The life cycles of plants and animals change in line with the seasons. Phenology is the study of these seasonal changes. They may be triggered by temperature, rainfall, or the length of daylight, but temperature is probably the single most important factor in Earth's temperate and polar regions, whereas rainfall is the key factor in the tropics. In 2003, climate change scientists Camille Parmesan and Gary Yohe proved that spring change is now happening earlier – a phenomenon called spring creep.

Season creep
For several decades, people have observed leaves and flowers appearing earlier in spring. In the past, these claims were often dismissed as lacking "hard science", such as facts, figures, or datasets.

The impact of seasonal changes on plants and animals

Plants **grow leaves**, produce **flowers** and **fruit**, and **shed their leaves**.

Mammals **breed and raise young**. Some mammals go into **hibernation** over the winter.

Seasonal changes in the weather

Birds **nest and breed**. Many birds (and some other animals) make long-distance **migrations**.

After hatching, amphibians, insects, and some other animals **metamorphose** from one body form to another.

All life forms respond to changes in weather brought about by the seasonal cycle. Migration, breeding, flowering, hibernation, and metamorphosis are some of the events affected by this cycle.

See also: Animal ecology 106–113 ▪ Animal behaviour 116–117 ▪ The foundations of plant ecology 167 ▪ Global warming 202–203 ▪ Endangered habitats 236–239 ▪ Halting climate change 316–321

When Camille Parmesan and Gary Yohe published evidence in 2003 – based on an analysis of more than 1,700 species – they demonstrated that change was very real. Their data showed that spring change was indeed taking place earlier – by an average of 2.3 days per decade. Studies by other scientists in recent years have supported their findings.

Many of the changes that take place in plants are governed by temperature, including growth spurts; the appearance of leaves, flowers, and fruit; and leaves dying in autumn. Most food chains start with plants, so these changes affect grazers and browsers, from rabbits to deer, and pollinators, including bees and butterflies. All of these are at the bottom of the food chain (primary consumers). If they struggle to find food, those that prey on them (secondary consumers) also suffer from the absence of prey.

Effects of climate change

A warmer Earth produces many effects. In most cooler parts of the world, the frost-free season is longer than before, providing a longer growing season for plants. As some regions become drier and some wetter, bouts of extremely heavy rainfall and flooding have become more common. Toxic algal blooms in lakes are occurring more frequently. Ice cover in polar regions is also decreasing. All these changes have affected and will continue to affect animal and plant behaviour.

Since 1993, the European Environmental Agency (EEA) has worked in earnest to pull together data from thousands of studies – dating back to at least 1943 – to create a picture of spring creep in Europe. The EEA's evidence shows earlier dates for plants producing pollen, frogs spawning, and birds nesting. According to their data, many insects whose life cycles are governed by air temperature (thermophilic insects, such as butterflies and bark beetles) now have a longer breeding season, enabling them to produce extra generations each year. For example, some butterflies that previously had two generations now have three.

The leaves of some oak species turn red shortly before they fall in autumn. Comparing the date on which this occurs from year to year can provide evidence for climate change.

In Spain, botanists studied data for 29 species of plants. They found that in 2003, leaves first appeared 4.8 days earlier on average than in 1943; flowers first bloomed 5.9 days earlier; trees produced fruit 3.2 days earlier; and leaves died 1.2 days later. In the UK, the »

Camille Parmesan

Born in 1961, Professor Camille Parmesan is an American academic who has established a reputation as one of the leading climate change scientists. She received her doctorate in biological sciences from the University of Texas at Austin in 1995 and her early research concerned the evolution of insect–plant interactions. For the best part of 20 years, she has focused on documenting the shifting geographical ranges of butterflies across North America and Europe, linking these to climate change. Parmesan has been a leading figure in the IPCC and her work has won her many accolades and has been cited in hundreds of academic papers. She is professor in Integrative Biology at the University of Texas at Austin and advises international conservation bodies.

Key works

2003 "A globally coherent fingerprint of climate change impacts", *Nature*
2015 "Plants and climate change: complexities and surprises", *Annals of Botany*

evidence was even more dramatic: across 53 plant species, leaves, flowers, and fruits appeared almost six days earlier in 2005 than they had done in 1976. Similarly, the fruiting season of 315 different kinds of fungi studied in Britain lengthened from 33 to 75 days in the second half of the 20th century.

Longer plant growing seasons sound like good news, but warmer temperatures create problems as well as advantages. Not all insects are welcome, and shorter, milder winters kill fewer dormant insects, some of which may consequently

Some bee species now emerge earlier in spring, in line with earlier flowering dates for the plants that they pollinate. Other bees, though, have not been able to synchronize their emergence.

undergo explosive population increases and produce damaging infestations. Warmer springs allow pine sawflies, whose larvae eat pine needles, to develop too rapidly for the birds and parasites that feed on them to keep their numbers in check. Out of control, the sawflies strip trees of their needles and stunt their growth.

Migration and hibernation

Birds that migrate in spring to reach rich food sources also face problems. Some have adjusted their flight schedules to benefit from the earlier abundance of insects. After making the long journey from sub-Saharan Africa, the first swallows arrive in the UK about 20 days earlier than they did in the 1970s, and the first sand martins reach

their destination 25 days sooner than previously. However, there is evidence that birds migrating from Central America to New England in the US have declined faster than birds that remain in New England all year. This is probably because the migrant birds have been unable to adjust their departure dates from Central America to arrive in time to benefit from the earlier abundance of insects the way local birds do.

Climate change also appears to have changed the behaviour of hibernating mammals. Zoologists at the Rocky Mountain Biological Laboratory found that yellow-bellied marmots living in Colorado emerged 38 days earlier in 1999 than they had done in 1975. In 2012, scientists at the University of Alberta found that in the last two decades, late snowfall has delayed the emergence of the Rocky Mountain ground squirrel from hibernation by 10 days. This has cut down the already short active period in which they mate, give birth, and feed to prepare for the next hibernation cycle.

Decoupling

Some organisms' survival could be threatened by the "decoupling" of interactions between species. This

We are now sure of what we only suspected years ago. Policy needs to catch up with science.
Camille Parmesan

A great tit feeds its chicks. If breeding takes place after the peak period for spring caterpillars, there will be less food for the young birds, and fewer will survive to breed.

could seriously upset the balance of ecosystems. If flowers appear earlier, the bees that pollinate them can respond in one of two ways: they, too, can emerge earlier; or they can move to a higher latitude to match later flower emergence further from the equator. Studies of 10 wild bee species in north-east North America have shown that their behaviour has, indeed, changed in line with earlier flowering. However, bumblebees in Colorado have not matched the changes and their population has fallen. If pollinators decline, so may the plants that they pollinate.

There is evidence that many primary consumers have adjusted to changed natural phenomena, but species higher in the food chain seem to find it harder to make the change. Although birds are now nesting earlier than they once did, the timing of insect emergence has advanced more rapidly. This is a problem for birds that depend on peaks in insect abundance. For example, pied flycatchers and great tits feed their chicks on caterpillars that are abundant for a short period in spring. Due to climate change, the caterpillar peak is now earlier, but the birds have not been able to advance their egg-laying dates sufficiently to take advantage of the glut of food. Studies show that fewer pied flycatcher and great tit chicks are surviving. Pied flycatcher numbers have declined in Dutch woodlands, possibly as a result of climate change.

Taking action

All of this disturbing evidence has prompted climate scientists worldwide to lobby governments and demand policy change. Spring creep has been used by scientists as a definitive piece of proof that climate change is occurring, and researchers have called upon policy makers to fight global warming to save the familiar species that find their very existence threatened by phenological changes. ∎

Wall butterflies and climate change

Climate change sometimes produces unexpected results. For example, in the UK, the life cycle of the wall butterfly has been disrupted by changing climatic conditions. Previously, the butterfly produced two generations every summer. The late-summer adults would mate, the females lay eggs, and the eggs then developed into caterpillars. In September, these caterpillars found sufficient food to grow large and sustain themselves in hibernation through winter. In spring, the caterpillars metamorphosed into pupae, and then became adults. Warmer weather has allowed a third generation to develop in autumn, with adults flying as late as mid-October. By the time the third generation caterpillars hatch there is little food, so most starve and die. Scientists call this a "developmental trap" and it is probably responsible for the decline in wall butterflies.

This butterfly I was studying shifted its entire range across half a continent – I said this is big ... Everything since then has just confirmed it.
Camille Parmesan

ONE OF THE MAIN THREATS TO BIODIVERSITY IS INFECTIOUS DISEASES

AMPHIBIAN VIRUSES

IN CONTEXT

KEY FIGURE
Malcolm McCallum
(1968–)

BEFORE
1989 The formerly common golden toad of Costa Rica is declared extinct. Various explanations are proposed.

1998 In the US, many poison-dart frogs die at the National Zoo in Washington DC. The chytrid fungus is implicated as a cause.

AFTER
2009 The Kihansi spray toad of Tanzania is declared extinct in the wild as a result of chytrid infection.

2013 A second species of chytrid fungus causes the near-extinction of fire salamanders in the Netherlands.

2015 The chytrid fungus is detected in amphibians in 52 out of 82 countries sampled.

Since the 1980s, hundreds of species of amphibians have suffered population crashes and localized extinctions – at a rate thought to be more than 200 times the natural, "background" extinction rate unaffected by modern human activity. This alarming phenomenon first attracted public attention in 1999, when American environmental scientist Malcolm McCallum published his findings about the dramatic increase in deformities in frogs. He went on to produce landmark studies on amphibian decline and extinction.

The causes of the problem are wide-ranging, and include habitat destruction and pollution, as well as competition from non-native species. But one of the most devastating causes is undoubtedly disease, with two particularly lethal culprits.

Chytrid and ranavirus

Chytridiomycosis is a disease caused by the chytrid fungus, and it has ravaged populations of frogs and toads in particular. The fungus affects amphibians' skin, such that

The North American bullfrog is resistant to the chytrid fungus, but acts as a deadly carrier of the infection to other species of amphibians.

they are not able to breathe, hydrate, or regulate their temperature. The exact origin of the fungus is not known, but the global trade in live amphibians for various uses, be it pets, food, fishing bait, or research, has been a major factor in its spread.

Ranaviruses evolved from a fish virus. They infect amphibians and reptiles, and have caused mass mortality in frogs since the 1980s. The common midwife toad ranavirus causes bleeding, skin sores, lethargy, and emaciation. It is notably virulent as it has the ability to "jump" from one species to another. ∎

See also: Biomes 206–209 ▪ Pollution 230–235 ▪ Endangered habitats 236–239 ▪ Deforestation 254–259 ▪ Overfishing 266–269

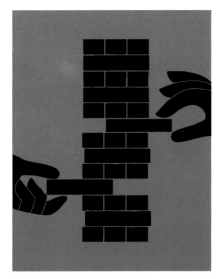

IMAGINE TRYING TO BUILD A HOUSE WHILE SOMEONE KEEPS STEALING YOUR BRICKS
OCEAN ACIDIFICATION

Adding carbon dioxide (CO_2) to the air not only triggers climate change but also makes the oceans more acidic. So far, the oceans have buffered the worst effects of global warming, absorbing up to half of the carbon dioxide added to the atmosphere by human activity. However, the gas alters the oceans' chemistry.

In 2003, American climate scientists Ken Caldeira and Michael E. Wickett investigated the effects of CO_2 pollution on the oceans. They took samples of seawater from around the world, and found that the acidity had increased measurably in the past 200 years of industrialization. They coined the term "ocean acidification" and predicted that this change could accelerate over the next 50 years, with damaging results.

Many sea creatures rely on the natural alkalinity of seawater to maintain carbonates for building their shells and skeletons. Even a slight decrease in alkalinity seriously disrupts growth, especially for sensitive creatures such as corals and plankton. Acidification might wipe out corals within decades; if they go, so do the reef ecosystems. Phytoplankton are the foundation of the ocean food web, and are vital to maintaining global oxygen levels.

Ocean acidification is far harder to reverse than the atmospheric effects of CO_2 emissions, and its devasting impact on biodiversity, fisheries, and food security remains a serious concern. ∎

Most carbon dioxide released into the atmosphere as a result of burning fossil fuels will be absorbed by the ocean.
Ken Caldeira and Michael Wickett

See also: Global warming 202–203 ▪ Pollution 230–235 ▪ Endangered habitats 236–239 ▪ Acid rain 248–249 ▪ Halting climate change 316–321

THE ENVIRONMENTAL DAMAGE OF URBAN SPRAWL CANNOT BE IGNORED

URBAN SPRAWL

IN CONTEXT

KEY FIGURE
Robert Bruegmann (1948–)

BEFORE
1928 British architect Clough
Williams-Ellis compares
London's growth to an octopus
devouring the countryside.

1950s With postwar prosperity
and increased car ownership in
the US, the middle classes leave
cramped city centres and move
to new, low-density areas in
the suburbs.

AFTER
2017 A housing crisis in the
UK prompts calls for the lifting
of restrictions on new building
on the greenbelts around major
UK cities.

2050 The date by which,
according to UN estimates
published in 2014, the urban
population of the world is
set to rise to 6.34 billion out
of a projected population of
9.7 billion.

Since the 1950s, the term "urban sprawl" has been widely used to describe the growth of low-density suburbs beyond high-density city cores. The term was first used by *The Times* newspaper in the UK in 1955 to describe the spread of London's suburbs. At this time, the British planning authorities were introducing "green belts" around cities, where new building was almost entirely banned. Green belts were designed to stop cities from spreading and merging with other towns.

The old city is submerged in a far-flung, multicentred, mostly low-density, highly heterogeneous urban region.
Robert Bruegmann

Modern definitions of urban sprawl vary, but it generally has negative overtones. At its most extreme, it has created megacities – defined by the United Nations as cities of more than 10 million people. Examples of such megacities include Tokyo-Yokohama (38 million), Jakarta (30 million), and Delhi (25 million).

Ecological upset

Some researchers claim urban sprawl is the most serious threat to biodiversity from any human activity. The new suburbs house relatively few people, yet require extensive and disproportionate levels of infrastructure, such as power and water supplies and transport networks. As cities swell, valuable farmland is covered in concrete and natural habitats are disrupted or lost entirely. Sprawl can also disturb local fauna and flora through the introduction of pets and invasive plants that threaten indigenous species. Limited public transport in low density areas also means that suburban populations tend to be multiple car owners, which adds to the levels of air pollution in cities – as do the wood- and coal-burning stoves of the poor in outlying shanty towns.

See also: Pollution 230–235 ▪ Endangered habitats 236–239 ▪ Deforestation 254–259 ▪ Depletion of natural resources 262–265 ▪ Amphibian viruses 280

Toluca was once a picturesque old town to the west of Mexico City. Now a city of more than 800,000 people, it is gradually merging into the sprawl of Mexico City – at a high ecological cost.

The area of the world currently covered in urban development is one-and-a-half times the size of France. Mexico City has expanded more than any other city in the West. Spreading far beyond its official boundaries to become the home of more than 21 million people, it has also grown disproportionately: in 1970–2000, the surface area of the city grew 1.5 times faster than its population. While 59 per cent of the city's territory is conservation land, illegal logging and urban sprawl continue to degrade urban forest, grassland, and water supplies.

It is estimated that 37 per cent of all urban growth by 2050 will occur in China, India, and Nigeria alone. In Beijing and other cities in China, densely populated *hutongs* (alleyways), where the urban poor used to live, are being demolished to make way for low-density luxury blocks, pushing the city limits – and the urban poor – far from city centres. The reliance on cars in the new neighbourhoods, and the lack of central hubs, means there is little opportunity for community life.

Aware of the problems caused by urbanization, the Chinese government is now trying to limit the population of Shanghai to 25 million and that of Beijing to 23 million by restricting land available for building and controlling the inflow of people, forcing out low-skilled workers. China is also building higher-density neighbourhoods with narrower streets, more intersections, and more public transport that will help the formation of communities. ▪

The endangered axolotl

One of the victims of the urban sprawl of Mexico City has been the tiny axolotl, a pale-coloured salamander that looks like a fish but is actually an amphibian, and is sometimes known as the Mexican walking fish. Capable of growing up to 30 cm (1 ft) long, the axolotl feeds on aquatic insects, small fish, and crustaceans, and has the ability to regenerate severed limbs – a quality that has made captive specimens an important subject of scientific research. The captive version is also a familiar pet in aquariums around the world.

Historically, the wild axolotl lived in the urban canals created by the Aztecs as they built their capital city in the 13th century, and in the network of lakes around the city that fed these canals. As Mexico City has expanded, these canals have been lost, and the wild axolotl has declined. In 2006, it was added to the list of critically endangered species and by 2015 it was thought that the creature may have been extinct. However, specimens have since been found in Lake Xochimilco in southern Mexico City.

OUR OCEANS ARE TURNING INTO A PLASTIC SOUP

A PLASTIC WASTELAND

IN CONTEXT

KEY FIGURE
Charles J. Moore (1947–)

BEFORE
1970s Scientists begin to research plastic litter at sea after reports in the journal *Science* describe large numbers of plastic pellets in the North Atlantic.

1984 The first International Marine Debris Conference, held in Hawaii, raises awareness of the growing problem of litter in the oceans.

AFTER
2016 The documentary *A Plastic Ocean*, directed by Australian journalist Craig Leeson, highlights the global effects of plastic pollution.

2018 The Earth Day Network, an organization committed to spreading the environmental movement worldwide, makes End Plastic Pollution the theme of Earth Day, on 22 April 2018.

When plastics were first mass produced in the early 20th century, the world marvelled at the versatility and durability of a material that could be moulded into any shape, used, and then thrown away. The problem with plastic, however, is that most of it never goes away. According to the British business publication *The Economist*, only 20 per cent of the 6.3 billion tonnes of plastic produced in the world since the 1950s has been burned or recycled. This means that 80 per cent – 5 billion tonnes – is in landfills or elsewhere in the environment.

Polluting the oceans

Microplastics – tiny fragments of plastic less than 5 mm (1/4 inch) across – are even harder to clean up than other plastics. Comprising 90 per cent of the plastics in the oceans, they surge through currents like a murky soup. The problem was first identified in 1997 by the American oceanographer Captain

A "seabin" is emptied in Sydney harbour. The Seabin Project, introduced in Australia in 2015, helps counteract plastic pollution by filtering surface water in ports and harbours.

Charles Moore, who highlighted it in his 2011 book *Plastic Ocean*. Sailing home from a yachting competition, Moore came across a vast patch of plastic debris in the Pacific Ocean. Now known to have a bigger surface area than France, Germany, and Spain combined, the Great Pacific Ocean Garbage Patch (GPOGP) comprises 79,000 tonnes of microplastics amassed by the swirling current known as the North Pacific Gyre.

The GPOGP is one of several oceanic garbage patches – there are others in the Atlantic and

See also: The food chain 132–133 ■ Humankind's dominance over nature 296 ■ Human devastation of Earth 299 ■ Man and the Biosphere programme 310–311

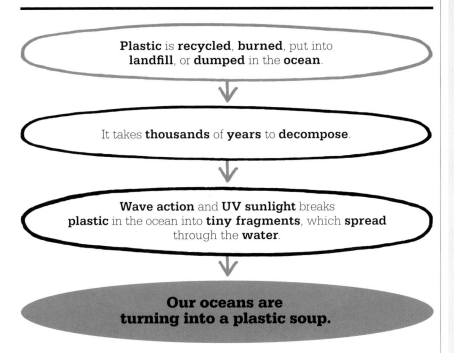

Plastic is **recycled**, **burned**, put into **landfill**, or **dumped** in the **ocean**.

It takes **thousands** of **years** to **decompose**.

Wave action and **UV sunlight** breaks **plastic** in the ocean into **tiny fragments**, which **spread** through the **water**.

Our oceans are turning into a plastic soup.

Effects on wildlife

Plastics pose a danger to wildlife in many ways. Larger items such as plastic carrier bags can choke or strangle birds and marine animals; if ingested, they can damage their digestive tracts or cause starvation by obstructing the stomach. If microplastics are ingested, toxins can pass into an animal's fatty tissues, a process that then passes up the foodchain.

According to Greenpeace, nine out of ten seabirds, one in three sea turtles, and more than half the population of whales and dolphins have eaten plastic. Even some of the crustaceans living in the western Pacific's Mariana Trench, the deepest point in the world's oceans, are known to have ingested plastic.

Companies are starting to take the need to reduce plastic use seriously. A brewer in Florida, for instance, has found a way to make six-pack rings from by-products of the brewing process, so that seabirds can chew them off if they become caught in them.

Indian Ocean as well as in smaller bodies of water such as the North Sea. Plastic microbeads, introduced by cosmetic companies in the 1990s, add to the problem. Used in personal care products such as soaps, facial scrubs and toothpastes, the beads travel from wastewater systems into rivers and oceans, where they are consumed by fish and other animals, with the same damaging effects as microplastics (see panel, far right).

Steps to limit plastic

Cleaning up plastic pollution is a gargantuan task. Breaking plastics down into their constituent chemicals requires huge amounts of energy, which also damages the environment. The best solution is to learn to live without plastic. Most countries have banned or are working towards phasing out the use of microbeads in beauty products, and many countries, following the lead of Bangladesh in 2002, are banning the provision of single-use plastic bags. Other measures include banning plastic straws and promoting the use of reusable water bottles and recyclable or compostable packaging. ■

The throwaway society cannot be contained – it has gone global. We cannot store and maintain or recycle all our stuff.
Charles J. Moore

A northern gannet is entangled in the plastic rings of a six-pack. Birds that scavenge along the shore such as seagulls are especially prone to being caught in such debris.

WATER IS A PUBLIC TRUST, AND A HUMAN RIGHT

THE WATER CRISIS

IN CONTEXT

KEY FIGURE
Maude Barlow (1947–)

BEFORE
1983–85 Droughts in Ethiopia, Eritrea, and Sudan cause 450,000 deaths.

1990 The desiccation of the Aral Sea is declared the world's worst ecological disaster of the 20th century by the UN Environment Programme.

2008 The United Nations estimates that around 42,000 people die every week from diseases related to bad water and poor sanitation.

AFTER
2011–17 California suffers one of its worst droughts on record. It impacts on agriculture, nature, and daily life.

2017 Water campaigner Maude Barlow reveals that half of China's rivers have disappeared since 1990.

Life requires access to clean water; to deny the right to water is to deny the right to life. The fight for the right to water is an idea whose time has come.
Maude Barlow

Indians queue for water in a slum area of Hyderabad in 2007. India suffered a severe water crisis in 2018, and demand is projected to be twice the available supply by 2030.

In 2008, Canadian activist Maude Barlow argued that water shortage had become the most pressing ecological and human crisis of the 21st century. Stressing that water is a "Commons" (a shared resource) and that access to water is a fundamental human right, she set out how wastage, pollution, and overconsumption meant that the water cycle – the constant exchange of water between Earth's surface and the atmosphere – could not be relied on to provide water for evermore. She said that shortage of water was already a crisis in the developing world, where the burden is borne particularly by women and children who collect water – and unless drastic action is taken, the rest of the world will be affected too.

About 1.1 billion people lack easy access to water, and 2.7 billion find water scarce for at least one month of the year. Although 70 per cent of Earth's surface is covered by water, almost all of it is saline ocean water. Only 0.014 per cent of the planet's water is both fresh and easily accessible. It is obtained mostly from rivers, lakes, and underground aquifers (rock containing groundwater). People use water to drink, wash, irrigate crops, and run industry, and since all plants and terrestrial animals require fresh water to live, all are affected by the water crisis.

Wasted water

A larger human population uses more water, and a large proportion of that is wasted, especially in developed countries, where people on average use about 10 times more than those in the developing world. Sources of fresh water have dried up (for example, much of the Rio Grande between Mexico and the US) or are becoming too polluted to

See also: The ecosystem 134–137 ▪ Pollution 230–235 ▪ Acid rain 248–249 ▪ Overpopulation 250–251
▪ Depletion of natural resources 262–265

Distribution of the world's water

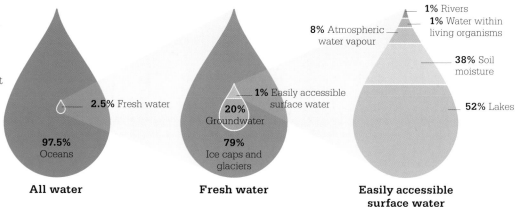

Easily accessible fresh water is a very fragile resource. Only a tiny fraction of the total amount of water available on our planet is immediately fit for human consumption.

2.5% Fresh water

97.5% Oceans

All water

1% Easily accessible surface water

20% Groundwater

79% Ice caps and glaciers

Fresh water

1% Rivers
1% Water within living organisms
8% Atmospheric water vapour
38% Soil moisture
52% Lakes

Easily accessible surface water

use. The Ganges in India and the Citarum in Indonesia are two of the most polluted rivers in the world. At the current rate of consumption this situation will deteriorate further. By 2030, two-thirds of the world's population may face shortages. Ecosystems will also suffer.

Increased demand
Human use of fresh water has tripled since about 1970, and demand is increasing by 64 billion cubic metres (2,260 billion cubic feet) – due in part to the population growing by 80 million people each year. The rise in demand has also been driven by changing lifestyles and eating habits that require more water per person. The production of biofuels has also risen sharply, with significant impact on water demand. Between 1,000 and 4,000 litres (260–1,060 gallons) of water are needed to make 1 litre (about ¼ gallon) of biofuel.

In the last century, half of Earth's wetlands have disappeared to make way for farmland or development, or because groundwater has been removed from aquifers faster than it has been replaced. A reduction in wetlands means plants and animals dependent on them have also gone. Nearly half of all drinking water comes from aquifers. About 1,000 cubic km (240 cubic miles) is taken from these every year. Two-thirds is used for irrigation, 22 per cent for »

Maude Barlow

Born in Toronto, Canada, in 1947, Maude Barlow is an activist and water policy critic. She is the author or coauthor of 18 books, including the bestseller *Blue Gold: The Fight to Stop the Corporate Theft of the World's Water*. Barlow formerly served as an adviser on water to the United Nations, and led moves to have water recognized as a basic human right. In 2012, she helped to found the Blue Planet Project, which campaigns for the right to water. Barlow chairs the Council of Canadians social action group, and was one of the "1000 Women for Peace" nominated for the 2005 Nobel Peace Prize. In 2008, she received the Citation of Lifetime Achievement, Canada's highest honour for environmentalism.

Key works

2002 *Blue Gold: The Fight to Stop the Corporate Theft of the World's Water*
2007 *Blue Covenant: The Global Water Crisis and the Coming Battle for the Right to Water*
2014 *Blue Future: Protecting Water for People and the Planet Forever*

A stranded ship on the dried-up bed of the Aral Sea. The loss of such a large body of water has had a devastating effect on agriculture, climate, and the local fishing industry.

The desiccation of the Aral Sea

The disappearance of most of the Aral Sea, once the world's fourth-largest lake, in Kazahkstan has been a huge ecological disaster. In the early 1960s, the two main rivers that fed the lake were diverted to irrigate millions of cotton plants across central Asia. In June 2004, the UN warned that the lake could dry up completely unless measures were taken to save it. It was then receiving only 10 per cent of the water that it once did, had divided into several smaller lakes, and contained only one-tenth of its 1960 volume of water. Large areas are now desert. Most of the lake's fish and other aquatic life disappeared with its water. Once fishermen here could catch Syr Darya sturgeon, but its numbers declined sharply when the lake shrank and became more saline. Efforts to replenish the waters have achieved an increase in surface area and depth, and fish populations are now increasing.

domestic use, and 11 per cent for industry. However, most aquifers replenish much more slowly than they can be emptied, so water yields reduce with use. If the water table falls, some lakes and rivers dry up. About half the total length of China's rivers has been lost since 1990. In North America, the Great Lakes are shrinking, Lake Winnipeg is threatened, and the massive Ogallala aquifer is being depleted. There are even water supply problems in Brazil, which is the most water-rich nation on Earth. As the situation worsens, it becomes a growing source of conflict.

Water scarcity

There are two types of water scarcity. Physical water scarcity affects regions that naturally do not have plentiful water, such as North Africa, the Arabian Peninsula, large areas of central and south Asia, northern China, and the southwest United States. In contrast, economic water scarcity occurs when water is available but the infrastructure does not exist to utilize it. This is the situation in much of sub-Saharan Africa and parts of Central America. People

Water stress around the world

KEY

- None
- 1–9 months
- 10–12 months
- Major cities experiencing water scarcity

This map illustrates the average exposure of water users to water stress – the ratio of total withdrawals to total renewable supply in a given area. A higher proportion of withdrawals means that more water users are competing for limited supplies.

London • New York • Los Angeles • Mexico City • Istanbul Beijing • Tokyo • Osaka • Shanghai Cairo • Delhi • Mumbai • Rio de Janeiro • Sao Paulo

> The world has not really woken up to the reality of what we are going to face in terms of the crises as far as water is concerned.
> **Rajendra Pachauri**
> *IPCC Chair*

living in these areas may have to spend hours each day walking to the nearest supply of water. Many children miss out on an education because they are collecting water.

Wildlife concerns

The water crisis is bad for humans and can mean extinction for some animals and reduction in numbers for others. Populations of the Amazon river dolphin, which lives in the Amazon and Orinoco river basins in South America, for example, have been much reduced, partly by the increase of heavy metal pollution from mining but also by the construction of dams, which restrict the migration of fish, the dolphins' food, to their spawning grounds. Elsewhere, in China, the world's largest amphibian, the Chinese giant salamander, has also become critically endangered by dams being built for water storage and hydro-electric power. Such engineering works change the natural flow of rivers, upsetting the creature's habitat.

A holistic view of ecosystem management is crucial to prevent the water crisis from getting much

worse. For example, a sewage treatment plant run on "clean" energy can provide the wastewater needed to fertilize biofuel crops, which in turn can be used to purify the water – without emitting greenhouse gases.

Drinkable waste water

New technologies can also convert wastewater directly into drinkable water – a process that has been energy-hungry in the past. The Intergovernmental Panel on Climate Change (IPCC) stresses that water management policies can lead to higher greenhouse gas emissions. However, that is not the case if the conversion is fuelled using solar energy, which is starting to take over from oil to power desalination plants in the Middle East. In parts of the world, there is seasonal heavy rain – for example, in countries with a monsoon – but it runs off into polluted rivers and cannot be used. Rainwater catchment and storage schemes would help.

Other helpful initiatives include reducing pollution, cutting irrigation and industrial wastage, providing new technological solutions for developing countries, and reaching international agreements – after all, water catchments do not stick to national boundaries. ∎

> There is no water-rich country in the world that is not facing problems.
> **Maude Barlow**

Salisbury Water

In Adelaide, South Australia, an innovative water recycling system in use in the suburb of Salisbury has reduced extraction from the Murray River and aquifers by about a half. Wastewater from the local sewage treatment works and rainwater from drains are treated, and then directed into a series of 50 wetlands. These contain reedbeds and other aquatic vegetation that further clean the water. The recycled, non-drinking water from the wetlands is then piped to the inhabitants of Salisbury to use for flushing toilets, watering gardens, washing cars, and filling ornamental ponds.

In addition to providing a more sustainable source of water, the system has boosted biodiversity within the newly established wetlands. Among the birds that are currently resident or visitors are ducks, spoonbills, herons, pelicans, cormorants, and migratory waders, along with species of amphibians and fish, and many aquatic invertebrates.

Salisbury's recycled water has environmental benefits including reduced demand on existing water resources and improved biodiversity through the newly created wetlands.

ENVIRONM
AND CONS

ENTALISM
ERVATION

Francis Bacon's work espouses the idea that man has **dominion over nature** – a view that is later termed "**imperial ecology**".

Written in a cabin in the woods, Henry David Thoreau's book **Walden** presents a **romanticized view** of the natural world.

The first working photovoltaic **solar cell** panel is built by inventor **Charles Fritts** in the US.

UNESCO launch their **Man and the Biosphere Programme** to encourage economic development that is **sustainable** and **eco-friendly**.

c.1620   **1854**   **1883**   **1971**

1789   **1864**   **1966**

Gilbert White's *Natural History and Antiquities of Selborne* records in great detail the **wildlife** around his rural home.

George Perkins Marsh warns of the **destructive impact** that **human action** is having on nature.

Lynn White argues that Western – largely Christian – **anthropocentric** worldviews have placed humankind in an **environmental crisis**.

Early in the 17th century, English philosopher and scientist Francis Bacon wrote of the need to control and manage nature. By the end of the 18th century, in contrast, English vicar Gilbert White was writing in favour of a peaceful coexistence between people and the natural world. Yet in his lifetime, powerful new steam engines unleashed the ravages of industrialization – the reaction against which would later provide a major impetus for the environmental movement.

Possibly the first systematic analysis of humanity's destructive impact was American diplomat George Perkins Marsh's 1864 book, *Man and Nature*. Marsh warned, among other things, that deforestation could lead to the creation of deserts, and he pointed out that resource scarcity was generally the result of human actions rather than natural causes.

Renewable and clean

Before the Industrial Revolution, most energy had been renewable – the energy of human and animal labour, wind- and watermills, and sustainable wood. From the mid-18th century there was a dramatic shift to coal. The most efficient fuel for firing furnaces and factories, it came at a price – choking pollution and the then-unknown rise in atmospheric greenhouse gases.

In the 1880s, however, the key to a new form of renewable energy was provided by American inventor Charles Fritts – a photovoltaic cell, which could convert solar power to electricity. German industrialist Werner von Siemens soon saw its potential for producing limitless energy, but it took a century for solar power to be widely adopted. "Clean" hydroelectric power was the first sustainable source capable of generating electricity on a large scale – joined in the late 20th century by modern wind power, and tidal, wave, and geothermal energy.

An environmental ethic

In 1937, following the devastating "Dust Bowl" caused by intensive farming in the US, President Franklin D Roosevelt wrote, "A nation that destroys its soils destroys itself". In 1949, American ecologist and forester Leopold Aldo articulated a recurring theme in environmental thought, by advocating a "land ethic", a responsible relationship between people and their local environment.

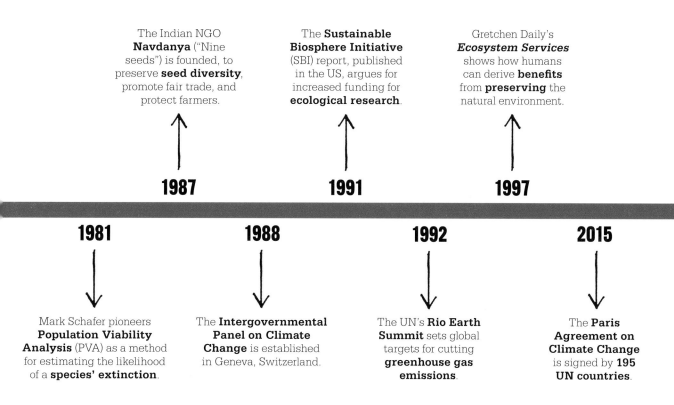

The Indian NGO **Navdanya** ("Nine seeds") is founded, to preserve **seed diversity**, promote fair trade, and protect farmers.

The **Sustainable Biosphere Initiative** (SBI) report, published in the US, argues for increased funding for **ecological research**.

Gretchen Daily's *Ecosystem Services* shows how humans can derive **benefits** from **preserving** the natural environment.

1987

1991

1997

1981

1988

1992

2015

Mark Schafer pioneers **Population Viability Analysis** (PVA) as a method for estimating the likelihood of a **species' extinction**.

The **Intergovernmental Panel on Climate Change** is established in Geneva, Switzerland.

The UN's **Rio Earth Summit** sets global targets for cutting **greenhouse gas emissions**.

The **Paris Agreement on Climate Change** is signed by **195 UN countries**.

The post-war period saw many governments legislating to ensure the quality of air and drinking water and establish national parks and other protected areas. In 1968, the world first found its collective voice, when UNESCO (the United Nations Educational, Scientific and Cultural Organization) held the Paris Biosphere Conference. This resulted, three years later, in the creation of the Man and Biosphere Programme.

Growing awareness

Public concern for the environment was marked by the establishment of major conservation organizations. The International Union for the Conservation of Nature had been established in 1948, and it was followed by the World Wildlife Fund (1961), Friends of the Earth (1969), and Greenpeace (1971). After the 1969 massive oil spill in Santa Barbara, California, US senator Gaylord Nelson proposed the idea of a national event to highlight the varied threats to the environment. On the first Earth Day, which took place on 22 April 1970, millions turned out on marches across the US. The scale of the event helped the passage of the Clean Air, Clean Water, and Endangered Species Acts and led to the creation of the United States Environmental Protection Agency (EPA).

In 1973, German economist Ernst Schumacher used the term "natural capital" in his best-seller *Small is Beautiful* to describe how ecosystems provide us with complex services. The concept inspired American environmentalist Gretchen Daily and others, who argued that ecosystems are capital assets which, when properly managed, provide a flow of vital goods and services.

International cooperation

Two UN agencies – the World Meteorological Organization and the UN Environment Programme – established the Intergovernmental Panel on Climate Change (IPCC) in 1988 to assess the risk of human-induced climate change.

The IPCC continues to monitor climate change. In 1992, the Rio Earth Summit, a UN initiative, was unprecedented in both its size and the scope of its concerns. It was the first of a number of international gatherings seeking, with much success, to get global agreement on greenhouse emissions. International cooperation is now seen as key to saving Earth's environment. ∎

THE DOMINION OF MAN OVER NATURE RESTS ONLY ON KNOWLEDGE

HUMANKIND'S DOMINANCE OVER NATURE

The Renaissance ("rebirth")
between the 14th and 17th
centuries is primarily
associated with the arts and
culture that flourished across
Europe as the Catholic Church's
authority began to be challenged.
It was also a time of extraordinary
scientific advances, which some
saw as the beginnings of a
"scientific revolution". Discoveries
in astronomy, physics, and medicine
gave rise to the idea that science
could tell humans everything about
the universe, and that knowledge
would make humans its masters.

Many scientists of the era believed
that humans had a privileged place
in a universe created by God for
humanity to inhabit. English
philosopher and scientist Francis
Bacon (1561–1626), a pioneer in the
development of scientific method,
reinforced this idea; the natural
world, in his view, existed to
provide for humans, and should
be conquered and exploited.

Bacon's view later became
known as "Imperial Ecology" – the
idea that humanity's knowledge of
science and technology should be
used to gain dominance over the
natural world. Imperial ecology
became the predominant ideology
throughout the Renaissance, the
Enlightenment – an 18th century
movement dedicated to the pursuit
of knowledge – and later the
Industrial Revolution of the 18th
and 19th centuries. ∎

Sir Francis Bacon sits for a portrait
in parliamentary robes. Bacon had
an illustrious political career; knighted
in 1603, he served as Lord Chancellor
of England from 1618 to 1621.

See also: Global warming 202–203 ▪ A holistic view of Earth 210–211
▪ Pollution 230–235 ▪ Environmental ethics 306–307

NATURE IS A GREAT ECONOMIST
THE PEACEFUL COEXISTENCE OF HUMANKIND AND NATURE

IN CONTEXT

KEY FIGURE
Gilbert White (1720–93)

BEFORE
4th century BCE Diogenes, a Greek philosopher, advocates forgoing the comforts of civilization in favour of a life "in accord with nature".

1773 American naturalist William Bartram starts his field studies of the wildlife of the southeast US, documented in his 1791 book, *Travels*.

AFTER
1949 American ecologist Aldo Leopold publishes *A Sand County Almanac*, exploring the idea of humans' "land ethic", or responsibilities towards nature.

1969 Friends of the Earth is founded in the US – initially as an anti-nuclear group – marking the beginning of the modern Green movement.

I n the late 18th century, rapid advances in science and technology – particularly in Britain – led to widespread industrialization and urbanization as people sought to control and exploit the natural world. There were, however, many in Britain who still lived and worked on the land. Among the educated rural class, some had a fascination for both science and nature. From this group, a new generation of naturalists emerged, suggesting that humans should learn from their scientific studies to live in harmony with the natural world rather than attempt to dominate it.

Arcadian ideology

In 1789, rural parson and naturalist Gilbert White published his *Natural History and Antiquities of Selborne*, which became a seminal work in what was later called "Arcadian Ecology". Educated at Oxford and a keen gardener and ornithologist, White closely observed the wildlife around his Hampshire village, and made meticulous notes from 1751

From reading White's *Selborne* I took much pleasure in watching the habits of birds, and even made notes.
Charles Darwin

onwards. The book was compiled from his correspondence about his findings with several like-minded naturalists, but it was more than simply a collection of data. White's engaging and often poetic style sent a persuasive message; his work rejected the "imperial" idea of conquering nature, and instead encouraged a balance between humans and the natural world – like that of the Ancient Greeks' mythical idyll of Arcadia, for which White's approach was named. ∎

See also: Romanticism, conservation, and ecology 298 ▪ Environmental ethics 306–307 ▪ The Green Movement 308–309 ▪ Halting climate change 316–321

IN WILDNESS IS THE PRESERVATION OF THE WORLD
ROMANTICISM, CONSERVATION, AND ECOLOGY

IN CONTEXT

KEY FIGURE
Henry David Thoreau
(1817–62)

BEFORE
1662 English diarist John Evelyn's work *Sylva*, advocating forest conservation, is presented to the Royal Society.

1789 Gilbert White publishes his *Natural History of Selborne*, inspiring a reaction against "imperial ecology".

AFTER
1872 A bill creating the first US national park, Yellowstone, is signed into law by President Ulysses S. Grant.

1892 In San Francisco, Scottish–American conservationist John Muir founds The Sierra Club.

1971 The UNESCO "Man and the Biosphere" project is launched.

I n many ways, Romanticism – a new cultural movement that emerged towards the end of the 18th century – was a reaction to the scientific rationalism of the Enlightenment. As industrialization took hold in urban areas, writers, artists, and composers began increasingly to glorify the natural world. The now prosperous middle classes were particularly inspired by Romantic portrayals of nature, and took up leisure pursuits such as hiking and mountaineering. The Romantic movement even affected scientific attitudes to nature by inspiring interest in the nascent field of ecology and the environmental movement.

The wild world
A key figure in the Romanticization of nature was Henry David Thoreau, an American writer from Concord, Massachusetts. His book *Walden* (1854) described his time living in a cabin in the woods by Walden Pond. Thoreau advocated preserving nature not for its own sake, but as a necessary resource in sustaining human life and a kind of spiritual enrichment. While Thoreau's "wilderness" was not far removed from modern life, his Romantic portrayal of the natural world significantly influenced the conservation movement in the US and helped to inspire its National Parks system. ∎

Thoreau's simple hut at Walden Pond appeared on the title page of this 1875 edition of *Walden*. Thoreau claimed he went to the wilderness to be free of the obligations of city life.

See also: Global warming 202–203 ▪ A holistic view of Earth 210–211 ▪ Urban sprawl 282–283 ▪ The Green Movement 308–309

MAN EVERYWHERE IS A DISTURBING AGENT
HUMAN DEVASTATION OF EARTH

IN CONTEXT

KEY FIGURE
George Perkins Marsh
(1801–82)

BEFORE
1824 Joseph Fourier, a French physicist, describes the greenhouse effect – later identified as a contributing factor in global warming.

1830s Scientists posit that the Dutch colonization of Mauritius in the 17th century caused the dodo to become extinct.

AFTER
1962 In the US, Rachel Carson's *Silent Spring* describes the harmful effect of pesticides on the environment.

1971 Greenpeace is founded by American environmentalists.

1988 The Intergovernmental Panel on Climate Change (IPCC) is set up to assess the "risk of human-induced climate change".

The widely held view that the natural world existed to be exploited by humankind saw a major rebuttal in the form of the 19th-century environmental movement. Arguments against the "imperial" attitude to nature, which had prevailed since the dawn of global exploration in the late 15th century, began with naturalists such as Gilbert White, and were echoed in the sentiments of Romanticism. Such ideas tended to focus on the idealization of nature, rather than examining the harm done by human conquests of the natural world.

In contrast to the emotive Romantic responses to modernism, American polymath George Perkins Marsh took a close look at humans' impact on the environment and suggested changes. Marsh was horrified by the destructive effects of human management of natural resources. In his book *Man and Nature, Or, Physical Geography as Modified by Human Action* (1864), he pointed in particular to the mass deforestation which had virtually desertified some areas of the US.

George Perkins Marsh in an engraving from 1882. As well as being an environmentalist, the Vermont native was also a skilled linguist, lawyer, congressman, and diplomat.

Marsh believed that people must be made aware of their destructive impact and find new ways of managing natural resources to preserve the natural equilibrium. An activist as well as writer, he helped to establish the principle of protected areas, and inspired the idea of sustainable resource management that became a core element of the 19th-century environmental movement. ∎

See also: Global warming 202–203 ▪ A plastic wasteland 284–285 ▪ Humankind's dominance over nature 296 ▪ Environmental ethics 306–307

SOLAR ENERGY
IS BOTH WITHOUT LIMIT AND WITHOUT COST

RENEWABLE ENERGY

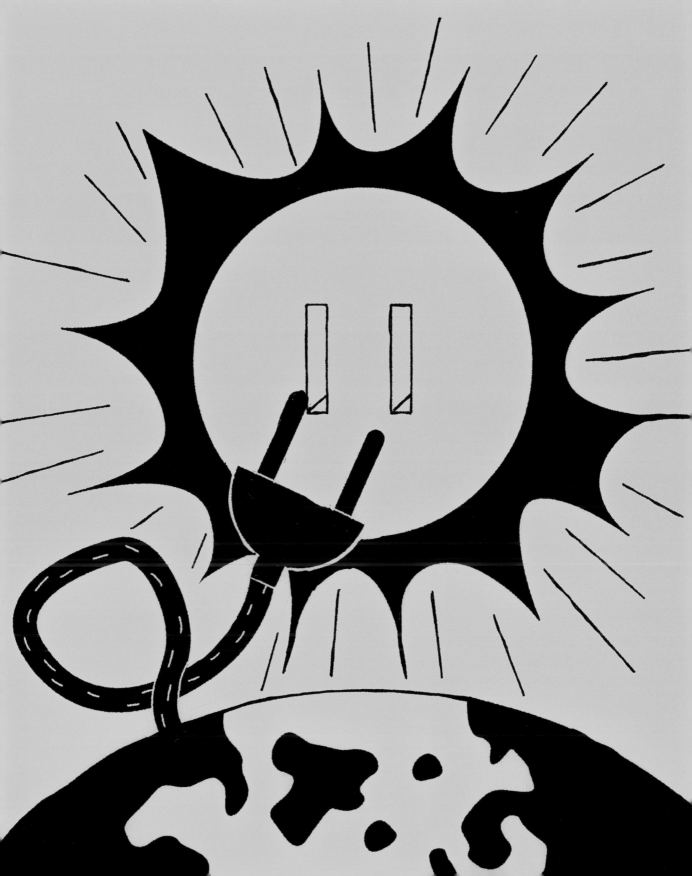

By the late 19th century, fears were already growing in industrial Europe that the world could not rely on fossil fuels forever. When the first working selenium solar cell panel was built in 1883 by American inventor Charles Fritts, the progressive German industrialist Werner von Siemens immediately recognized its huge potential for renewable energy. He declared: "the supply of solar energy is both without limit and without cost." Yet, because no one at the time understood exactly how selenium created photoelectricity, and Siemens's calls for more experiments went unheeded, solar cells were not developed until the 1950s. Today, solar power is the fastest growing source of new energy and predicted to dominate future growth in renewables.

Renewables v fossil fuels

Human civilizations have drawn on renewable energy for millennia – from burning firewood to harnessing the wind to propel sailing ships. Renewable sources such as sunlight or tidal power are not at all depleted by use.

By contrast, fossil fuels – such as coal, oil, and gas – have taken thousands of years to form, and when exhausted, cannot be replaced. Natural gas is an abundant fossil fuel, but its extraction can cause environmental problems, such as earth tremors and water contamination. Nuclear power, although sustainable for a long period of time, is not considered renewable because its production requires a rare type of uranium ore.

Energy sources such as solar power, wind, and water are also generally "clean" – unlike fossil fuels, they produce zero or very low greenhouse gas emissions. However, not all renewables are clean. People have burned wood and animal dung for heat and light for hundreds of thousands of years. Trees can be replanted and animals produce more dung, so the practice is sustainable, but burning such fuels also emits carbon dioxide (CO_2), which is one

The Ivanpah solar plant in the Mojave Desert, California, generates enough concentrated solar power to serve more than 140,000 homes at peak hours of the day.

See also: Global warming 202–203 ▪ Pollution 230–235 ▪ Ozone depletion 260–261 ▪ Depletion of natural resources 262–265 ▪ Waste disposal 330–331

Solar energy is derived from **solar radiation**.

→

Solar radiation can be used to **produce energy** on Earth.

↓

The **supply** of this energy will not stop as long as the **Sun exists**.

←

Solar energy is without limit.

reason why, unlike other forms of renewable energy, they are not classed as "alternative" sources.

Renewable, clean energy will have huge long-term benefits for populations and ecosystems. It reduces pollution, mitigates against global climate change, builds sustainability, and increases the energy security of countries. If it can be provided cheaply enough, it will also pull many people out of poverty. In some 30 countries, renewable energy now makes up more than 20 per cent of the supply.

Solar energy
The Sun's power could supply the world's energy needs several times over. The International Energy Agency (IEA) believes that – in the short term – it has the greatest potential of all the renewables. Its radiation can be converted directly into electricity via photovoltaic cells (as with solar panels on buildings) or indirectly by using lenses or mirrors to create heat, which can be converted to electricity. This is called concentrated solar power.

Solar panels on a roof can heat domestic water. Sunlight can be employed to desalinate water

through an evaporative process, first adopted by 16th-century Arab alchemists and used on an industrial scale in Chile in the late 19th century. In the developing world, solar disinfection is bringing safe drinking water to more than two million people; the process involves using solar heat and ultra-violet light to kill pathogens.

Wind power
For more than 2,000 years, people have built windmills to pump water and grind grain. Today, wind farms onshore and offshore account for around 9 per cent of renewable energy consumption. A wind turbine's huge blades turn around a rotor attached to a main shaft, which spins a generator to produce electricity. Wind power is now the leading area of energy growth in Europe, the US, and Canada. Almost 50 per cent of Denmark's energy comes from the wind, and in Ireland, Portugal, and Spain the figure is 20 per cent. Its global potential is thought to be around five times its present level.

It is only economic to build wind farms where there is regular wind, however, so the potential is not »

Artificial photosynthesis

Since the early 1970s, scientists have been working to develop the technology to mimic the process of photosynthesis and create liquid fuels from carbon dioxide, water, and sunlight. As all three are plentiful, if the process can be replicated it could produce an endless, relatively inexpensive supply of clean fuel and electricity.

There are two crucial steps: to develop catalysts that use solar energy to split water into oxygen and hydrogen, and to create other catalysts that convert hydrogen and carbon dioxide into an energy-dense fuel, such as liquid hydrogen, ethanol, or methanol. Scientists at Harvard University recently used catalysts to split water into oxygen and hydrogen, then fed the hydrogen, plus carbon dioxide, to bacteria. The bio-engineeered bacteria converted the carbon dioxide and hydrogen into liquid fuels. The next challenge is to transfer a successful lab experiment into something commercially viable.

This solar fuel generator mimics the way plants turn sunlight and carbon dioxide in the air into energy and oxygen.

Hot-dry-rock energy

Natural rock fractures that bring hot water to the surface from deep underground have been described as the "low-hanging fruit" of geothermal energy because they are easy to exploit. However, they are rare in most parts of the world. The vast majority of geothermal energy locked beneath Earth's surface is in dry, non-porous rock.

The enhanced geothermal system (EGS), a similar process to fracking for natural gas and oil, aims to overcome this problem by fracturing rock strata and injecting water into it at a great depth. The water is heated by contact with the rock, then returns to the surface through production wells. Depending on the economic limits of drill depth, the technology might be feasible across many parts of the world, but there are risks. Like fracking, EGS can cause small earth tremors, and so should not be conducted near populated areas or power stations.

… the wind and the sun and the earth itself provide fuel that is free, in amounts that are effectively limitless.
Al Gore
American environmentalist and former US Vice President

evenly spread around the globe. Offshore wind is generally stronger and more regular than onshore. Floating turbines can generate wind energy far offshore, unlike seabed-anchored wind turbines, which have to be sited in shallow water close to the coastline.

Geothermal energy

The heat in Earth's interior is derived both from the original formation of the planet and from the radioactive decay of materials within it. People have bathed in hot pools, where geothermally heated water reaches the surface, since Paleolithic times. Ancient Romans made use of it to heat their villas. Today it is employed to generate electricity in at least 27 different countries, with the United States, the Philippines, and Indonesia the world's leading producers.

Geothermal heat is also utilized directly to heat homes and roads in Iceland. Technology is now being developed that will use geothermal hot water to operate desalination plants. The only drawback of this renewable energy source is that it is concentrated near tectonic plate boundaries, where Earth's mantle heat rises close to the surface. The potential is much greater, but drilling for deep resources is very expensive.

Water power

Since water is 800 times denser than air, even a slow-moving flow can yield considerable amounts of energy if harnessed, for instance, by dams or tidal barrages that drive

China's Three Gorges Dam, the world's largest hydroelectric dam, was completed in 2012. Critics point to its ecological impact on the Yangtze River's habitat and biodiversity, and the risk for local people of flooding and landslides.

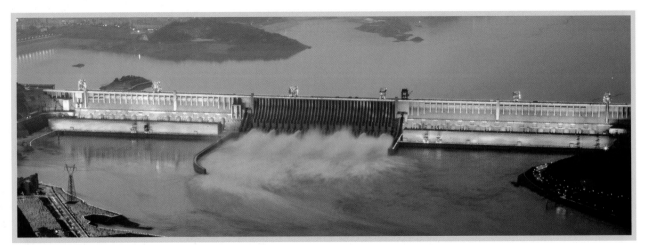

turbines connected to generators. China is the biggest producer of hydroelectric power (HEP), with 45,000 small installations in addition to "big-dam" schemes such as the Three Gorges project, whose 32 giant turbines have the capacity to produce 22,500 megawatts of electricity. A downside of big HEP schemes is that reservoirs created upstream of the dam can flood good farmland, forcing people to relocate and destroying ecosystems. Despite this, the IEA has estimated that by 2023, hydropower will be meeting 16 per cent of the global demand for electricity.

Tidal power is based on the same principle: moving water turns turbines, which drive electricity generators. The source of energy from a tidal scheme is reliable, generating power each time the tide ebbs and flows, but such schemes are expensive to construct. At present, the largest is the Sihwa Lake Tidal Power Station in South Korea, which was completed in 2011 and has reduced the annual amount of CO_2 the nation generates by 286,000 tonnes (315,000 tons). Wave power involves the capture of wave energy through a converter. The first commercial wave power scheme began off the west coast

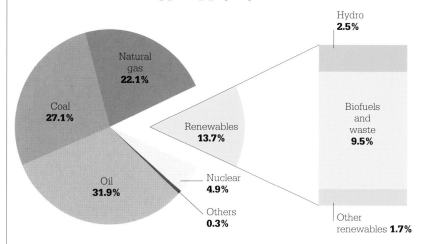

World energy supply by source in 2016

A pie chart illustrating the sources for the total energy produced and supplied throughout the world in 2016, according to data published by the IEA. "Others" includes non-renewable wastes and other sources not included elsewhere such as fuel cells.

… someday, renewable energy will be the only way for people to satisfy their energy needs.
Hermann Scheer
President, European Association for Renewable Energy

of Scotland in 2000, and the first multi-generator wave farm opened at Aguçadoura in Portugal in 2008.

Biomass
Organic matter from plants or animals is known as biomass. It contains stored energy because plants absorb the solar power they need for growth via photosynthesis, and creatures absorb that energy either from the plants they eat or from what their prey consumes. Creating a renewable fuel from plant, animal, and human waste products such as straw, dung, and garbage may seem an attractive option, and some coal-fired power stations have been converted to wood-burners. Burning biomass produces heat, electricity, and transport fuels, such as ethanol and biodiesel. However, biomass energy is not necessarily "clean". Burned as a fuel, biomass releases CO_2, and creates air and particle pollution. Clearing prime forest for its wood or to cultivate biomass crops, such as

grain for biofuels, can also damage the environment. Perhaps because of this, biomass is a more common fuel in nations that cannot afford other renewable options. According to the IEA, the majority of solid biofuel supply in 2016 took place in Africa, accounting for 33.2 per cent.

The future
As growth in renewables increases, the advantages of each type must be balanced against their adverse effects – from biomass pollution to the reported role of wind turbine blades in the deaths of migrating birds. In 2014, the IEA predicted that renewables would provide 40 per cent of global energy needs by 2040. In 2018, the IEA further predicted that renewables would account for almost a third of all world electricity by 2023, with solar power taking the biggest share. Energy from ocean currents could also generate huge amounts of electricity, as could large arrays of solar panels in space or floating on the seas. ∎

THE TIME HAS COME FOR SCIENCE TO BUSY ITSELF WITH THE EARTH
ENVIRONMENTAL ETHICS

IN CONTEXT

KEY FIGURE
Aldo Leopold (1887–1948)

BEFORE
1894 In *The Mountains of California*, Scottish–American naturalist John Muir describes his travels through wild places in California, evoking the deep spirituality and adventure he feels when in the wilderness.

1909 Gifford Pinchot's *The ABC of Conservation* argues that future generations should be able to utilize Earth's natural resources.

AFTER
1968 US academic Paul R. Erlich and his wife, Anne, publish *The Population Bomb,* warning of the dangers of human population growth.

1970 On 22 June, the first Earth Day is celebrated in the US. It becomes an annual global celebration of environmental education and reform.

A t its heart, the discipline of environmental ethics extends the boundaries of ethics beyond humans, and into the natural world. It forces humans to question their role in the environment, their responsibility to the planet itself, and their duty to future generations.

The field of environmental ethics grew out of an urgent sense of encroaching crisis, expressed in both popular and academic writings. In 1962, the book *Silent Spring*, written by US biologist and conservationist Rachel Carson,

A thing is right when it tends to preserve the integrity, stability, and beauty of the biotic community. It is wrong when it tends otherwise.
Aldo Leopold

documented the serious impact of pesticides on the environment, and brought these issues to the forefront of American public debate. Six years later, US ecologist Garett Hardin's article "The Tragedy of the Commons" outlined the danger of overusing shared resources and allowing the human population to grow unchecked.

Other writers viewed the impending crisis from a more philosophical perspective. Aldo Leopold's "land ethic", outlined in *A Sand County Almanac* (1949), placed human beings on an equal footing with others in a wider ecosystem. As one part of a larger whole, our ethical concerns should be with the healthy functioning of the entire ecosystem, rather than simply the advancement of human health and happiness.

In his seminal 1966 lecture "The Historical Roots of Our Ecologic Crisis", later published as an article, the US historian Lynn White claimed that the environmental crisis was the fault of Western society's worldview. In particular, he blamed the Christian thinking that promoted anthropocentrism – the idea that humans are superior to all other

See also: Endangered habitats 236–239 ▪ Pesticides 242–247 ▪ Depletion of natural resources 262–265 ▪ Ecosystem services 328–329

The remote, subalpine Mineral King Valley has survived the threat of development. It remains an ecosystem that aims to benefit all – following Aldo Leopold's "land ethic" principle.

creatures, leading to the view that nature was created for humanity's use and exploitation.

Ethical dilemmas

Environmental ethics questions the moral imperatives behind sustainability and stewardship by asking if the motivations are grounded in anthropocentrism, or in the protection of the natural world because it inherently deserves protection. These questions have played out not only in philosophical arenas, but also in the legal and political spheres.

In 1969, the Sierra Club, an environmental lobbying group, challenged a US Forest Service permit allowing Walt Disney Enterprises to survey the Mineral King Valley in California – Disney

wanted to build a ski resort there. The Valley had no official protected designation beyond that of a game refuge, but the Sierra Club argued that the area should be preserved in its original state for its own sake. The suit went to the Supreme Court, which in 1974 ruled in favour of the Forest Service and Disney. By then, however, Disney's interest had waned; today the Valley is part of Sequoia National Park.

The battle between those who follow anthropocentric ethics and those who argue for ecocentric approaches has continued. It often takes place in political arenas, particularly with the increased prominence of globally sensitive issues such as climate change. Sustainable development has generally been an anthropocentric endeavour, to ensure future generations have their needs met. Environmental ethicists tend to argue that sustainability is only viable if it preserves the future of all members of the ecosystem. ▪

Aldo Leopold

Born in 1887, Aldo Leopold grew up in Burlington, Iowa. He received his degree from the Yale School of Forestry, after which he took a job with the US Forest Service. While there he was instrumental in the proposal to manage the Gila National Forest as a wilderness area, and in 1924 it became the first official Wilderness Area in the US. Leopold then moved to Wisconsin to continue his work in the Forest Service, and in 1933 became a Professor of Game Management at the University of Wisconsin. Leopold died in 1948, while helping to fight a grass fire. Most of his many essays on natural history and conservation were published posthumously in collections, such as *A Sand County Almanac*, that greatly influenced the emerging environmental movement.

Key works

1933 *Game Management*
1949 *A Sand County Almanac*
1953 *Round River: From the Journals of Aldo Leopold*
1991 *The River of the Mother of God: and Other Essays*

THINK GLOBALLY, ACT LOCALLY
THE GREEN MOVEMENT

IN CONTEXT

KEY FIGURES
David Brower (1912–2000),
Petra Kelly (1947–92)

BEFORE
1892 The Sierra Club is
founded in San Francisco,
California, by the Scottish–
American conservationist
John Muir.

1958 Environmentalists
protest against proposals
for a nuclear power plant
at Bodega Bay, California.

AFTER
1970 On 22 April, the first
Earth Day is held across the US.

1972 Environmentalist
candidates stand for election
in Tasmania, New Zealand,
and Switzerland.

1996 Ralph Nader stands as
candidate for President of the
US on the Green Party ticket.

The roots of the modern
"green movement"
developed in organizations
established in the late 19th and
early 20th centuries, such as the
Sierra Club. Faced with the threat
of increasing urbanization and
industrialization, the Sierra Club
sought to protect the natural
environment for people's enjoyment.

A greater awareness of humans'
relationship with the environment
led to the emergence of a more
politically active environmental
movement in the second half of the
20th century. This took off in the
1960s, when the Cold War was at
its height and the Cuban Missile
Crisis of 1962 brought the US and
the Soviet Union to the brink of
nuclear war, galvanizing calls for
nuclear disarmament among
many campaigners.

In this atmosphere, the idea
of conserving particular natural
sites, as in the national parks
system in the US and the UK,
gave way to a much broader
concept of environmentalism.
Several organizations emerged with
a strong activist agenda involving
mass protests and direct action.

Organized protest
One of the first of the activist
organizations was Friends of
the Earth. It was founded in the US
in 1969 by a group that included
conservationist David Brower,
a former leader of the Sierra Club,
with the aim of preventing the
building of nuclear power plants.
Politically active from the outset,
Friends of the Earth continues to
lobby governments across the world
and campaigns on a broad range of
environmental issues, emphasizing
the importance of sustainable
economic development. In 1971,
a small group of activists in North
America formed the Don't Make a

Only through care
for the environment can
the livelihoods of those
most dependent on it
be sustained.
Petra Kelly

See also: Citizen science 178–183 ▪ Pesticides 242–247 ▪ Humankind's devastation of Earth 299 ▪ Halting climate change 316–321

Activists in a dinghy patrol in front of two ships from the UK carrying illegal toxic substances as part of regular Greenpeace protests.

Wave Committee to protest against nuclear bomb testing by the US on the island of Amchitka, Alaska. The organization favoured direct action rather than political lobbying and chartered a boat to sail to the island in protest. The publicity generated by the group swayed public opinion and halted the tests. This was the first action of what was to become Greenpeace, an organization that continues to use direct action to challenge those engaged in environmentally damaging activities.

Green politics
During the 1970s, political parties with dedicated environmentalist manifestos emerged in several countries. For example, The Ecology Party was established in the UK in 1975, and the Green Party

formed in Germany in 1979. As the movement gained momentum, many smaller parties began to coalesce to form national, unified Green Parties.

In recent years, as issues such as pollution and climate change have risen up the news agenda, other established political parties have adopted environmentally friendly policies. ▪

We have everything
we need, save perhaps,
political will. But, you know
what … political will is a
renewable resource.
Al Gore

Petra Kelly

Born Petra Lehmann in Günzburg, West Germany, in 1947, Kelly later adopted the surname of her stepfather, an American army officer. When she was 12, the family moved to the US, where Kelly studied political science in Washington, D.C.

In 1970, Kelly returned to Europe. While working at the European Commission in Brussels, she joined Germany's Social Democratic Party, but grew disillusioned with traditional politics. She joined Germany's newly formed Green Party in 1979, and in 1983 was one of 28 members to be elected to parliament. Kelly campaigned on issues of environmentalism and human rights. In 1992, she and her companion, Green politician Gert Bastian, were found dead at her home in Bonn, apparently the result of a suicide pact.

Key works

1984 *Fighting for Hope*
1992 *Nonviolence Speaks to Power*
1994 *Thinking Green: Essays on Environmentalism, Feminism, and Nonviolence*

THE CONSEQUENCES OF TODAY'S ACTIONS ON TOMORROW'S WORLD
MAN AND THE BIOSPHERE PROGRAMME

IN CONTEXT

KEY ORGANIZATION
UNESCO

BEFORE
1925 The International Institute of Intellectual Cooperation – which aims to exchange intellectual ideas and improve quality of life – is set up in Paris, France.

1945 The United Nations Conference establishes the constitution of UNESCO.

AFTER
1983 First International Biosphere Reserve Congress takes place in Minsk, Belarus.

1995 Statutory framework of the World Network of Biosphere Reserves is agreed.

2015 The UN launches its 17 Sustainable Goals initiative.

2017 The US withdraws 17 sites from the UNESCO World Network of Biosphere Reserves, but 23 new sites are added elsewhere.

During the second half of the 20th century, there was an increasing global awareness of the importance of the relationship between humans and the natural world. This led, in 1971, to the United Nations Educational, Scientific, and Cultural Organization (UNESCO) launching the Man and the Biosphere Programme (MAB). This is an intergovernmental programme devoted to encouraging environmentally sustainable and equitable economic development, while protecting natural ecosystems.

UNESCO was founded after World War II with the aim of fostering "the building of peace, the eradication of poverty, sustainable development and intercultural dialogue through education, the sciences, culture, communication and information". As such, it was in a unique position to examine carefully the relationship between people and the environment.

Global network
The organization began by setting up a number of internationally recognized protected sites, known

---

Humankind is **altering the environment** with processes such as **deforestation** and **urban sprawl**.

→ Such **actions** have **consequences**.

↓

The MAB programme predicts the consequences of today's actions on tomorrow's world.

← **Data** gathered from **global MAB reserves** helps to **generate a picture** of what these consequences could be.

See also: Human activity and biodiversity 92–95 ▪ The ecosystem 134–137 ▪ The peaceful coexistence of humankind and nature 297 ▪ Renewable energy 300–305 ▪ Environmental ethics 306–307 ▪ Sustainable Biosphere Initiative 322–323

as the World Network of Biosphere Reserves (WNBR). These set out to show how human cultural and biological diversity are mutually beneficial and encourage the balanced integration of people with their natural environment. They also sought to find ways to manage natural resources efficiently for the benefit of the environment as well as its inhabitants.

There are now over 650 sites around the world, providing a platform for collaborative scientific and cultural research in a range of marine, coastal, and terrestrial ecosystems. Through the network, the programme monitors the effects of human activity on the biosphere, particularly examining climate change, and fosters the exchange of information.

Local knowledge

The MAB programme recognizes three interconnected functions of a biosphere reserve: conservation; sustainable development; and support though education and training. These objectives are

achieved by zoning areas within the reserve to protect core locations, whilst simultaneously providing places for appropriate and sustainable development by local inhabitants.

To this end, communities are encouraged to participate in the management of the reserve, and use their local knowledge of the area to make the best use of natural resources. The idea of educating people about the environment and

Moroccan women gather the health-giving fruits of the argan tree. These trees in the Arganeraie Biosphere Reserve are carefully sustained by the local population.

sharing knowledge across the World Network is key to the success of the project as a whole.

Conflicting opinions

The sites of the WBNR, as well as being of international scientific significance, are often culturally important to the host state. They are not nominated by UNESCO, but by national governments, and they remain under the jurisdiction of the states they are in. International recognition of their status does not impinge upon the rights of those states over the Biosphere Reserves.

In recent years, some states have chosen to manage certain sites as national rather than international reserves and have withdrawn them from the programme. Nevertheless, there has been a steady increase in sites nominated for the programme from governments around the world. ▪

UNESCO

UNESCO, an agency of the UN based in Paris, France, was founded in 1946 to promote international collaboration for peace and security. It was established in line with the United Nations Charter, through education, science, and culture. Today, the organization has 195 member states.

UNESCO continues the work begun by the League of Nations International Committee on Intellectual Cooperation in the 1920s, which was interrupted by

the outbreak of World War II. Today, members aim to achieve their objectives by sponsoring international educational and scientific programmes. These include dedicated projects that promote and protect human rights and sustainable development, while encouraging cultural diversity.

The organization is perhaps best known for establishing internationally recognized World Heritage Sites, which aim to preserve as many aspects as possible of the world's diverse cultural and natural heritage.

PREDICTING A POPULATION'S SIZE AND ITS CHANCES OF EXTINCTION

POPULATION VIABILITY ANALYSIS

IN CONTEXT

KEY FIGURE
Mark L. Shaffer (1949–)

BEFORE
1964 The IUCN publishes its first Red List of threatened mammal and bird species.

1965 In *The Destruction of California*, ecologist Raymond Dasmann charts the rapid loss of flora and fauna in the state.

1967 *The Theory of Island Biogeography* by Robert MacArthur and Edward O. Wilson explores island patterns of immigration and extinction.

AFTER
2003 Population viability analysis (PVA) of the Fender's blue butterfly is used to guide conservation in the US.

2014 PVA studies in the Sonoran Desert, US, help assess the response of birds and reptiles to climate change.

Population viability analysis (PVA), or extinction risk assessment, is a process used to estimate the probability that a population of a target species has the ability to sustain itself for a specific time, be it 10, 30, or 100 years. A key feature of PVA is the definition of minimum viable population sizes and minimum habitat areas – information which can then inform decisions on conservation priorities.

A tool for conservationists
PVA combines both statistics and ecology to calculate the fewest organisms required for a species to survive long-term in its preferred

Fender's blue butterfly was not seen after the 1930s and was deemed extinct until it was rediscovered in 1989. It is endangered, but small populations live in northwest Oregon, US.

habitat. This minimum number also dictates the amount of suitable habitat that the species needs. PVA is a useful tool for conservationists when lobbying governments and developers to give protected status to an area. Armed with a PVA, they can explain precisely why reducing a stretch of forest, heathland, or reedbed will threaten certain flora or fauna. Protecting an area that is extensive enough to support a large species also benefits many smaller organisms sharing the same environment.

A number of creatures can only survive in environments where human disturbance is minimal. This is especially true for those that live in specialist habitats, such as certain owls in old-growth forest, reptiles on acid heathland, or amphibians in fast-flowing, unpolluted streams. However, as the human population grows, there is a constant demand for land for building, agriculture, leisure, roads, or forestry. This pressure is a particular threat to species that cannot easily adapt and move elsewhere. Where they are already confined to "islands" of suitable

habitat, it takes no more than a low level of environmental damage or human disturbance to nudge them towards extinction.

Counting grizzlies

In 1975, grizzly bear numbers were shrinking in Yellowstone National Park. Only an estimated 136 of the bears were left, and this isolated population was considered to be endangered. As part of his doctoral research, Mark L. Shaffer began to study the long-term sustainability of this geographically isolated grizzly bear population.

Shaffer, a pioneer of population viability analysis, applied four factors that he considered would decide their fate. The first was demographic stochasticity: irregular, unpredictable fluctuations in numbers, age, gender, and birth and death rates. For example, if the overwhelming majority of animals in a population are males, breeding success will be poorer than in a »

Uncertainty is just about the only certainty in PVA.
Steven Beissinger
American conservation biologist

Vulnerability of small populations

A minimum viable population has to be of a sufficient size not only to maintain itself under average conditions but also to endure extreme events. Mark Shaffer likened this to a reservoir built to withstand the type of flood that occurs only once in 50 years, but not a devastating once-in-a-century flood.

Small populations are especially vulnerable to multiple threats occurring successively. The heath hen in New England, US, had been widespread in colonial times, but relentless hunting for food and sport caused a dramatic decline in heath hen numbers by 1908. In that year, the last surviving population on the island of Martha's Vineyard was given protected status. However, a catastrophic wildfire during the 1916 breeding season, severe winters, inbreeding, disease, and heavy predation by birds of prey, all combined to push the heath hen population below a viable level. By 1927, only two females remained, and the species was extinct by 1932.

A female grizzly and her cubs forage in Yellowstone. In a grizzly's lifetime, a female's home range is 775–1,400 sq km (300–550 sq miles), while a male's is as much as 5,000 sq km (2,000 sq miles).

more evenly balanced population, and will influence its chances of survival. The second consideration was environmental stochasticity: unpredictable fluctuations in environmental conditions, such as habitat and climate changes, which may affect the availability of food and shelter. The third was natural catastrophes, such as forest fires, or floods. The fourth of Shaffer's factors was genetic changes, including problems created by inbreeding. For each of these, statistical modelling can determine a range of possibilities.

Since Shaffer's initial research in the 1970s and '80s, and subsequent new management and conservation strategies, grizzlies have extended their habitat by more than 50 per cent within the

extensive Greater Yellowstone Ecosystem – an area of 89,031 sq km (34,375 sq miles) which has the national park at its core. In 2014, the US Geological Survey estimated that around 757 bears lived in the ecosystem, based on 119 sightings of grizzly sows and cubs. However, the population had dropped to around 718 in 2018, and population

Technology is increasingly allowing scientists and policymakers to more closely monitor the planet's biodiversity and threats to it.
Stuart L. Pimm
American–British biologist

modellers have suggested that Yellowstone may have reached its maximum carrying capacity – the largest number of animals an area of suitable habitat can support. In 2017, grizzlies were briefly removed from the threatened species list, but their protections were restored by a federal judge in 2018.

How studies are devised
PVA studies are now conducted in several ways. The simplest type is the time-series PVA, which looks at the entirety of a population over a period of time in order to calculate a rough average growth trend and any variations. In such studies, all individuals are treated as identical.

Demographic PVAs tend to be more precise and detailed. They are based on estimated reproductive and survival rates for different age bands within the population. Such analyses require much more data, but can provide extra information on the needs and vulnerability of different sections of the population,

A **population is identified** as being **at risk**.

A **population viability analysis** is conducted to **assess the situation**.

A **management solution** is found to **combat the threat** to the population.

The population has a chance to recover.

The island foxes of the Channel Islands, off California, numbered fewer than 200 in the late 1990s. By 2015, there were more than 5,000, but on one island, a subpopulation is still at risk.

fuelling a case for conservation where protection is required. As reliable information on age ranges and breeding rates is often not available for small, threatened populations, ecologists sometimes use data from other populations of the same species – or a different but similar species – to conduct a PVA. However, the results are variable, even in populations of the same species in the same area. In a 2015 study of three colonies of California sea lion in the Gulf of California,

"surrogate" data from one colony was used to make forecasts on the other two; they proved valid for one colony, but not the other.

Making a difference

Methods are still being refined, but PVA has now become a cornerstone of conservation biology. PVAs have been applied to populations as varied as island foxes in California, sea otters in Alaska, Fender's blue butterflies and northern spotted owls in Oregon, and bottlenose

dolphins off the coasts of Argentina and Australia. With the development of increasingly efficient computer programs incorporating ever more variables, PVA will undoubtedly be used even more effectively in the future. It is impossible to predict every extinction, but PVA provides tools for identifying endangered populations and determining the management actions likely to be most effective in improving population viability, and preserving a species at risk. ∎

A Japanese study

The Japanese rock ptarmigan lives in the Japanese Alps at an altitude of around 2,500 m (8,200 ft). Its population of some 2,000 birds is divided into several small communities on mountain peaks. When a combination of climate warming and predators moving further up the mountains prompted fears for its survival, ecologist Ayaka Suzuki and his team set out to find the minimum viable population size for the birds on Mount Norikura. The team

collected population growth data, including the number of female offspring that survived to the next breeding season and the annual survival rate of all birds. Their calculations included variables for a range of offspring from each pair.

Their findings indicated that there was a relatively low risk of extinction in the next 30 years, even if the starting population was only 15. One potential conclusion is that the Mount Norikura population is strong enough to supplement declining populations on other mountains.

Population viability analysis can indicate how urgently recovery efforts need to be initiated in specific populations.
William F. Morris
American biologist

CLIMATE CHANGE

IS HAPPENING HERE.
IT IS HAPPENING NOW

HALTING CLIMATE CHANGE

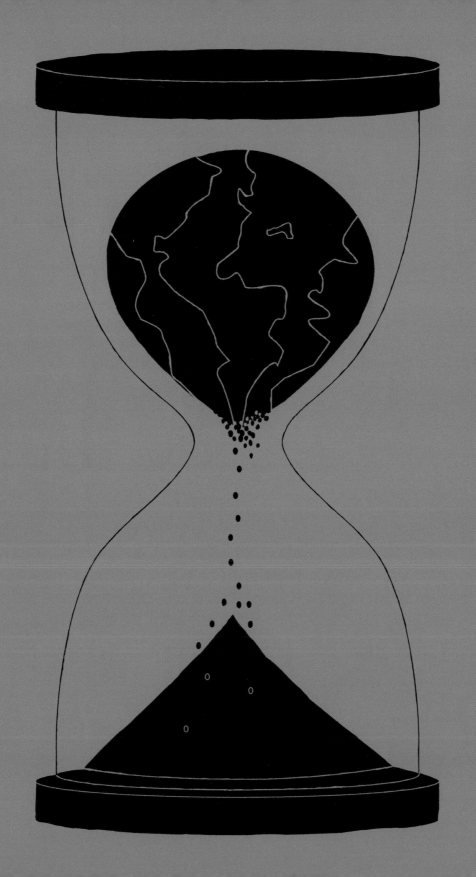

IN CONTEXT

KEY FIGURES
Bert Bolin (1925–2007),
**Intergovernmental Panel
on Climate Change** (1988–)

BEFORE
1955 American scientist
Gilbert Plass concludes that
higher concentrations of
carbon dioxide (CO_2) will lead
to higher temperatures.

1957 American scientist Roger
Revelle and Austrian physical
chemist Hans Suess jointly
publish a report proving that
the oceans will not absorb the
excess CO_2 in the atmosphere.

1968 British glaciologist
John H. Mercer theorizes a
catastrophic rise in sea levels
in the next 40 years due to the
collapse of Antarctic ice sheets.

AFTER
2020 Plans created by the
Paris Agreement to combat
climate change are due to
be implemented.

S ince the Industrial
Revolution, humans have
been altering Earth's natural
environment through increased
carbon dioxide (CO_2) emissions.
Societies have become more
technologically advanced, but this
technology – from coal-powered
trains, ships, and factories, to oil-
fuelled cars and planes – has had
an adverse impact on the natural
world and the species inhabiting
it. As scientists have become more
aware of the human causes of
climate change, global research
groups have formed to study the
phenomenon and suggest ways
in which humanity can halt, if not
reverse, the damage.

The effects of climate change
are varied. As more CO_2 in the
atmosphere creates global warming,
this causes the polar ice caps to
melt, the oceans to warm and rise,
and species that are unsuited to
warmer oceans to die out. Global
weather patterns are also changing:
hurricanes in the North Atlantic
region have increased in intensity,
leaving devastation and death in
their wake. Fires and droughts have
become more frequent in dry areas,
winters are more severe in colder
climates, and areas of the world
already susceptible to extreme

Firefighters battle flames from the
"Holy Fire" that ravaged Orange
County, California, in 2018. Higher
temperatures led to an extended and
difficult forest fire season.

weather-related catastrophes,
such as those impacted by tropical
monsoons, are seeing the most
severe repercussions, especially
in terms of loss of life and habitat.

Global cooperation

Scientists have been aware
that human actions contribute
to climate change since 1896, when
Swedish scientist Svante Arrhenius
suggested that people burning

Natural causes:
• Volcanic eruptions
• Shifting plate tectonics
• Ocean currents

Human causes:
• Deforestation
• Farming practices
• Fossil fuel burning
• Industrial emissions

**... lead to increased amounts of
carbon dioxide in the atmosphere,
causing climate change.**

Brainse Fhionnglaise
Finglas Library
Tel: (01) 834 4906

fossil fuels were adding to global warming. It was not until the 1970s, however, that governments began to act upon this knowledge. Around this time, the general public had begun to be made aware of the reality of climate change due to news articles and broadcasts that shared the bleak outlooks of climate scientists with the wider world.

International efforts to halt or delay climate change began with the first United Nations conference on the environment, which was held in Stockholm, Sweden, in 1972. The conference paid little attention to the issue of climate change compared to other environmental issues – such as pollution and renewable energy – but did create the United Nations Environment Programme (UNEP), an agency to oversee environmental policies and programmes such as

An Inconvenient Truth, a 2006 documentary on climate change by former US Vice President Al Gore, aimed to educate the public on the causes and effects of climate change.

> ... human beings are now carrying out a large scale geophysical experiment of a kind that could not have happened in the past...
> **Roger Revelle and Hans Suess**

ecosystem management, natural disaster relief, and anti-pollution activities. UNEP later became responsible for coordinating UN efforts against climate change.

In 1987, UN members also agreed to the Montreal Protocol, pledging to protect Earth's ozone layer by ending the use of ozone-depleting substances. Although it was not specifically designed to combat climate change, the

agreement, which was ratified by all UN member states, did reduce greenhouse gas emissions.

Creation of the IPCC

In 1988, the Intergovernmental Panel on Climate Change (IPCC) was established in Geneva, Switzerland, by two United Nations organizations: UNEP, and the World Meteorological Organization (WMO). Swedish meteorologist Bert Bolin – who served on the Advisory Group on Greenhouse Gases that the IPCC supplanted – was the panel's first chairman.

The IPCC was created to serve as a globally coordinated response to climate change linked to human activity. It issues reports based on scientific research in support of the main international treaty on climate change: the UN Framework Convention on Climate Change (UNFCCC), which was signed at the Earth Summit in Rio de Janeiro, Brazil, in 1992. The IPCC's work also involves issuing the Summary for Policymakers (SPM), which provides summaries of climate »

change research to governments around the world to help them to understand the threats to humans and the environment as a result of climate change.

The Kyoto plan

Nine years after the creation of the IPCC, in 1997, UN members signed the Kyoto Protocol, which sought to improve regulation of global carbon emissions. This protocol was the first agreement between nations to mandate country-by-country reductions in greenhouse gas emissions, aiming to reduce them to levels that would stop humans from negatively impacting the world's ecosystems.

Although signed in 1997, the Kyoto Protocol did not take effect until 2005. At the end of the first commitment period in 2012, all signatory nations had achieved their target reduction except for Canada, which withdrew from the protocol because it could not meet its targets. Australia also failed to reduce emissions, but in the initial period, their target was set as an 8 per cent increase. Most nations are

on track to meet their target for 2020, except for Norway, which had set a very high target (a 30–40 per cent reduction from 1990 levels).

Paris and the future

The Kyoto Protocols set targets for nations to meet from 2005 to 2020. After 2020, signatory nations will begin to abide by a new protocol: the Paris Agreement. In November 2016, after decades of calls for a

An underwater cabinet meeting is held in the Maldives in 2009 to call for action against climate change. Rising sea levels could mean that the nation is eventually swallowed by the ocean.

more aggressive global resolution climate change, the Agreement was signed by 195 UNFCCC member countries at the UN headquarters in New York City. Like Kyoto, the primary aim of the Paris Agreement is to cut greenhouse gas emissions to agreed-upon levels.

With the decision of Syria to sign the Paris Agreement in 2017, the United States became the only country in the world not to take part in the agreement. Although the US initially signed the agreement under Barack Obama's presidency, his successor, Donald Trump, has rejected the agreement, claiming that it asked too much of the United States and too little of other nations. This decision struck a blow to the other signatories; as well as having plenty of wealth to fund climate research, the US is also the world's second-largest greenhouse gas emitter. President Trump has since clarified his position by saying

Climate change denial

Despite the consensus by the majority of scientists around the world that climate change is a human-caused phenomenon and requires urgent intervention, climate change denial persists in many of the world's most powerful nations. Several scholars have termed the opposition to the facts of climate change a "denial machine", in which conservative media and industries benefiting from lax environmental regulations create an environment of uncertainty and scepticism about climate change science.

Some scepticism comes from those who suggest scientists' estimations are too alarmist, and that global warming is happening more slowly than predicted. Others see the idea of climate change as a human phenomenon as a hoax, instead claiming that global warming is a natural cycle for the planet and not a product of human behaviour. Whatever the reason, denial of climate change amongst some policymakers and business leaders is a position that that the IPCC and scientists continue to disprove.

that he believes climate change to be a natural phenomenon from which the world can "come back" without significant changes to human behaviour.

Other nations have voiced their own concerns with the Paris Agreement. The government of Nicaragua, which joined the accord in 2017, criticized the Agreement for not going far enough and argued that it will not reduce carbon emissions quickly enough to avert global climate disaster. The Paris Agreement also lacks a mechanism to ensure that countries that have signed it comply with its terms.

Desperate measures

According to the terms of the Paris Agreement, countries must work together to limit the increase in the global average temperature to below 2°C (3.6°F) above pre-industrial levels. The Agreement also seeks to go further, suggesting that the increase should be limited to only 1.5°C (2.7°F). In a study published in the journal *Earth System Dynamics* in 2016, climate scientist Carl-Friedrich Schleussner and his co-researchers argued that while an increase of 1.5°C would

We have presented governments with pretty hard choices. We have pointed out the enormous benefits of keeping to 1.5°C.
Professor Jim Skea
Co-Chair, IPCC working group III

The burden of change

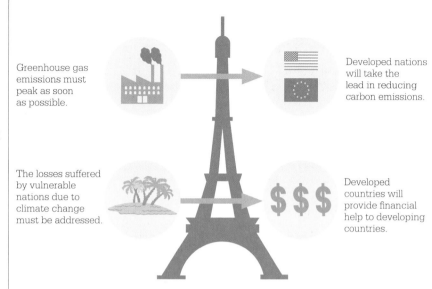

Greenhouse gas emissions must peak as soon as possible.

Developed nations will take the lead in reducing carbon emissions.

The losses suffered by vulnerable nations due to climate change must be addressed.

Developed countries will provide financial help to developing countries.

The Paris Agreement was signed by 195 member countries of the UNFCCC. It placed the responsibility on developed nations to assist those who lacked the funds or resources to combat climate change alone.

create a global environment mirroring the current highest temperatures experienced, a 2°C increase would usher in a "new climate regime" unlike anything humans have seen before.

Subsequent research has shown that this 1.5°C target will prove difficult to meet. In 2018, the IPCC produced a Special Report on global warming, as it had been tasked to do by the Paris Agreement. Its findings were alarming. Rather than being on track for the 1.5°C target, the world is now headed closer to 3°C above pre-industrial levels. To recover and hit the target of 1.5°C would require nations to take unprecedented and drastic measures. Global human CO_2 emissions would need to drop 45 percent from 2010 levels by 2030, and in 2050, would need to reach "net zero", meaning that humans

create no emissions without also removing an equivalent amount of CO_2 from the atmosphere.

The IPCC's 2018 report also appealed to individuals to do their part to lower CO_2 emissions. Land use, energy, cities, and industry are the major areas in which the IPCC suggests change is necessary: people should embrace electric cars; walk and cycle more; and fly less often, as planes produce a significant proportion of greenhouse gases. The IPCC also encouraged people to buy less meat, milk, cheese, and butter, as reduced demand for these products should lead to lower emissions by the meat and dairy processing industries. While global agreements such as Kyoto and Paris have dominated the conversation, it is now clear that any and all methods to lower CO_2 emissions must be pursued. ∎

THE CAPACITY TO SUSTAIN THE WORLD'S POPULATION
SUSTAINABLE BIOSPHERE INITIATIVE

IN CONTEXT

KEY FIGURE
Jane Lubchenco (1947–)

BEFORE
1388 England's Parliament makes it illegal to throw waste into public watercourses such as ditches and rivers.

1970s British scientist James Lovelock and American microbiologist Lynn Margulis develop the Gaia hypothesis.

AFTER
1992 Canadian ecologist William Rees introduces the concept of the "ecological footprint" to describe human impact on the environment.

2000 Dutch Nobel laureate Paul Crutzen popularizes the idea that the world has entered a new geological epoch known as the Anthropocene, or "Age of Man". This era recognizes the monumental and often dangerous ecological impacts humans make on the planet.

T he Sustainable Biosphere Initiative (SBI) emerged in 1988 due to the efforts of the Ecological Society of America (ESA) to establish what scientific research should be prioritized given the limited funding available. At this time, the field of ecology was undergoing a transition towards applied science – using knowledge to develop practical solutions relevant to contemporary environmental issues. American environmentalist Jane Lubchenco led the SBI, and paved the way for the ESA (and others) to promote

The SBI has stimulated improvements in understanding and in advancing connections between ecological knowledge and society.
Jane Lubchenco

useful ecological knowledge as scientists raced to combat environmental degradation.

Prioritizing the planet
The scientists of the SBI set out a new path for the field of ecology, and determined which research areas would be the most important in the years to come. They sought to prioritize three fields of research: global change, biological diversity, and sustainable ecological systems. Studies of global change look at the atmosphere, climate, soil and water (including changes due to pollution), and patterns of land- and water-use. Research into biological diversity focuses on the conservation of endangered species and the study of natural and manmade changes in genetic and habitat diversity. Finally, studies of sustainable ecological systems analyse the interactions between humans and ecological processes in order for scientists to find solutions to the stresses they detect in ecosystems.

The SBI stressed the need for funding for such research, and also highlighted the importance of sharing findings with those outside the scientific community. It set out a process for applied ecological

See also: Chaotic population change 184 ▪ The ecosystem 134–137 ▪ The Gaia hypothesis 214–217 ▪ Overfishing 266–269 ▪ Halting climate change 316–321 ▪ The economic impact of climate change 324–325 ▪ Waste disposal 330–331

research that included not only acquiring new knowledge, but communicating it and helping to incorporate it into real-world policy changes.

The future of research

Lubchenco and her colleagues created the SBI as both a mission statement and an argument for why ecological research deserved more funding and attention. Their report was published in 1991 in the journal *Ecology* as "The Sustainable Biosphere Initiative: An Ecological Research Agenda". It was well received within the scientific community, and has been adapted for use at a global level – first in the International Sustainable Biosphere Initiative that was developed in Mexico in 1991, and then in Agenda 21, an action plan adopted in 1992 at the United Nations Earth Summit in Rio de Janeiro, Brazil.

Since 1991, the SBI and its report have influenced a generation of ecologists, opening up new avenues of funding and collaboration, forming committees, putting on

workshops, and creating reports to advance its agenda. The SBI has brought ecology into the public eye, and today ecologists sit on advisory boards, influencing both corporate and government policies.

Despite such improvements, Lubchenco still believes that the changes that have been made have not kept pace with the growing dangers the planet faces. New campaigns such as the ESA's Earth Stewardship Initiative, created in

Wind turbines are explained to young students. The SBI advocates ecological education in schools and universities so that people can learn how to manage and sustain the biosphere.

2013, build on the work of the SBI. They hope to effect greater change in the next two decades, so that sustainable development can satisfy humans' current needs without compromising the needs of future generations. ▪

Jane Lubchenco

An acclaimed environmental scientist and marine ecologist, Jane Lubchenco grew up in Denver, Colorado. She earned a bachelor's degree in biology at Colorado College, followed by a Master's in zoology. She went on to gain a PhD in marine ecology at Harvard. Her research centres on the interaction between humans and the environment, with an emphasis on biodiversity, climate change, and oceanic sustainability.

From 2009 to 2013, she served as Under Secretary of Commerce for Oceans and Atmosphere, and Administrator of the National Oceanic and Atmospheric Administration (NOAA). She was the first female and the first marine ecologist head of NOAA. In 2011 Lubchenco oversaw the creation of Weather-Ready Nation, a project to prepare the public in case of extreme weather.

Key works

1998 "Entering the century of the environment: a new social contract for science", *Science*
2017 "Delivering on science's social contract", *Michigan Journal of Sustainability*

WE ARE PLAYING DICE WITH THE NATURAL ENVIRONMENT

THE ECONOMIC IMPACT OF CLIMATE CHANGE

IN CONTEXT

KEY FIGURE
William Nordhaus (1941–)

BEFORE
1993 In *Reflections on the Economics of Climate Change*, William Nordhaus summarizes the issues surrounding climate change and the economy, highlighting uncertainties and potential solutions.

AFTER
2008 In *Common Wealth: Economics for a Crowded Planet*, Jeffrey Sachs argues that although humanity faces daunting economic crises – including that of climate change – we have the knowledge to address them.

2013 *The Climate Casino: Risk, Uncertainty, and Economics for a Warming World*, by William Nordhaus, explains how global warming relates to the world's economy, and provides ideas for reducing its impact.

C limatology is an uncertain science. Future projections will change, based on human actions and new technology, as well as natural cycles. However, it is vitally important to assess the financial impacts of climate change. Once potential costs are understood, we can explore ways in which to mitigate its direct impacts. It is necessary to consider not only the direct costs – such as damage to property from flooding or fire – but also the costs associated with broader effects, such as a decline in biodiversity, habitat destruction, shifts in growing seasons, and enforced human migration.

Counting the cost

The social cost of carbon (SCC) is a monetary estimate of the damage to human society caused by every additional tonne of carbon dioxide released into the atmosphere.

Protesters in Lamu, Kenya, in 2018, opposing the construction of a coal-fired power plant. Growing awareness of ecological damage has seen an increase in public disapproval.

See also: Renewable energy 300–305 ▪ Man and Biosphere programme 310–311 ▪ Halting climate change 316–321

These damages include reductions in agricultural productivity, harm to infrastructure, energy costs, and impacts on human health. The SCC provides a starting point for energy policy. For example, if the SCC is factored into proposals for a new power plant, the cost of building it becomes much higher. This may also make the cost of alternative forms of energy – such as solar or wind power – more financially viable. However, it is extremely difficult to calculate the SCC.

Forecast models

Economists use several models in order to calculate the SCC. In 1999, William Nordhaus developed RICE (Regional Integrated Climate-Economy model) – a variant of his own preceding DICE (Dynamic Integrated Climate-Economy model), which weighed the costs and benefits of slowing down global warming. The RICE model integrates carbon emissions, carbon concentrations in the atmosphere, climate change, damages, and controls that are in place to reduce

emissions. The model divides the world into distinct regions for its analysis. It predicts that the combined SCC in 2055 will be between \$44 and \$207 per tonne (\$40–\$188 per short ton) of carbon dioxide released, depending on the rate of warming and the mitigation policies enacted.

Economic models incorporate assumptions, such as the discount rate. Discount rates prioritize the present over the future, because the future cannot be predicted perfectly. The rate is selected based on how the balance between present and future priorities is weighted. Higher discount rates indicate that future populations will be wealthier, and prepared to deal with climate change. Lower discount rates suggest that the disruption caused by climate change will make people in the future poorer than we are today. Nordhaus suggests a 3 per cent discount rate, meaning that if the monetary damages from climate change will be \$5 trillion in the year 2100, we could invest \$382 billion today to avoid it. ▪

Analysing the costs of reducing carbon dioxide

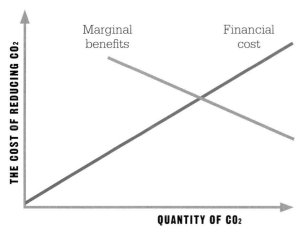

Marginal benefits

Financial cost

THE COST OF REDUCING CO_2

QUANTITY OF CO_2

The cost of reducing CO_2 increases in line with the quantity, but this is offset by the benefits gained. The lines intersect at the point of equilibrium, where maximum benefits are achieved at the lowest cost.

William Nordhaus

Born in New Mexico, America in 1941, Nordhaus is a leader in the field of the economics of climate change. He stumbled upon this field of research through sharing an office with a climatologist. Nordhaus's economic theories – the DICE and RICE models – are widely used to analyse policy decisions. Nordhaus is principally concerned with placing a realistic price on carbon. Today, the social cost of carbon is generally agreed to be around \$40 per ton, but Nordhaus's models show that it should be higher to account for the impacts of climate change. Nordhaus is Sterling Professor of Economics at Yale University, and serves on the Congressional Budget Office Panel of Economic Experts and the Brooking Panel on Economic Activity. In 2018, Noordhaus was awarded the Nobel Prize in economics.

Key works

1994 *Managing the Global Commons: The Economics of Climate Change*
2000 *Warming the World: Economic Models of Global Warming*

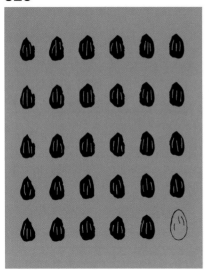

MONOCULTURES AND MONOPOLIES ARE DESTROYING THE HARVEST OF SEED

SEED DIVERSITY

IN CONTEXT

KEY FIGURE
Vandana Shiva (1952–)

BEFORE
1966 A new high-yielding strain of rice known as IR8 leads to a big increase in production in several rice-growing countries. First developed in the Philippines, it is also called "miracle rice".

AFTER
1994 The World Trade Organization introduces the Trade Related Aspects of Intellectual Property Rights (TRIPS) agreement.

2004 After protests by farmers who developed the crop, the Monsanto company's patent on an Indian strain of wheat known as Nap Hal is revoked.

2012 Indian initiative Navdanya International launches its worldwide Seed Freedom campaign to protect food sovereignty and safety.

In 1987, Indian environmental campaigner Vandana Shiva launched a movement to protect native seed diversity in response to changes in agriculture and food production. She founded Navdanya, a non-governmental organization, to protect agricultural biodiversity from the combined threat of genetic engineering and patents.

Agro-biodiversity

Agricultural biodiversity (also known as agro-biodiversity) has resulted from the selective breeding, over thousands of years, of plants and animals taken from the wild. These practices led to the extraordinary genetic diversity of different breeds of crops and domesticated animals.

Californian rice production is high yield but there are problems with soil salinity. Although salt tolerance can be genetically introduced, traditional rice varieties can be naturally salt-resistant.

For example, a grass in the genus *Oryza* was first cultivated for rice in Asia between 8,200 and 13,500 years ago; today, there are more than 40,000 varieties of this rice in existence. Intrinsic to agro-biodiversity are the many non-harvested species that support production. These include microorganisms in the soil, species that feed on pests, and pollinators. Through the ages, the skills and knowledge of millions of farmers have shaped this biodiversity.

See also: Human activity and biodiversity 92–95 ▪ Pesticides 242–247
▪ Humankind's dominance over nature 296 ▪ Ecosystem services 328–329

> Seed patents threaten
> the very survival and
> freedom of peasants …
> and farmers …
> **Vandana Shiva**

From the late 1960s, a technology transfer to the developing world included high-yield varieties of cereals in association with chemical fertilizers, pesticides, and herbicides, mechanization, and more efficient irrigation. Known as the "Green Revolution", this transformation shifted the focus of agriculture in the developing world away from biodiversity to higher crop yields. New Green Revolution crops such as "miracle rice" (IR8) boosted production, but there was a downside. As more emphasis was placed on fewer productive strains, the genetic base of traditional seed varieties for grains, potatoes, fruits, vegetables, and cotton declined.

The United Nations Food and Agriculture Organization estimates that 75 per cent of crop biodiversity has been lost from the world's fields. Some environmentalists have argued that traditional varieties are more compatible with local farming conditions, cheaper for farmers to use, and more environmentally sustainable than new, high-yield varieties. Additionally, many of the new strains are patented by the companies that created them. Trade deals impose regulations on who can use what. These work against small-scale farmers but favour the powerful agricultural corporations that produce the seed.

Seed sovereignty

Shiva argues that rural farms are threatened if the appropriate seed is no longer available. Traditionally, most small-scale farmers routinely save their seed from one harvest to the next. Now, when farmers buy in seed – especially if it is genetically modified – they often have to agree not to save it. Having to buy seed from a company every year can leave them worse off financially.

Shiva criticized the practice of corporations patenting seed varieties as "biopiracy" and set up Navdanya to support "seed sovereignty". It campaigns for agro-biodiversity via a network of seed-keepers and organic producers and has helped to found more than 100 community seed banks, effectively gene banks, where seeds of crops and rare plant species are stored for future use. ▪

Fertilizers have hugely increased food grain production in India – whose population of 1.3 billion people makes food security paramount – but the chemicals also destroy soil fertility.

Vandana Shiva

Environmental campaigner Vandana Shiva was born in northern India. Her mother was a farmer, and her father a forester. She studied in India and Canada, obtaining a doctorate in the philosophy of physics. After returning to India, in 1982 she founded the Research Foundation for Science, Technology, and Ecology. Following the Bhopal pesticide plant disaster in 1984, her interest in agriculture grew and three years later she founded Navdanya to protect biodiversity and native seeds. Shiva campaigns against the World Trade Organization's Trade Related Aspects of Intellectual Property Rights (TRIPS) agreement, which broadens patents to include plants and animals. *TIME* magazine hailed Vandana Shiva as an Environmental Hero in 2003.

Key works

1989 *The Violence of the Green Revolution*
2000 *Stolen Harvest: The Hijacking of the Global Food Supply*
2013 *Making Peace with the Earth*

NATURAL ECOSYSTEMS AND THEIR SPECIES HELP SUSTAIN AND FULFIL HUMAN LIFE
ECOSYSTEM SERVICES

IN CONTEXT

KEY FIGURE
Gretchen Daily (1964–)

BEFORE
c.400 BCE Greek philosopher
Plato is aware of the human
impact on nature, noting that
deforestation can cause soil to
erode and springs to run dry.

1973 German statistician and
economist E.F. Schumacher
coins the term "natural capital"
in his book *Small is Beautiful*.

AFTER
1998 The UN Environment
Programme, NASA, and the
World Bank release a study
on how protecting the planet
serves human needs.

2008 A study at the University
of California, Berkeley, shows
that ecological destruction by
the world's richest countries
means they owe the world's
poorest countries more than
the developing world's debt.

The benefits that humans derive from ecosystems are referred to by ecologists as ecosystem services. Some of the natural processes most important to the continuation of human life can be classified as ecosystem services, such as pollination of crops, decomposition of waste, and the availability of clean drinking water. Ecologists argue that because the enormous contributions of ecosystem services to human life are not readily quantifiable, humans drastically undervalue these services while exploiting the natural world's resources for profit.

As a sacred mountain, Mount Fuji supplies a cultural ecosystem service for the people of Japan, while the surrounding rich volcanic soil provides a service to the local tea plantations.

Although the idea that humans benefit from nature has a long history, it was not until the 1970s that the balance between nature and human needs came to the forefront of ecological debate. The term "ecosystem services" first appeared in the mid-1980s, and in 1997 the concept was developed in two key articles: "Ecosystem Services: Benefits Supplied to Human Societies by Natural

See also: Human activity and biodiversity 92–95 ▪ Ecological resilience 150–151
▪ The Gaia hypothesis 214–217 ▪ Human devastation of Earth 299

> If current trends
> continue, humanity will
> dramatically alter virtually
> all of Earth's remaining
> natural ecosystems
> within a few decades.
> **Gretchen Daily**

Ecosystems", edited by Gretchen Daily, and "The Value of the World's Ecosystem Services and Natural Capital", edited by American ecological economist Robert Costanza. In 2001, UN Secretary General Kofi Annan launched the Millennium Ecosystem Assessment (MEA), which helped popularize the concept of ecosystem services in 2005, when they published a wide-ranging appraisal of how humans impact the environment.

The four types of service
The MEA's 2005 report detailed four categories of ecosystem services: supporting, provisioning, regulating, and cultural. Supporting services, such as soil formation and water purification, allow for the existence of all other services. Provisioning services consist of fresh water; food, such as crops and livestock; fibres, including wood, cotton, and other materials used for human essentials such as building and clothing; and natural medicines, and plants used in pharmaceuticals. Regulating services include

nature's ability to control pests – as opposed to humans' use of pesticides – and the atmosphere's capacity to cleanse itself naturally, as well as the control of weather hazards through natural buffers such as wetlands and mangrove forests. Pollination is another important regulating service, one that is endangered by the global decline of pollinators such as bees. Cultural services involve the ways that humans assign cultural or spiritual significance to elements of ecosystems such as sacred trees, animals, rivers, and mountains. The aesthetic or recreational value of a natural landscape is another type of cultural service.

At its heart, the concept of ecosystem services allows humans to see how inextricably connected they are to nature, and how without the natural world human existence would be impossible. Ecologists use the concept to illuminate how precious these systems are for basic life conditions and to convince industries, businesses, and governments of the necessity for ecological preservation. ▪

> Plans to protect air
> and water, wilderness and
> wildlife are in fact plans
> to protect man.
> **Stewart Udall**
> *American politician and
> conservationist*

Gretchen Daily

Born in 1964 in Washington, DC, Gretchen Daily developed a passion for ecology at a young age. After her family moved to West Germany in 1977, she witnessed a national crisis over acid rain, and saw people protesting in the streets over environmental degradation. Daily went on to earn two degrees and then her doctorate in biology at Stanford University, where she is now the Bing Professor of Environmental Science.

Daily studies biodiversity within the framework of "countryside biogeography", or the portions of nature that have not been used for human development, but whose ecosystems are still impacted by human activity. She is a cofounder of the Natural Capital Project, which aims to incorporate environmentalism into business practices and public policy.

Key works

1997 *Nature's Services: Societal Dependence on Natural Ecosystems*
2002 *The New Economy of Nature: The Quest to Make Conservation Profitable*

WE ARE LIVING ON THIS PLANET AS THOUGH WE HAVE ANOTHER ONE TO GO TO
WASTE DISPOSAL

IN CONTEXT

KEY FIGURE
Paul Connett (1940–)

BEFORE
1970 The first Earth Day takes place in the US to raise awareness of clean waste disposal and recycling.

1988 The Resin Identification Code is introduced in the US to encourage the recycling of plastic goods.

1992 At the Rio Earth Summit, 105 heads of state pledge their commitment to sustainable development.

AFTER
2010 The United Nations launches its Global Partnership on Waste Management to promote resource conservation and efficiency.

2012 Goals outlined at the UN Conference on Sustainable Development include waste reduction and eco-friendly production methods.

More than 65,000 people from at least 180 nations travelled to Johannesburg, South Africa, in 2002 to attend the United Nations World Summit on Sustainable Development. Its final resolutions included a call to minimize waste and maximize reuse and recycling, and to develop "clean" waste disposal systems.

In the last decades of the 20th century, it had become clear that refuse was reaching unmanageable proportions. Industrialization, the growth of large urban populations, and increasing use of plastic were all adding to the world's garbage heap. Traditionally waste had been burnt or buried – both options now associated with toxic greenhouse gas emissions and, in the case of landfills, the potential for poisoning ground water. The answer to the world's growing waste heap had to be found elsewhere.

Pollution is nothing but the resources we are not harvesting. We allow them to be dispersed because we've been ignorant of their value.
R. Buckminster Fuller
American inventor and architect

The recycling revolution
Recycling for reuse is not a new concept, but its use as a way of reducing mountains of public waste that would otherwise go into landfill has its origins in the 1960s and 1970s, when organizations such as Greenpeace made the public more aware of environmental issues. Recently, campaigners such as Paul Connett, author of *Zero Waste* (2013), have renewed the global call to reduce consumption, and reuse or recycle items, rather than discard them.

Since the 1970s, many US states and most European countries, as well as Canada, Australia, and New Zealand, have introduced kerbside collections of recyclable items sorted into bins. Sweden has been especially active. In 1975, Swedes recycled only 38 per cent of their rubbish, but today they lead

See also: Global warming 202–203 ▪ Pollution 230–235 ▪ Urban sprawl 282–283 ▪ A plastic wasteland 284–285 ▪ Renewable energy 300–305

Refuse plastic bags and excessive packaging. Buy products in **large containers** or without packaging.

Rethink as you shop. Do you **really need** what you are buying?

Individual actions can reduce waste – households in the developed world add a tonne of waste to landfill each year.

Reuse what you can or **pass it on** to someone else who can use it.

Recycle what can't be used so that it can be turned into **new products**.

Methane from landfill

After carbon dioxide, methane is the most critical greenhouse gas. Although its atmospheric concentrations are lower than CO_2, methane is 25 times more powerful at trapping heat in the atmosphere. Atmospheric methane comes from various natural sources, including the decay of vegetation in habitats such as bogs and wetlands, but also from livestock rearing, from the use of fossil fuels, and from the decomposition of rubbish in landfill sites.

In many places, including the UK and US, a number of landfill sites are now trapping and collecting methane to produce energy. Landfill gas contains up to 60 per cent methane, depending on the composition of the waste and the age of the site. Vertical and horizontal pipelines are placed through the landfill to collect the methane, which is then processed and filtered. Most of it is used to generate electricity, but it may also be used in industry. After further processing, it can be turned into fuel for vehicles, too.

the world, recycling 99 per cent of household waste. About 50 per cent of this waste is burnt in recycling plants that generate heat for the nation's homes. Sweden also imports waste from other countries to process in its 32 incineration plants. In 2015, it imported some 2.3 million tonnes (2.5 million tons) of waste from Norway, the UK, Ireland, and other nations.

"Mining" electronics
The fastest-growing type of waste is discarded electronics. E-waste from mobile phones, computer hard drives, TVs, and other electrical goods reached almost 42 million tonnes (46 million tons) in 2014 – almost 25 per cent more than in 2010. E-waste often contains precious metals, such as the gold,

silver, copper, and palladium used in circuit boards. It has been shown that "mining" landfill sites to extract the metals can be more cost-effective than mining natural mineral deposits. However, e-waste also includes toxic metals, such as cadmium, lead, and mercury. In countries that both generate and import e-waste, landfill scavenging for metals can be polluting. While Europe now has an e-waste reprocessing industry, relatively few efficient schemes exist elsewhere.

There are many new initiatives, but the world is still very far from Connett's zero waste ideal. A huge challenge remains for individuals and governments: how to cut consumption and recycle global refuse that will soon reach 2 billion tonnes ($2\frac{1}{4}$ billion tons) a year. ▪

Methane is extracted at the Payatas landfill, Manila – the first in the Philippines to have the gas converted to energy, as part of a United Nations programme.

DIRECTO

RY

DIRECTORY

In addition to the scientists covered in the preceding chapters of this book, many other men and women have made significant contributions to the development of ecology. They have ranked among the greatest scientific thinkers of their time. Some have excelled in academia, while some came from other walks of life but pioneered new approaches to advance. Still more have been formidable campaigners. Although they worked in a range of disciplines, all have contributed to our understanding of Earth's biosphere, how it has evolved, and humanity's place in it. Crucially, their work continues to show what needs to be done to preserve the natural world and to protect Earth from the destructive consequences of human behaviour.

SAMUEL DE CHAMPLAIN
1574–1635

A French explorer, cartographer, soldier, and naturalist, Champlain explored and mapped much of Canada. He founded the city of Quebec and established the colony of New France. As a keen observer and chronicler, he documented animals and made notes about plants, including details of leaves, fruits, and nuts, and enquired about how the Native American people used them.
See also: Classification of living things 82–83

JAMES AUDUBON
1785–1851

The pioneer of North American ornithology, Audubon grew up in Haiti and France before emigrating to the US in 1803. He developed an interest in nature, especially birds, and was a talented artist. His artistic technique was unusual: after shooting a bird, he held it in a "natural pose", using fine wire, and painted it with a backdrop of the animal's natural habitat. Between 1827 and 1838 he published *The Birds of America* in a series of instalments. It included 435 coloured prints of 497 species, six of which are now extinct. Audubon also discovered 25 previously

undescribed species and used yarn to "ring" birds – meaning he tied it to their legs, allowing him to identify each individual bird – to find out more about their movements.
See also: Animal ecology 106–113

MARY ANNING
1799–1847

In 2010, the Royal Society named Anning as one of the 10 British women who have most influenced the history of science. She found fame as a fossil collector and palaeontologist, and her extraordinary fossil finds, from Jurassic strata in the cliffs of the Dorset coast, included the first correctly described ichthyosaur, two relatively complete plesiosaurs, and the first pterosaur from outside Germany. Her finds helped to change views about Earth's history, providing strong evidence for extinction.
See also: Mass extinctions 218–223

CATHERINE PARR TRAILL
1802–99

A botanist and prolific author, Traill was born in the UK, and emigrated to what is now Ontario, Canada, after she married in 1832. There, she wrote about life as a settler in Canada. She also wrote about the natural environment,

notably in *Canadian Wild Flowers* (1865) and *Studies of Plant Life in Canada* (1885). Her many albums of plant collections are housed in the National Herbarium of Canada, at the Canadian Museum of Nature in Ottawa.
See also: Endangered habitats 236–239

KARL AUGUST MÖBIUS
1825–1908

A German pioneer, Möbius was primarily interested in the ecology of marine ecosystems. After studying at the Natural History Museum of Berlin, and gaining a doctorate at the University of Halle, he opened a seawater aquarium in Hamburg in 1863. While a professor of zoology at the University of Kiel, his work on the viability of commercial oyster production in the Bay of Kiel led him to recognize the various dependent relationships between organisms in the oyster bank ecosystem.
See also: The ecosystem 134–137

ERNST HAECKEL
1834–1919

Haeckel was a biologist, physician, and artist who popularized Charles Darwin's ideas in Germany (while also rejecting many of them) and introduced the word "ecology" in 1866. Born in

Potsdam, he studied at several universities before becoming a zoology professor at the University of Jena in 1861. Haeckel was the first biologist to propose the kingdom *Protista* – for organisms that are neither animal nor vegetable – and he researched and painstakingly recorded tiny deep-sea protozoans called radiolaria.
See also: Evolution by natural selection 24–31

WILLIAM BLAKE RICHMOND
1842–1921

Best known as a British artist, sculptor, and designer of stained glass and mosaics, London-born Richmond became an environmental activist after having to endure the poor light and smoky air produced by London's winter coal fires. In 1898 he founded the Coal Smoke Abatement Society (CSAS) to lobby politicians for clean air. The CSAS was instrumental in the introduction of the UK's Public Health (Smoke Abatement) Act in 1926 and the Clean Air Act in 1956.
See also: Pollution 230–235

THEODORE ROOSEVELT
1858–1919

To deal with severe childhood asthma, Roosevelt became a keen sportsman and outdoorsman, developing a lifelong passion for nature. When, in 1900, he stood as William McKinley's running mate in the US presidential election, he did so on a ticket of peace, prosperity, and conservation. Roosevelt became the 26th President when McKinley was assassinated in 1901, and went on to establish the US Forest Service, five new national parks, 51 bird reserves, and 150 national forests.
See also: Deforestation 254–259

JÓSEF PACZOSKI
1864–1942

Paczoski was a Polish ecologist, born in what is now Ukraine. He studied botany at the University of Kiev and went on to pioneer phytosociology, the study of natural plant communities, first using the term in 1896. In the 1920s Paczoski established the world's first institute of phytosociology, at the University of Poznan, where he was professor of plant systematics. An accomplished botanist, he published works on central European flora, including that of the Białowieza Forest, which he managed as a national park.
See also: Organisms and their environment 166

JACK MINER
1865–1944

Also known as "Wild Goose Jack", Miner moved with his family from the US to Canada in 1878. He was illiterate until the age of 33 but embarked on local conservation projects, such as building winter feeding stations for bobwhite quail. He was one of the first people in North America to put aluminium rings on birds' legs to track their movements. A duck ringed by him, and later seen in South Carolina, was the first ringing recovery made in North America. Miner is thought to have ringed more than 90,000 wildfowl, helping to establish a huge database of migration routes.
See also: Citizen science 178–183

JAMES BERNARD HARKIN
1875–1955

Sometimes referred to as the "father of Canadian national parks", Harkin had a passion for politics and conservation. In 1911, he was appointed the first commissioner of the Canadian National Parks Agency and oversaw the establishment of Point Pelee, Wood Buffalo, Kootenay, Elk Island, Georgian Bay Islands, and Cape Breton Highlands national parks. Harkin realized the commercial value of the parks, and his policy of encouraging road-building to attract tourists was not universally popular. He was a prime mover behind legislation to regulate the hunting of migrant birds in 1917.
See also: Endangered habitats 236–239, Deforestation 254–259

MARJORY STONEMAN DOUGLAS
1890–1998

A formidable campaigner for the protection of the Florida Everglades, Douglas was also a successful journalist and author, suffragist, and campaigner for civil rights. Her 1947 book *The Everglades: River of Grass* was influential in building an appreciation of the Florida wetlands, and in 1969 she founded the Friends of the Everglades to defend the area from draining for development. Douglas remained active well into her second century, and at the age of 103 she was awarded the Presidential Medal of Freedom.
See also: Citizen science 178–183

BARBARA MCCLINTOCK
1902–92

In 1983 McClintock became the first solo woman to win the Nobel Prize in Physiology or Medicine, and the first American woman to win any unshared Nobel Prize. The award recognized her discovery – more than 30 years before – of transposable genetic elements, or "jumping genes", which sometimes create or reverse mutations. As a cytogeneticist concerned with how chromosomes relate to cell behaviour, she also discovered the first genetic map for maize – linking physical traits with regions of the chromosome – and the mechanism by which chromosomes exchange information.
See also: The role of DNA 34–37

JACQUES COUSTEAU
1910–97

French undersea explorer Cousteau was well known as the presenter of several documentaries on the aquatic world. After inventing underwater breathing apparatus called the Aqua-Lung in 1943, he worked with the French Navy to clear underwater mines after World War II. He later converted the *Calypso*, a former minesweeper, into a research vessel from which he explored the oceans, writing several books and

making hours of television. The *Calypso* was badly damaged in 1996, but Cousteau died suddenly in 1997 before he could afford to replace it.
See also: A plastic wasteland 284–285

PIERRE DANSEREAU
1911–2011

Dansereau was a French Canadian plant ecologist who pioneered the study of forest dynamics and is considered one of the "fathers of ecology". Born in Montreal, he gained his doctorate in plant taxonomy at the University of Geneva in 1939. He later helped to set up the Montreal Botanical Garden and wrote numerous papers on botany, biogeography, and the interaction of humans and the environment. In 1988 he was appointed Professor Emeritus at the University of Montreal, a post he held until he retired, aged 93, in 2004.
See also: Biogeography 200–201

MARY LEAKEY
1913–96

London-born Mary Leakey, one of the world's foremost paleoanthropologists, experienced her first archeological excavation at the age of 17, when she was hired as an illustrator at a "dig" in Devon. In 1937 she married paleoanthropologist Louis Leakey, and the couple moved to Africa to work in the Olduvai Gorge – a site rich in fossils, in what is now Tanzania. In 1948, Mary found the fossil skull of an 18-million-year-old ancestor of apes and humans, *Proconsul africanus*. More breakthroughs in understanding human ancestry followed, including the discovery in 1960 of *Homo habilis*, a 1.4–2.3-million-year-old hominid who used stone tools.
See also: Evolution by natural selection 24–31

MAX DAY
1915–2017

An ecologist and entomologist, Day developed an interest in wildlife, particularly insects, as a boy in

Australia. He graduated from the University of Sydney with a degree in botany and zoology in 1937, and then studied at Harvard University, gaining a doctorate for his work on termites. After World War II, he returned to Australia, where he became the first head of the Commonwealth Scientific and Industrial Research Organisation's Division of Forest Research in 1976. Particularly known for his work on myxomatosis and its use in controlling rabbit populations, Day published his first paper in 1938, and his last – on moths – 74 years later.
See also: Thermoregulation in insects 126–127 ▪ Invasive species 270–273

JUDITH WRIGHT
1915–2000

Principally a poet, Wright was also renowned in her native Australia for her campaigning on Aboriginal land rights and environmental issues. She was born in Armidale, New South Wales, and studied at the University of Sydney, before publishing her first book of poetry in 1946. Between 1967 and 1971, along with artist John Busst and environmentalist Len Webb, she built an alliance of conservation groups, trade unions, and concerned citizens to fight Queensland state government's plans to open up the Great Barrier Reef to mining. The campaign, detailed in her book *The Coral Battleground* (1977), eventually succeeded.
See also: The Green Movement 308–309

EILEEN WANI WINGFIELD
1920–2014

As a young Aboriginal woman in Australia, Wingfield herded cattle and sheep with her father and sister. In the early 1980s she lay down in front of bulldozers at Canegrass Swamp in opposition to construction of the Olympic Dam uranium mine. Later, Wingfield teamed up with Eileen Kampakuta Brown and other Aboriginal elders to campaign against the government's proposals to dump nuclear waste in South Australia. The

women toured the country, speaking at meetings to highlight the dangers of the dump, which they feared could grow as foreign governments and corporations saw an opportunity to dispose of their radioactive waste.
See also: Pollution 230–235

EUGENIE CLARK
1922–2015

Known as the "Shark Lady" for her research on shark behaviour, Clark was a Japanese-American marine ecologist and a pioneer in the use of scuba diving for scientific research – she undertook many dives around Florida's Cape Haze Marine Laboratory, where she worked alongside other female ecologists such as Sylvia Earle. Clark made several key discoveries about sharks and fish, and was a major advocate of marine conservation. In 1955, she founded the Mote Marine Laboratory, which works to protect shark species, preserve coral reefs, and found sustainable fisheries.
See also: Animal behaviour 116–117

DAVID ATTENBOROUGH
1926–

British naturalist and television producer Attenborough served as a controller for the BBC before stepping down to dedicate more time to writing and producing documentaries. He wrote and narrated a series of nature programmes, notably the *Life* series, beginning with *Life on Earth* (1979). Attenborough's work has been credited with renewing public interest in nature and conservation in Great Britain.
See also: A plastic wasteland 284–285

PETER H. KLOPFER
1930–

Berlin-born Klopfer is an ecologist whose main area of interest is ethology, studying animal behaviour in a natural environment. His influential 1967 book *An Introduction to Animal Behaviour: Ethology's First Century* acted as a survey and synthesis of past and

present ethological theories. In 1968, he began teaching in the Department of Zoology at Duke University, North Carolina, where he was instrumental in starting its primate centre.
See also: Animal behaviour 116–117

DIAN FOSSEY
1932–85

Most of what is known about the lives and social structures of wild mountain gorillas in Africa derives from the work of primatologist and conservationist Fossey. The daughter of a San Francisco fashion model, she graduated and worked as an occupational therapist before visiting Africa, where she met, and was inspired by, Mary and Louis Leakey. In early 1967 Fossey founded the Karisoke Research Centre in the Rwandan mountains, where she studied gorillas. Her best-selling 1983 book about her experiences – *Gorillas in the Mist* – was later adapted for the screen. Fossey was murdered at her camp in December 1985, probably because of her anti-poaching stance.
See also: Animal behaviour 116–117

TOMOKO OHTA
1933–

Ohta is a Japanese population geneticist who in 1973 proposed the evolutionary Nearly Neutral Theory, which included the idea that mutations that are neither neutral nor harmful play an important part in evolution. After graduating from the University of Tokyo in 1956, Ohta worked on the cytogenetics (how chromosomes relate to cell behaviour) of wheat and sugar beet, and now works at Japan's National Institute of Genetics.
See also: The selfish gene 38–39

STANLEY C. WECKER
1933–2010

An American animal behaviourist, Wecker was an influential researcher into animal population and community ecology, especially the study of what determines where animals choose to live. His 1963 paper on habitat selection by prairie deer mice demonstrated that instinct and experience both play a role in how the mice select their habitat.
See also: Ecological niches 50–51

SYLVIA EARLE
1935–

An American marine biologist, author, and conservationist, Earle is an expert on the impact of oil spills. In 1991, she assessed the damage caused by the destruction of Kuwaiti oil wells during the Gulf War. Earle undertook similar work after the Exxon Valdez, Mega Borg, and Deepwater Horizon oil spills. In 2009, Earle launched Mission Blue, which, by 2018, had established nearly 100 marine protected areas around the world.
See also: Pollution 230–235

ROBERT E. SHAW
1936–

Shaw is an American pioneer of ecological psychology, which looks at how perception, action, communication, learning, and evolution in humans and animals are determined by the environment. In 1977 he coedited the book *Perceiving, Acting, and Knowing: Toward an Ecological Psychology*, which effectively launched this new area of study. In 1981 Shaw was the founding president of the International Society for Ecological Psychology and is now an emeritis professor in the Department of Psychological Sciences at the University of Connecticut.
See also: Using animal models to understand human behaviour 118–125

DAVID SUZUKI
1936–

Canadian scientist Suzuki graduated from the University of Chicago with a degree in zoology in 1961, and two years later became a professor in the genetics department at the University of British Columbia. Since the mid-1970s, he has also been a TV and radio broadcaster and the author of books on nature and the environment. He cofounded the David Suzuki Foundation in 1990 to investigate sustainable ways for people to live in harmony with the natural world.
See also: Environmental ethics 306–307

DANIEL B. BOTKIN
1937–

Botkin, a prominent American author and environmentalist, gained his doctorate in plant ecology in 1968 at Rutgers University. He writes and speaks on all areas of the environment, from forest ecosystems to fish populations, and also advises agencies, corporations, and governments. After decades spent researching climate change, Botkin has questioned how far it is impacted by human activity. He is a research scientist at the Marine Biological Laboratory, near Boston, and is involved in environmental studies programmes at several American universities.
See also: Halting climate change 316–321

EILEEN KAMPAKUTA BROWN
1938–

In the early 1990s, the Australian government revealed plans to build a nuclear waste dump near Woomera, in the South Australian desert. Together with Eileen Wani Wingfield, Brown, an Aboriginal elder, established a *kungka tjuta* (women's council) in the town of Cooper Pedy to fight the plans. They were aware of the birth defects, cancer, and other health issues following the British military's nuclear tests in the desert in the 1950s and 1960s, and feared that radiation could seep into the groundwater. The plans were abandoned and Brown and Wingfield won the 2003 Goldman Environmental Prize.
See also: Pollution 230–235

LYNN MARGULIS
1938–2011

American biologist Margulis attended university in Chicago aged only 15 and gained her doctorate at the University

of California, Berkeley, in 1965. The next year, at Boston University, she proposed that cells within nuclei had evolved as a result of the symbiotic merger of bacteria. This idea, although not gnerally accepted until the 1980s, transformed the understanding of cell evolution.
See also: The Gaia hypothesis 214–217

PAUL F. HOFFMAN
1941–

Canadian scientist Paul Hoffman's discovery of "cap carbonates" – evidence for ancient glaciation in Precambrian sedimentary rocks in Namibia – revived the "Snowball Earth" hypothesis in climate change studies in 2000. The term was first used by American geologist Joseph Kirschvink in 1992, although there had been speculation since the late 19th century that Earth's surface was almost entirely frozen more than 650 million years ago.
See also: Ancient ice ages 198–199

SIMON A. LEVIN
1941–

Levin, an American ecologist, specializes in the use of sophisticated mathematical modelling, alongside field and lab observation, to understand the workings of ecosystems. He also researches the relationships between ecology and economics. Levin gained a doctorate in mathematics from the University of Maryland in 1964 and taught at Cornell University from 1965 to 1992. After moving to Princeton, he was appointed director of the university's Center for BioComplexity, which investigates the mechanisms that generate and maintain complexity in the living world.
See also: Predator–prey equations 44–49

JAMES A. YORKE
1941–

An American mathematician and physicist based at the University of Maryland, Yorke is best known for his work on chaos theory. In his 1975 paper "Period Three Implies Chaos", written with Chinese mathematician Tien-Yien Li, he argued that above a certain rate of growth, population forecasts become totally unpredictable, a discovery with major ecological implications.
See also: Population viability analysis 312–315

IAN LOWE
1942–

Lowe, an Australian environmentalist who studied engineering and science at the University of New South Wales and gained his doctorate in physics at the University of York, advises the UN's Inter-governmental Panel on Climate Change. He is outspoken on the need for renewable energy, arguing that it is "quicker, less expensive, and less dangerous than nuclear power". In |1996, Lowe chaired the expert group responsible for the first report on the state of Australia's environment. Lowe is now Emeritus Professor of Science, Technology, and Society at Griffith University, Brisbane.
See also: Renewable energy 300–305
▪ Halting climate change 316–321

AILA KETO
1943–

Keto spent much of her youth exploring the Great Barrier Reef and surrounding rainforests. She studied biochemistry at university and went on to work at the University of Queensland. In 1982, with her husband Keith, she founded the Australian Rainforest Conservation Society, which did much to save Australia's Wet Tropics area.
See also: Biomes 206–209

BOB BROWN
1944–

After studying medicine at the University of Sydney, Brown practised as a doctor in Australia and the UK. He moved to Tasmania in 1972 and soon became involved in the environmental movement. In the early 1980s, he was one of the leaders of a successful campaign to prevent the building of the Franklin Dam, which would have destroyed key habitats. In 1996, Brown was elected to the Australian Senate as a Green Party representative. On retirement in 2012, he set up the Bob Brown Foundation to campaign for the protection of Australian habitats.
See also: The water crisis 286–291

BIRUTE GALDIKAS
1946–

German-born anthropologist and primatologist Galdikas has pioneered the study of orangutans in the wild. Along with Jane Goodall and Dian Fossey, she was one of "The Trimates", chosen by Louis Leakey to study great apes. Leakey persuaded her to support the establishment of an orangutan research station in Borneo, which she moved to in 1971. For more than 30 years, Galdikas studied the great apes, advocated protection for them and their rainforest habitat, and undertook the rehabilitation of orphaned orangutans.
See also: Animal behaviour 116–117

BRIAN A. MAURER
1954–2018

Maurer's 1989 paper "Macroecology: The Division of Food and Space Among Species on Continents" – written with James H. Brown – was the first clear articulation of the idea that there is value in studying ecological patterns and processes over large areas and long time frames. In his last years he researched the dynamics of the spread of exotic birds and species diversity among mountain-dwelling mammals in North America.
See also: Macroecology 185

NANCY GRIMM
1955–

Based at Arizona State University, Grimm is a climate change ecologist and sustainability scientist, whose research concentrates on the

interaction of climate change, human activity, and ecosystems. Her work has particularly focused on the movement of water and chemicals through ecosystems. Grimm is a past president of the Ecological Society of America and a senior scientist on the US Global Climate Change Research Program.
See also: Ecosystem services 328–329

TIM FLANNERY
1956–

One of Australia's most prominent environmentalists, Flannery gained a doctorate in 1984 from the University of New South Wales for his work on kangaroo evolution. He later built up a reputation as a mammalogist, discovering several new species, and as an expert on climate change. He was chief commissioner of the Climate Commission, an Australian government body, and champions renewable energy.
See also: Renewable energy 300–305

SUSAN KAMINSKYJ
1956–

From her laboratory at the University of Saskatchewan, Canada, Kaminskyj – a cell biologist and mycologist – has pioneered the use of fungi to clean oil-contaminated sites, in a process known as bioremediation. Kaminskyj and her team found that when seeds are treated with a fungus named TSTh20-1, plants can establish in the substrate of such land and clean the soil as they grow.
See also: The ubiquity of mycorrhizae 104–105 ▪ Pollution 230–235

ROSEMARY GILLESPIE
1957–

Scottish-born Gillespie studied zoology at the University of Edinburgh before moving to the US to gain her doctorate at the University of Tennessee. She is known particularly for investigations into what drives biodiversity at species level, concentrating her evolution research on "hotspot archipelagos"

such as the Hawaiian chain, where the date of each island is already known with some accuracy. Most of her work is focussed on the evolution of spider species. Gillespie is based at the University of California, Berkeley, where she runs the EvoLab, a research group that focuses on arthropods, such as spiders and insects.
See also: Thermoregulation in insects 126–127 ▪ Island biogeography 144–149

HARVEY LOCKE
1959–

Born in Calgary, Canada, Locke trained and practised as a lawyer before switching to full-time conservation in 1999. He is committed to areas of ecology known as large landscape and connectivity conservation, which involve the connection of all lands, whether urban or wild, across a wide network. Locke was a founder of the Yellowstone to Yukon Conservation Initiative, which campaigns to create a continuous wildlife corridor between those two areas of North America. In 2009, Locke also cofounded the Nature Needs Half movement, which advocates for the protection of half of Earth's land and water area by 2050. Locke argues that this policy is necessary to avoid a sixth mass extinction on Earth.
See also: Mass extinctions 218–223

MAJORA CARTER
1966–

When her dog led her through a degraded brownfield site to the banks of the Bronx River, in her native New York City, Carter realized the potential for the regeneration of this area. She won funding from the city council to develop Hunts Point Riverside Park on the site, providing a natural retreat and river access for locals. Subsequently, her organization Sustainable South Bronx (SSBx) advocated and won support for "green" urban renewal in disadvantaged communities elsewhere in New York. SSBx also campaigns to improve air quality and food choices.
See also: The Green Movement 308–309

SARAH HARDY
1974–

Hardy is an American marine biologist and polar explorer who studies the effect on the environment of deep-ocean mining. She argues that to protect marine communities and biodiversity it is important to develop a systematic approach to the zoning of the oceans – with deep-sea marine protection areas a priority. Hardy studied marine biology at the University of California and gained her doctorate in oceanography at the University of Hawaii in 2005.
See also: A plastic wasteland 284–285

KATEY WALTER ANTHONY
1976–

Based at the University of Alaska, Walter Anthony is an aquatic ecosystems ecologist specializing in polar environments. She has studied carbon dioxide and methane emissions from lakes in the North American Arctic. In 2017, she discovered that unusually large amounts of methane were escaping from an Arctic lake, where the gas was seeping into the water from greater depths than previously discovered. If replicated elsewhere, such emissions from reserves deep in the permafrost could produce a dramatic increase in the amount of methane in the atmosphere.
See also: The Keeling Curve 240–241

AUTUMN PELTIER
2004–

Peltier, a member of the Wikwemikong First Nation who lives in Ontario, Canada, is a campaigner for clean drinking water, arguing that humanity should treat water with greater respect. In 2018, at the age of 13, she was one of the youngest people ever to speak to the UN General Assembly. Here, she advocated the policy that "No child should grow up not knowing what clean drinking water is, or never know what running water is."
See also: The water crisis 286–291

GLOSSARY

Abiotic Non-living; often used to refer to the non-living components of an ecosystem (such as climate and temperature).

Abundance The number of a given species within an ecosystem; an abundant species is strongly represented within the wider population.

Acid rain Any form of precipitation with high levels of acidity, causing damage to the environment; may occur naturally or as a result of human activity.

Anthropogenic Originating in, or influenced by, human activity.

Apex predator A predator that is not prey for any other species.

Atmosphere The layer of gases surrounding Earth. It also protects organisms from ultraviolet radition.

Autotroph A producer; an organism that makes its own food from sources such as light, water, and chemicals in the air.

Behavioural ecology The study of animal behaviour and how ecological pressures influence this.

Biodegradable Usually used in reference to waste products, meaning something that can be broken down by natural processes.

Biodiversity The variety of ecological life within a given geographical area, encompassing variety between and within species.

Biogeography The study of how plants and animals are distributed geographically, and the changes to this distribution over time.

Biological community A collection of living organisms within one location; when combined with their environment, they make an ecosystem.

Biomass The total quantity of a given organism within a habitat, generally expressed as weight or volume. Also a type of fuel made from organic matter, usually burnt to generate electricity.

Biome An area of Earth that can be classified according to the species of plant and animal life within it.

Biosphere The layer of Earth in which life can exist, situated between the atmosphere and lithosphere; the sum of all ecosystems on the planet.

Botany The scientific study of plant life.

Carnivore An organism which eats only meat.

Catastrophism The theory that changes in Earth's crust were caused by dramatic and unusual events, as opposed to gradual change over time.

Cells The smallest structural and biological unit that can survive on its own; the "building blocks" of all life on Earth.

Citizen science Scientific research carried out by amateurs, typically involving large-scale data collection.

Climate change A shift in the world's interconnected weather patterns; a gradual natural process exacerbated by human actions.

Climax A biological community or ecosystem that has reached a stable point, so that populations of organisms will remain steady. This is the end result of succession, in which the type of species and population sizes that make up a community change over time.

Climax species A plant species that will not change as long as its environment remains stable.

Clutch size The number of eggs laid in one birthing.

Community ecology The study of how species interact within a given geographical space.

Competitive exclusion principle The idea that multiple species reliant on exactly the same resources cannot exist together without one population rising and the other falling, as one will always have an advantage over another.

Coniferous Describes trees with seed cones which mostly do not shed their needle-like leaves during winter.

Conservation The protection and preservation of animal life, plant life, and natural resources.

Consumer A species that eats other organisms to obtain its required nutrients; this term can apply to any organism that is not at the very bottom of the food chain

Deciduous Describes trees that shed their leaves in the autumn.

Decomposers Organisms, primarily bacteria and fungi, that break down dead organisms and waste matter to obtain energy.

Deforestation The cutting down of a large area of trees, carried out for a range of purposes, including farming, industry, and construction.

Detritivores Organisms that feed on waste matter.

Diatom Any of a large group of microscopic algae that often play an important role in stabilizing an ecosystem and facilitating the existence of a range of life forms.

Diversity A measure of the variety of species within a biological community or ecosystem.

DNA Deoxyribonucleic acid. A large molecule in the shape of a double helix that carries genetic information in a chromosome.

Ecology The scientific study of the relationships between living organisms and their environment.

Ecosystem A community of organisms in a given environment that interact with and affect one another.

Ecosystem services The benefits humans receive from an ecosystem; a term highlighting the importance of the environment to humanity.

Endangered Describes a species whose population is so small that it is at risk of dying out completely.

Epidemiology The study of how diseases spread through populations, and the impact this has on the wider ecosystem.

Ethology The scientific study of the evolution of animal behaviour as an adaptive trait, with a particular focus on observing animals in their natural habitat.

Evolution The process by which species change over time as traits are passed down over generations.

Extinction The permanent dying out of an entire species.

Extirpation Extinction of a species on a local level – when a species dies out within a specific geographic area but still exists elsewhere on the planet.

Feedback loop The effect that one part of an ecosystem has on the rest, and how this change feeds back into the system as a whole.

Fertilizers Substances, which can be either natural or chemical, that are added to soil to increase its nutrient content and help plants grow more successfully.

Fieldwork Studies undertaken in the wild, rather than under controlled laboratory conditions.

Food chain A series of predators and prey, in which each organism is dependent on the preceding lifeform for food.

Food web A collection of food chains within an ecosystem and the connections between them, illustrating how communities interact on a wider scale to survive.

Fossil The remains of a prehistoric organism, preserved and solidified in sedimentary rock or amber.

Fossil fuel Non-renewable fuels formed over millions of years from plant and animal remains.

Fracking A process by which oil or gas can be extracted from the ground. Fracking involves drilling down and injecting liquid into the rock at a high pressure in order to force the oil and gas to the surface.

Fungi A group of organisms, including mushrooms, that produce spores and feed on organic matter. Unlike plants, fungi do not utilize sunlight for growth.

Gene The most basic unit of heredity; part of a DNA molecule that transmits characteristics from a parent to its offspring.

Genome The complete set of an organism's genes.

Geology The scientific study of Earth's physical formation and structure. Geologists examine our planet's history, and the ongoing processes that are acting upon it.

Global warming A gradual increase in the temperature of Earth's atmosphere caused by the accumulation of greenhouse gases.

GMO Genetically Modified Organism – any life form that has been artificially and chemically altered by engineering techniques that modify its DNA.

Greenhouse effect The way in which gases in Earth's atmosphere trap heat. The buildup of these gases leads to global warming.

Greenhouse gas Gases such as carbon dioxide and methane that absorb energy reflected by Earth's surface, stopping it from escaping into space.

Green Movement A political ideology that encourages a greater focus on the importance of the environment, and asks people to take action to prevent damage to Earth's natural habitats.

Groundwater Water found below Earth's surface, such as in spaces in the soil, sand, or rock.

Habitat The area in which an organism naturally lives.

Herbivore An organism that eats only plants.

Homeostasis The regulation of conditions within an organism, such as temperature, water, and carbon dioxide, to maintain a stable internal state.

Hypothesis An idea or assumption, used as the starting point for a theory, which is then tested through scientific experimentation.

Inheritance The passing on of genetic qualities and behavioural predispositions to offspring, through both genetic information and parental nurture.

Invasive species A non-native species that has been introduced to an ecosystem and spreads rapidly, damaging the ecological balance of the area.

Irrigation The controlled application of water to areas of land, usually through the creation of channels, to help crops grow.

Keystone species A species that plays a centrally important role in an ecosystem, often disproportionate to its biomass, and whose removal would alter or endanger the entire ecosystem.

Kin selection An evolutionary strategy whereby individuals pursue the best tactic for their relatives' survival, even at the expense of their own safety, wellbeing, or reproduction.

Mass extinction The widespread and rapid dying out of an abnormally large number – at least half – of all species; this sharp change in biodiversity usually marks a shift to a new geological era in our planet's history.

Metabolism The chemical processes that occur within the cells of an organism to keep it alive, such as the processes that enable the digestion of food.

Metacommunity A set of independent communities that interact and are connected by the movement of some species between those communities.

Metapopulation A collection of smaller populations of a given species that are linked by the movement of some individuals.

Microorganism An organism, invisible to the human eye, that can only be seen with a microscope, such as a bacterium, virus, or fungus; also known as a microbe.

Migration A large-scale movement of a species from one environment to another; often occurs seasonally.

Monoculture Using land for the cultivation or growth of only one type of plant or animal. This often has damaging effects on the land, as it can decrease its mineral value.

Morphology The study of the external structure of organisms.

Mutation A change of structure within an organism's DNA, which may result in a genetic transformation giving it uncharacteristic traits. One example of a mutation is albinism, a lack of pigmentation.

Mutualism A situation in which two or more organisms depend on each other for survival.

Mycorrhizae Types of fungi that grow among the roots of plants and exist in a symbiotic relationship with these plants.

Natural selection The process by which characteristics that increase an organism's chances of reproducing are preferentially passed on.

Niche The specific space and role that a species occupies within an ecosystem.

Omnivore An organism that feeds on both animals and plants.

Organism General term for any living thing, from single-cell bacteria to complex, multicellular lifeforms such as plants and animals.

Ornithology A branch of biology that concerns the study of birds.

Overfishing The depletion of the fish population in a given area as a result of fishing too intensively.

Ozone layer Part of the upper level of Earth's atmosphere, with a high concentration of ozone (O_3) molecules; provides protection from ultraviolet radiation.

Palaeontology The study of fossils and biology of Earth's geological past. Palaeobotany is the branch studying plant fossils.

Parasite An organism that lives on or in another organism, and obtains nutrients from its host.

Pesticides Chemicals used to kill certain types of pest in order to protect cultivated plants. They can, however, also kill non-target species, and damage the wider ecosystem.

Photosynthesis The process by which plants and algae transfer the Sun's light energy into chemical energy as glucose, allowing it to be passed along the food chain. The process absorbs carbon dioxide and releases oxygen.

Physiology A branch of biology that focuses on the everyday functioning of organisms.

Pollination The transfer of pollen from a male plant part to a female one – by birds, insects, and other animals, or by the wind – enabling fertilization and seed production.

Pollution The introduction of harmful contaminants to the natural environment, inducing changes to the atmosphere.

Predator A species that hunts other animal species for food.

Prey A species that is hunted by another species.

Primary producer Any organism that makes its own food from nonorganic sources, namely light and/or chemical compounds such as carbon dioxide and sulphur, and thus sustains the animals that feed on it.

Primary vegetation The vegetation that has prevailed in a given area since the start of its current climatic conditions.

Recycling The process of converting waste into new objects or materials, or burning it to generate energy.

Renewables Fuel sources that are not finite, such as solar power, hydropower, and wind power.

Species A group of organisms capable of exchanging genes with one another through reproduction.

Stochasticity Unpredictable fluctuations in environmental conditions that affect populations and ecological processes.

Succession The process by which a biological community evolves over time, from a few simple species to a complex ecystem, through species' impact on the environment.

Taxonomy The science of naming and classifying different organisms.

Tectonic plates Pieces of Earth's crust and uppermost mantle that gradually shift over time, causing seafloor spreading, continental drift, and mountains, rift valleys, volcanoes, and earthquakes at plate boundaries.

Thermoregulation The internal processes that occur within an organism to ensure it maintains a stable temperature, a function that is crucial for survival.

Transmutation The process of evolutionary divergence by which one species transforms into an entirely new one.

Trophic cascade The impact that the removal of a trophic level of a food chain with at least three levels has on the wider ecosystem as a whole.

Trophic level The place of an organism within an ecosystem's hierarchy; organisms which are on the same level of the food chain are on the same trophic level.

Tropics The region of Earth that surrounds the equator, between the lines of the Tropic of Cancer and the Tropic of Capricorn, and does not experience the same seasonal changes as the rest of Earth.

Urbanization The process which occurs when rural areas are build upon intensively, almost always with negative consequences for the natural environment.

Urban sprawl The outward growth of a previously concentrated urbanized area, often with negative consequences for the environment.

Variation Differences within a species, caused either by genetic or environmental factors.

Vascular plant A type of plant with conductive tissue for the movement of water and minerals throughout, such as a fern or a flowering plant.

INDEX

D

E

F

G

Q R

S

QUOTE ATTRIBUTIONS

ACKNOWLEDGMENTS

Dorling Kindersley would like to thank Professor Fred D. Singer for his help in planning this book, Monam Nishat and Roshni Kapur for design assistance, and Anita Yadav for DTP assistance.

PICTURE CREDITS

The publisher would like to thank the following for their kind permission to reproduce their photographs:

(Key: a-above; b-below/bottom; c-centre; f-far; l-left; r-right; t-top)

21 Alamy Stock Photo: The Picture Art Collection (tr); The Natural History Museum (bl). 22 Alamy Stock Photo: North Wind Picture Archives (br). 26 Rex by Shutterstock: Granger (bl). 29 Alamy Stock Photo: Kamal Bhatt (tl); Laurentiu Iordache (cr). 30 Alamy Stock Photo: Blickwinkel (t). 31 Alamy Stock Photo: Cultura RM (crb). Dorling Kindersley: Frank Greenaway / Natural History Museum, London (cb). 33 Alamy Stock Photo: Pictorial Press Ltd (tr). Dreamstime.com: Gordana Sermek (bc). 35 Alamy Stock Photo: Alexander Heinl / Dpa Picture Alliance / Alamy Live News (tl). Science Photo Library: A. Barrington Brown © Gonville & Caius College (cla). 36 Science Photo Library: Pascal Goetgheluck (clb). 37 Alamy Stock Photo: BSIP SA (tl). 39 SuperStock: Animals Animals (cla); Guillem López / Age fotostock (bl). 46 Alamy Stock Photo: Historic Collection (bl). 47 Getty Images: Adam Jones (ca). 49 Science Photo Library: Nigel Cattlin (tr). 51 Getty Images: Pete Oxford / Minden Pictures (cla). 53 iStockphoto.com: Stefonlinton (cla). 54 Alamy Stock Photo: Suzanne Long (bc). 57 Ardea: © Gregory G. Dimijian M.D. / Scie (cla). Science Photo Library: Gilbert S. Grant (t). 59 Depositphotos Inc: Andaman (b). 62 Alamy Stock Photo: Richard Ellis (b). 63 Alamy Stock Photo: Kevin Schafer (tl). 64 Courtesy of National Park Service, Lewis and Clark National Historic Trail: (bl). 65 Alamy Stock Photo: Nick Upton (tl). 67 Alamy Stock Photo: Avalon / Photoshot License (tr). Getty Images: Roger Tidman (bl). 69 Alamy Stock Photo: GL Archive (br); Pictorial Press Ltd (cra). 70 Alamy Stock Photo: M.Brodie (tl); David speight (tr). 73 Alamy Stock Photo: PF-(bygone1) (tr). Getty Images: Fritz Polking (bl). 75 Alamy Stock Photo: Nigel Cattlin (bl). Getty Images: Visuals Unlimited, Inc. / Anne Weston / Cancer Research UK (b). 77 Alamy Stock Photo: David Lester (cla). 83 Getty Images: Douglas Klug (cra); DEA Picture Library (b). 85 Alamy Stock Photo: Art Collection 3 (bl); Science History Images (cra). 87 Alamy Stock Photo: ART Collection (tr); Florilegius (bl). 89 Alamy Stock Photo: Jeff J Daly (clb). 90 Getty Images: Shawn Walters / EyeEm (br). 91 Alamy Stock Photo: Henri Koskinen (crb). 94 Getty Images: Bettmann (tl); Education Images (br). 95 Alamy Stock Photo: De Luan (tl). 96 Alamy Stock Photo: Marka (br). 97 Getty Images: Denver Post (tr). 103 Alamy Stock Photo: BSIP SA (bl); Historic Images (tr). 104 Science Photo Library: Dr. Merton Brown, Visuals Unlimited (cra). 105 Alamy Stock Photo: Blickwinkel (crb). 109 Alamy Stock Photo: Wildlife GmbH (crb); DP Wildlife Invertebrates (ca). 110 Alamy Stock Photo: Vince Burton (clb). Getty Images: Universal History Archive (tc). 111 Alamy Stock Photo: Ingo Oeland (br). 112 Science Photo Library: Wim Van Egmond (bl). 113 Alamy Stock Photo: Biosphoto (t). 114 Dreamstime.com: Bernard Foltin (br). 115 SuperStock: Minden Pictures (cla). 116 Alamy Stock Photo: Austrian National Library / Interfoto (cra). 117 Getty Images: Rolls Press / Popperfoto (bl). 121 Dreamstime.com: Mark Higgins (cra). Getty Images: CBS Photo Archive (bl). 122 Getty Images: Michael Nichols (br). 123 naturepl.com: Anup Shah (br). 124 Getty Images: Dr Clive Bromhall (br); Dan Kitwood / Staff (tl). 125 Alamy Stock Photo: Terry

Whittaker Wildlife (br). 127 Alamy Stock Photo: Oliver Christie (cla). Getty Images: Alastair Macewen (br). 133 Getty Images: Wildestanimal (cra). 135 Alamy Stock Photo: The Picture Art Collection (bl). iStockphoto.com: Vlad61 (cra). 136 Alamy Stock Photo: A.P.S. (UK) (tr). 137 Getty Images: Olaf Protze (br). 139 Alamy Stock Photo: The Natural History Museum (tl). Science Photo Library: Ted Kinsman (crb). 141 Alamy Stock Photo: Danita Delimont (cra). 142 Alamy Stock Photo: Dennis Frates (bl). 143 Alamy Stock Photo: World History Archive (tl). Getty Images: Fine Art (crb). 146 Alamy Stock Photo: Mark Lisk (tr). 147 Courtesy of Marlboro College: www.marlboro.edu (tr). 149 Alamy Stock Photo: age fotostock (tl). Getty Images: Universal History Archive / UIG (bl). 151 Alamy Stock Photo: Jason Bazzano (cla). naturepl.com: Paul Williams (br). 153 Alamy Stock Photo: Bill Crnkovich (crb). 155 Alamy Stock Photo: Blickwinkel (ca). 156 North Carolina State University: Rebecca Kirkland (c). 163 Alamy Stock Photo: Greg Basco / BIA / Minden Pictures (cb); Pictorial Press Ltd (cra). 165 Alamy Stock Photo: GL Archive (tr). Science Photo Library: Solvin Zankl / Visuals Unlimited, Inc. (t). 166 NASA: Jeff Schmaltz, MODIS Rapid Response Team, NASA / GSFC (cra). 168 Alamy Stock Photo: Emmanuel Lattes (cr). 169 Alamy Stock Photo: RWI Fine Art Photography (br). 170 Alamy Stock Photo: Robert K. Olejniczak (cra). 173 Dreamstime.com: Anton Foltin (cra). 175 Dreamstime.com: Claudio Balducelli (cla). 176 Alamy Stock Photo: All Canada Photos (br). 180 U.S.F.W.S: (tr). 181 Alamy Stock Photo: Everett Collection (tr); Ian west (b). 182 Dreamstime.com: Yuval Helfman (tl). 183 Alamy Stock Photo: Natural History Archive (tr). 185 naturepl.com: Mary McDonald (crb). 187 naturepl.com: Jussi Murtosaari (cra). 188 Rex by Shutterstock: Antti Aimo Koivisto (tl). 189 Alamy Stock Photo: David Hall (tl); Genevieve Vallee (br). 191 Alamy Stock Photo: Mauritius images GmbH (bl). Getty Images: Danita Delimont (tl). 193 Alamy Stock Photo: Adam Burton (tl). 198 Getty Images: Philippe Lissac / GODONG (bc). 199 Alamy Stock Photo: Rolf Nussbaumer Photography (crb). Depositphotos Inc: swisshippo (cla). 201 Dreamstime.com: Rvo233 (cla). Getty Images: Hulton Archive / Stringer (tr). 202 IPCC : FAQ 1.3, Figure 1 from Le Treut, H., R. Somerville, U. Cubasch, Y. Ding, C. Mauritzen, A. Mokssit, T. Peterson and M. Prather, 2007: Historical Overview of Climate Change. In: Climate Change 2007: The Physical Science Basis. Contribution of Working Group I to the Fourth Assessment Report of the Intergovernmental Panel on Climate Change [Solomon, S., D. Qin, M. Manning, Z. Chen, M. Marquis, K.B. Averyt, M. Tignor and H.L. Miller (eds.)]. Cambridge University Press, Cambridge, UK and New York, NY, USA (br). 203 Getty Images: Wolfgang Kaehler / LightRocket (crb). 205 Alamy Stock Photo: Sputnik (tr). iStockphoto.com: Totajla (tl). 207 Alamy Stock Photo: Suzanne Long (tr). SuperStock: Wolfgang Kaehler (bl). 209 Depositphotos Inc: Pawopa3336 (tl). Getty Images: DEA / C.DANI / I. JESKE (crb). 210 Getty Images: R A Kearton (cr). 211 Alamy Stock Photo: ClassicStock (br). 212 Dorling Kindersley: Dorling Kindersley: Colin Keates / Natural History Museum, London (cra). 213 Alamy Stock Photo: AustralianCamera (tr). 215 Alamy Stock Photo: Ancient Art and Architecture (bc). Getty Images: Terry Smith (tr). 216 Alamy Stock Photo: Iuliia Bycheva (tr). 217 iStockphoto.com: Zhongguo (br). 220 Science Photo Library: Detlev Van Ravensnaay (tr). 221 Alamy Stock Photo: Pictorial Press Ltd (tr). 223 Getty Images: The Washington Post (cla). 225 Alamy Stock Photo: Arterra Picture Library (cla). Getty Images: Magnus Kristensen / AFP (cra). 233 Alamy Stock Photo: Chronicle (br). Getty Images: Sonu Mehta / Hindustan Times (bl). 234 Alamy Stock Photo: Design Pics Inc (b). Unicef: (tr). 235 UNSW: Aran Anderson (tl). 237 Alamy Stock Photo: Archive Pics (tl). iStockphoto.com: 4kodiak (cla). 238 Alamy Stock Photo: Paul Kennedy (bc). 239 Alamy Stock Photo: ImageBroker (crb); Huang

Zongzhi / Xinhua / Alamy Live News (tl). 241 Alamy Stock Photo: Arctic Images (br). Science Photo Library: Simon Fraser / Mauna Loa Observatory (cla). 244 Alamy Stock Photo: Walter Oleksy (bl). Science Photo Library: CDC (tr). 247 iStockphoto.com: Harry Collins (br). 248 Alamy Stock Photo: Christopher Pillitz (bc). 249 Gene E. Likens: On Location Studios, Poughkeepsie, NY (tr). 250 Alamy Stock Photo: North Wind Picture Archives (bc). 251 Getty Images: Peter Charlesworth (crb). Dr Max Roser: Esteban Ortiz-Ospina (2018) "World Population Growth". Published online at OurWorldInData. org. Retrieved from: https://ourworldindata.org/world-population-growth [Online Resource] (cla). 252 Alamy Stock Photo: Renault Philippe / Hemis (cr). 253 Alamy Stock Photo: Danita Delimont (cr). 256 Getty Images: Brazil Photos (tr). 257 Getty Images: Antonio Scorza / Staff (tr). 258 Getty Images: Michael Duff (b). 259 Getty Images: Micheline Pelletier Decaux (br). MongaBay.com: Rhett Butler / rainforests.mongabay.com (tl). 261 Getty Images: Orlando / Stringer (crb). NASA: Jesse Allen (cla). 263 Getty Images: Orjan F. Ellingvag / Corbis (cla); Fairfax Media (tr). 264 Dreamstime.com: Oliver Förstner (crb). 265 Alamy Stock Photo: IanDagnall Computing (tr). 267 Alamy Stock Photo: Poelzer Wolfgang (bl). 268 John M. Yanson: (tl). 269 Alamy Stock Photo: Science History Images (br). Getty Images: Barcroft Media (bl). 271 Getty Images: Scott Tilley (cla). NSW Department of Primary Industries: Dr Steven McLeod (cra). 272 123RF.com: Stephen Goodwin (crb). 273 Alamy Stock Photo: Jack Picone (tr). 277 Alamy Stock Photo: Gay Bumgarner (cra). Camille Parmesan: Marsha Miller, University of Texas at Austin (bl). 278 Getty Images: Bianka Wolf / EyeEm (br). 279 Alamy Stock Photo: Andrew Darrington (tl). 280 iStockphoto.com: ca2hill (cr). 282 Alamy Stock Photo: Christian Hütter (tr). 283 Getty Images: Hector Vivas (tr). 284 Getty Images: Peter Parks / AFP (br). 285 Ardea: Paulo de Oliveira (br). 288 Getty Images: AFP / Stringer (tr). 289 Getty Images: Jim Russell (b). 290 Alamy Stock Photo: ImageBroker (br). Mesfin Mekonnen and Arjen Hoekstra: (2016) http://advances.sciencemag.org/content/2/2/e1500323 (b). 291 Alamy Stock Photo: Russotwins (cra). 296 Alamy Stock Photo: Granger Historical Picture Archive (bc). 298 Alamy Stock Photo: The Granger Collection (bc). 299 Getty Images: DEA / Biblioteca Ambrosiana / De Agostini (cr). 302 Alamy Stock Photo: Jim West (br). 303 Getty Images: Bloomberg / reprinted with permission from Joint Center for Artificial Photosynthesis - California Institute of Technology (crb). 304 Getty Images: TPG (tr). 305 IEA: © OECD/IEA [2016], Renewables Information, IEA Publishing. Licence: www.iea.org/t &c<http://www.iea.org/t&c> (tl). 307 Alamy Stock Photo: Jonathan Plant (cla). Rex by Shutterstock: AP (tr). 309 Alamy Stock Photo: Steve Morgan (cla); Friedrich Stark (tr). 311 Alamy Stock Photo: Flowerphotos (tr). 313 Alamy Stock Photo: Rick & Nora Bowers (br). Ardea: © USFWS / Science Source / Science S (cla). 314 Getty Images: Design Pics / Richard Wear (tl). 315 Rex by Shutterstock: Chuck Graham / AP (tr). 318 Getty Images: Digital First Media / Inland Valley Daily Bulletin via Getty Images (tr). 319 Rex by Shutterstock: Eric Lee / Lawrence Bender Prods. / Kobal (b). 320 Rex by Shutterstock: Mohammed Seeneen / AP (tr). 323 Getty Images: Hero Images (cra). NOAA: (bl). 324 Getty Images: Tony Karumba (br). 325 Getty Images: Paul J. Richards / Staff (tr). 326 Alamy Stock Photo: Inga Spence (cr). 327 iStockphoto.com: pixelfusion3d (tr). Rex by Shutterstock: AGF s.r.l. (cr). 328 Getty Images: Amana Images Inc (cr). 329 Stanford News Service: Linda A. Cicero (tr). 331 Getty Images: Ted Aljibe / Staff (crb).

All other images © Dorling Kindersley
For further information see: **www.dkimages.com**